Ultrasonic Communication by Animals

The badge of No. 360 Squadron. This joint Royal Navy/Royal Air Force Squadron provides electronic countermeasures training for users of Air Defence radar systems. For the centrepiece of their badge, presented in September 1973 after Royal approval, they adopted *Melese laodamia*, a moth that uses acoustic countermeasures against the ultrasonic 'radar' of bats (see Chapter 4). This photograph was kindly provided by Commander G. Oxley, RN.

The first species of animal to be discovered by means of its ultrasound. The 'inaudible' song of this bush cricket was first detected (by J.D.P.) and recorded near Ibadan, Nigeria. The captured specimens did not belong to any known species and were later named *Anepitacta egestoides* (see Chapter 5). (Photograph by A. Howard and R. Reed, printed by M. Coughtrey).

Ultrasonic Communication by Animals

Gillian Sales
Research Associate in Zoology
David Pye
Reader in Zoology,
King's College, University of London

CHAPMAN AND HALL
London
A HALSTED PRESS BOOK
JOHN WILEY & SONS
New York

First published 1974
by Chapman and Hall Ltd.
11 New Fetter Lane London EC4P 4EE

Printed in Great Britain by
Richard Clay (The Chaucer Press), Ltd.,
Bungay, Suffolk

Library of Congress Cataloging in Publication Data

Sales, Gillian.
Ultrasonic communication by animals.

Includes bibliographical references.
1. Animal communication. 2. Animal sounds. 3. Ultrasonics in biology. I. Pye, David, 1932 – joint author. II. Title. [DNLM: 1. Animal communication. 2. Behavior, Animal. 3. Ultrasonics.
QL776 S163u 1974]
QL776.S24 1974 591.5′9 73-15213
ISBN 0-470-74985-7

Contents

Preface

In recent years there has been a rapid increase in the understanding of communication between animals and this is perhaps especially true of bio-acoustics. In the last 35 years a completely new branch of bio-acoustics, involving ultrasounds, has been made possible by technical developments that now allow these inaudible sounds to be detected and studied. This subject has a personal fascination for the authors, perhaps because of the novelty of 'listening in' to these previously unknown signals, perhaps because of the wide variety of ways in which different animals use them.

Many studies of different aspects of animal ultrasound have now been published and a review of them all seems to be timely. Ultrasound is defined in human terms and is biologically arbitrary; other animals may produce similar signals at lower frequencies for similar purposes. This book attempts to be comprehensive but the limits of the subject are rather difficult to define. It should be read in conjunction with other books on audible bio-acoustics. Each chapter has been written and may be read as a separate entity, although there is considerable cross-referencing. Chapters 1 and 2 form a common introduction and may help in understanding the later sections. The Appendix is not essential but is included for those who may be interested in the quantitative aspects of the echo-location phenomena described in Chapters 3 and 8.

It is sincerely hoped that readers initially concerned with one particular topic will also be interested in the other chapters and that many will come to share the enthusiasm of the authors for this subject.

We are grateful to many people who have helped to produce the book. Professor Kenneth Roeder, Dr David Ragge and Miss Jane

Smith kindly read parts of the text but the authors accept full responsibility for any errors of judgment or of fact that may be found. Dr Ann Brown has advised on parts of the literature on hearing. Figures or permission for reproduction have been kindly provided by A. M. Brown, F. W. Darwin, B. Dumortier, D. R. Griffin, John Huxley, B. Masterton, A. Michelsen, L. A. Miller, G. Neuweiler, H. Nocke, K. Norris, E. E. Okon, G. Oxley, K. H. Reid, L. H. Roberts, C. Scott Johnson, N. Suga, F. A. Webster and P. J. Whitfield. Photographic production work for the book was carried out by R. Reed and A. Howard of King's College and M. Coughtrey of Queen Mary College.

The authors' own studies in this field have been supported by the Science Research Council, the United States Office of Scientific Research and the Royal Commission for the Exhibition of 1851. We are indebted to King's College London for generous facilities and to many members of its staff, especially Professor Don Arthur, for providing much help, encouragement and stimulating discussion. The Entomology Department of the British Museum (Natural History) kindly identified many specimens, including some badly damaged experimental subjects.

Permission for the publication of illustrations has been granted by: The American Foundation for the Blind (Plate IVb, c); American Scientist (Plate Vb); The Biological Bulletin (Plate XIII and dust jacket); Journal of the Acoustical Society of America (Fig. 9.1); Journal of Auditory Research (Fig. 6.3); Journal of Experimental Zoology (Plate Va); Journal of Insect Physiology (Figs. 4.2, 4.4, 5.8 and Plate VIIIb); Journal of Morphology (Plate VIIIa); National Geographic Society (Plate IVa); Nature, London (Fig. 7.19b); Science, New York (Fig. 4.3 and Plate XIa) Copyright 1970 and 1971 respectively by the American Association for the Advancement of Science; Springer-Verlag, Berlin, etc. (Figs. 3.9, 6.2, 6.4); The Zoological Society of London (Fig. 7.8).

June 1973 — G. D. S.

King's College,* — J. D. P.

University of London

* The authors were at King's College when the text was written, but moved to Queen Mary College, University of London in October 1973.

List of Abbreviations

In general, standard units are used but the reader's attention is drawn to the following:

s^{-1}	per second	Hz	Hertz (one cycle per second)
in s^{-1}	inches per second	kHz	kiloHertz (one thousand cycles per second)
cm^{-2}	per square centimetre	dB	decibel (see p. 6)
μm	micrometre (10^{-6}m)	S.P.L.	sound pressure level
μbar	microbar (1 dyne per cm^2)		

CHAPTER ONE

What is Ultrasound?

Animals communicate with each other through all their senses: touch, vision, smell, hearing and in some fishes by electric fields. In recent years the study of communication by animals has increased rapidly and some fascinating phenomena have been revealed through the development of methods for studying both the communication signals of the animals and also the capabilities of their sense organs. In addition the rapid development of techniques for human communication has stimulated interest in the methods used by animals. The parallels between human and animal communication systems are often remarkably close and many animals rival or surpass the performance of human technology.

The study of sound communication, or bio-acoustics, has benefited tremendously from electronic technology. Quite recently such studies were barely scientific and almost entirely anecdotal; a bird song could only be described as 'a shrill twitter' or be transcribed as 'twit twit twit tweewhooo'. Now it is possible to describe songs accurately in terms of sound frequency, intensity and harmonic patterns. The rapid development of knowledge of animal acoustics has been summarized in a number of books, notably 'Acoustic Behaviour of Animals' edited by Busnel (1963a).

It has been found that animals may communicate by sound vibrations inaudible to man, either by ultrasonic waves 'above' our hearing or by infrasonic ones 'below' it (the old term 'supersonic' has been restricted to speeds faster than sound). Ultrasonic communication now provides examples of all the functions of communication: intraspecific communication for a variety of social purposes, interspecific communication between members of different species and autocommunication for the purposes of navigation.

THE DISCOVERY OF ANIMAL ULTRASOUND

The first discovery of ultrasound emitted by any animals appears to have been made in the nineteen-thirties by Pierce who was then Professor of Physics at Harvard University. As a physicist he invented the now well-known crystal-controlled oscillator and discovered much about the vibration of solids at high frequencies under electrical and magnetic forces. He was able to transmit ultrasonic waves through air and to detect them with a special receiver. Using this equipment, Pierce soon discovered that some bush-crickets emit ultrasounds. Entomology became a hobby and in retirement he wrote a delightful book called 'The Songs of Insects' (1948) which describes his ingenious methods and many important discoveries in this field. But further studies of the subject followed slowly and many entomologists continued to work only within the range of human hearing. Later studies of ultrasound production in insects and of their ability to hear high frequency sounds will be reviewed in Chapters 5 and 6.

In 1938 Griffin was a graduate student at Harvard University and was interested in the navigation of birds and bats. He took some bats to Pierce and, using the ultrasound detector, they together demonstrated the emission of ultrasonic pulses by flying bats (Pierce and Griffin, 1938). A series of papers by Griffin and Galambos soon confirmed the suggestion that had been made by Hartridge in 1920, that bats might navigate by listening to echoes of their own cries at ultrasonic frequencies. In addition Griffin showed that flying insects are detected and intercepted by the same means. This amazing phenomenon attracted a great deal of attention and the initial rapid progress was summarized by Griffin in a book called 'Listening in the Dark' (1958). The current position on the study of echo-location by bats will be described here in Chapter 3.

Other fields of investigation into animal ultrasound appeared gradually. In the nineteen-fifties Roeder and Treat discovered that noctuid moths can detect the ultrasonic cries of predatory bats and so take avoiding action. This interaction between moths and bats was then studied in detail and has been reviewed by Roeder in 'Nerve Cells and Insect Behaviour' (1963, revised 1967). Investigations into a number of other acoustic counter-measures against bats followed in different nocturnal insects and these will be described in Chapter 4.

Ultrasonic signals from rodents were apparently found independently by several workers from Anderson (1954) onwards but none of these observations were followed up until Noirot started a systematic study of the ultrasonic calls of mouse pups in 1966. She introduced the present authors to this subject which has expanded steadily as will be shown in Chapter 7.

Many other animals are now known to emit ultrasonic signals and these are briefly described in Chapter 8. The ultrasonic echo-location of dolphins was investigated by several groups of workers and has been described by Kellogg in his book 'Porpoises and Sonar' (1961). Two cases of echo-location by birds will be discussed for comparison, although neither is ultrasonic. An echo-location function has also been suggested for the cries of a number of other vertebrates, although these are less well substantiated. Finally Chapter 9 will summarize the present position comparatively and suggest where further examples might be found.

THE PROPERTIES OF ULTRASOUND

Human hearing has been intensively studied and the subject has been reviewed by Hirsch (1952). The 'average' ear is most sensitive to sounds at a frequency of about two thousand cycles (waves) per second, usually abbreviated to two kiloHertz or 2 kHz. On either side of this frequency the sensitivity of the ear declines steadily (Fig. 8.4) until it disappears completely at about twenty cycles per second (20 Hz) where tactile sensation may take over for intense sounds, and at about 17–20 kHz beyond which we have no sensation. Merely amplifying a higher frequency to increase its intensity cannot normally make it audible provided that it is still faithfully reproduced, although the human cochlea can respond to much higher frequencies if the sound is transmitted through the skull (Pumphrey, 1956; Deatherage *et al.*, 1954; Corso, 1963; Corso and Levine, 1963). Actually the upper limit of hearing varies greatly between individuals and generally declines with age; in many people and particularly the elderly it is as low as 10 kHz. So while engineers often arbitrarily define ultrasound as sounds above 20 kHz, it is safer to make instruments work down to 10–15 kHz, thus ensuring that there is no gap before our unaided ears can take over.

Many animals, especially small mammals and the Cetacea (whales and dolphins) among mammals (Brown and Pye, 1974) and many insects, do not share the human upper frequency limit and can hear to much higher frequencies (Fig. 1.1). To these animals ultrasound is entirely natural and it would be surprising if they did not exploit this band for communication, if only because the lower frequency bands can get congested, like some bands of radio frequencies, by too many users.

The only really important property of ultrasound that distinguishes it from 'ordinary sound' is therefore that the frequency is too high for human hearing. The distinction has no physical or general biological basis but because of it all investigations must rely on instruments

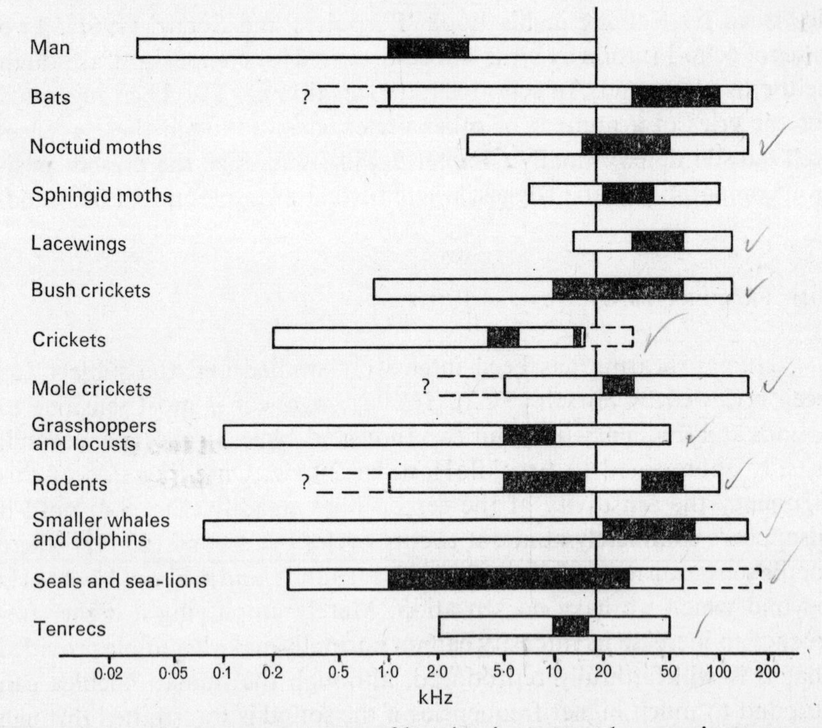

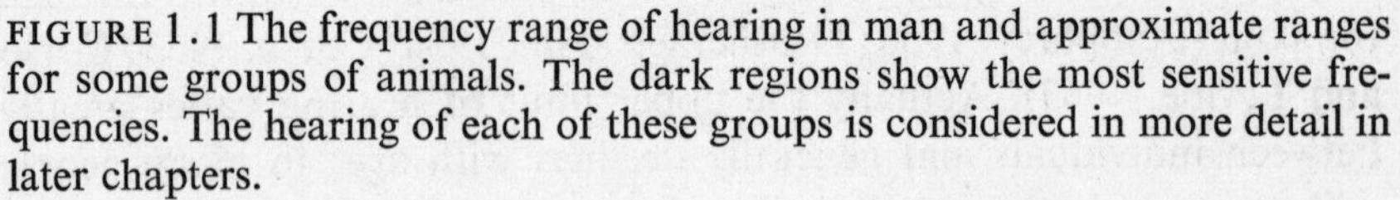

FIGURE 1.1 The frequency range of hearing in man and approximate ranges for some groups of animals. The dark regions show the most sensitive frequencies. The hearing of each of these groups is considered in more detail in later chapters.

capable of detecting the signals and analysing their characteristics. This limitation, however, produces one of the principal attractions of studying animal ultrasound for, with the proper equipment, it is still relatively easy to listen to signals that have never been 'heard' before. It is like entering a previously unsuspected world. The equipment and techniques that have been used will be briefly reviewed in the next chapter and although that section is not essential reading, it may help in understanding the results given in later chapters. It is hoped that many readers will be encouraged to build or acquire similar equipment and a more detailed account of principles, construction and operation will be given elsewhere (Pye, in preparation).

Most school science books describe the propagation of sound waves as alternate compressions and rarefactions of the medium. These waves share most of the properties of all other waves except that they have no movements transverse to the direction of propagation, as do waves on water surfaces, electromagnetic waves or waves travelling along a wiggled rope. Because sound waves consist only of alternate 'pushing

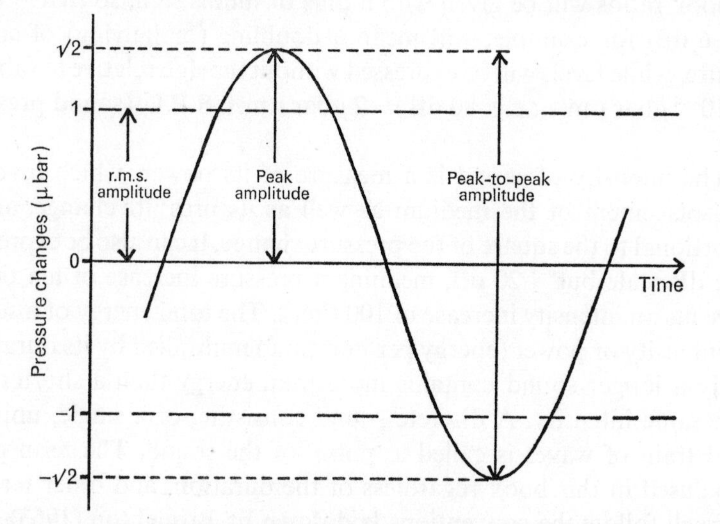

FIGURE 1.2 A sinusoidal pressure wave of 74 dB *re* 2×10^{-4} μbar r.m.s. The r.m.s. pressure change is 1 μbar, the peak change is $\sqrt{2} = 1{\cdot}4$ times this and the peak-to-peak is double this or 2·8 μbar. An r.m.s. value of 2·8 μbar would be 9 dB higher, or 83 dB *re* 2×10^{-4} μbar r.m.s.

and pulling' in the direction of propagation, they cannot show any of the phenomena due to polarization.

The strength of a sound wave is defined by the pressure changes that occur in the medium, above and below its mean pressure on alternate half cycles. The usual unit of pressure is the microbar (μbar) which is equivalent to 1 dyne cm^{-2} of the old system of units (or 0·1 Nm^{-2} of the new International System, although the μbar is retained for sound pressure). It may be used to express the maximum (peak) pressure change or more commonly the standard deviation of pressure changes, called the root mean square (r.m.s.) value (Fig. 1.2). Sound levels are also often expressed in the decibel scale which is a logarithmic scale of ratios such that the addition of one decibel (+1 dB) is equivalent to an increase in pressure by about 12·20%. Thus +6 dB is 12·20% six times by compound interest, which amounts to a doubling of the pressure, while +20 dB gives an increase of ten times in pressure. Since the decibel only expresses a pressure ratio (1·1220:1), it cannot be used to describe a pressure level unless another reference level is stated. The most common reference level for sounds in air is 2×10^{-4} μbar r.m.s. (roughly the threshold of human hearing at its most sensitive frequency) and other levels are commonly expressed in dB relative to this level. Throughout this book ratios will be given with a plus or minus sign, so that +6 dB (or −6 dB) for example, will mean a doubling (or halving) of sound pressure, while levels will be expressed without the sign relative to (above) 2×10^{-4} μbar r.m.s.; e.g. 80 dB = 2 μbar r.m.s. S.P.L. (sound pressure level).

The intensity of sound is a measure of its power, which involves the displacement of the medium as well as its pressure change, and is proportional to the square of the pressure change. It can also be expressed in the dB scale but +20 dB, meaning a pressure increase of ten times, represents an intensity increase of 100 times. The total energy of a sound is its intensity or power (energy per unit time) multiplied by its duration. Clearly a longer sound contains more total energy than a shorter one of the same intensity. A discrete sound consisting of a single, uninterrupted train of waves is called a 'pulse' of the sound. The term pulse will be used in this book regardless of the duration, and other terminology will follow the conventions laid down by Broughton (1963).

Ultrasound shows two important properties that are extensions of effects shown by audible sound but which become more marked at high frequencies. First, higher frequencies travelling through air are absorbed more rapidly than are lower frequencies. All sounds spread out as they travel, producing an effectively spherical wave-front which expands geometrically. Thus all the sound energy passing through a given area at a fixed distance from the source must occupy four times (2 × 2) the area at twice that distance and nine times (3 × 3) the area at three times the distance (Fig. 1.3). This gives rise to the inverse square law which states that the intensity of any sound decreases with the square of the distance travelled. In addition a certain amount of the sound energy is lost by warming the air through which it travels. This atmospheric attenuation proceeds steadily with the distance travelled and must be added to the effect of the inverse square law if the intensity at a given distance is to be calculated. Higher frequencies are attenuated more rapidly than lower ones and although the difference is quite slight over the band audible to man, it becomes increasingly significant at ultrasonic frequencies. Thus in dry air the total energy of a 1 kHz sound is reduced to half in about 60 m but at 10 kHz it falls to half in 15 m and at 150 kHz in only 1 m. The effect is considerably complicated, however, by humidity in the air. The presence of water vapour greatly reduces the attenuation of low frequencies but increases the loss at high frequencies even further. The whole subject has been reviewed in detail by Griffin

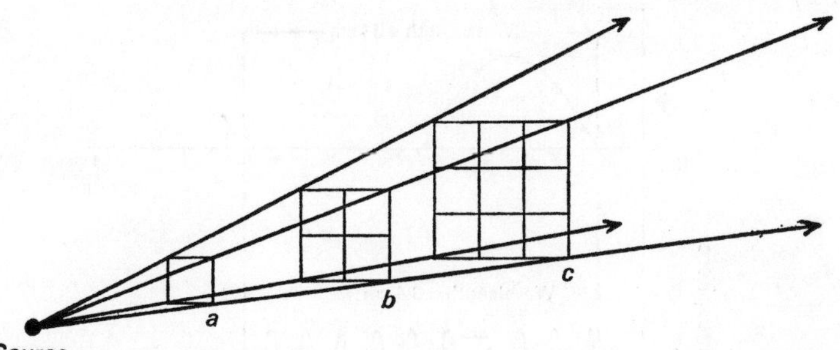

FIGURE 1.3 A diagram to illustrate the inverse square law. The energy passing a unit square at *a* expands to four times the area at twice the distance *b* and nine times the area at three times the distance *c*. If there is no absorption, the intensity is inversely proportional to the square of the distance.

(1971) who discussed its implications for the detection of distant targets by bats. Underwater losses are also greater at higher frequencies although all losses are less than in air. In seawater the total energy of a signal of 1 kHz falls to half after 1000 km, of 10 kHz after 10 km and of 100 kHz after 100 m (Tucker, 1966).

The second special characteristic of ultrasound depends entirely on the fact that the wavelengths are short. The velocity of sound in air is almost independent of frequency and is about 340 m s^{-1} at 16°C (increasing by 0·54 ms^{-1} for every 1°C rise in temperature). Since all sound waves travel at the same speed, it follows that if a greater number pass a fixed point in a given time their peaks must be closer together. This is expressed by saying that the velocity, which is constant, is equal to the frequency times the wavelength. The wavelength of 1 kHz is 34 cm, that of 10 kHz is 3·4 cm and that of 100 kHz is 3·4 mm (Fig. 1.4).

On a pond, or even in the bathtub, it can easily be seen that ripples of short wavelength are reflected from an object larger than a wavelength whereas longer waves pass right round the object, leaving almost no 'shadow'. The same applies to sound waves, for not only are they reflected better by larger objects (see Appendix) but larger surfaces can form more directional emitters and receivers. Since the effect depends on size relative to the wavelength, quite small objects can strongly influence the propagation and reflection of higher frequencies. A human shout needs a cliff or at least a large wall to give a good echo but a post-card acts as a specular surface (a mirror) to a sound of say 50 kHz. Again this is not a special property of ultrasound itself but is a direct consequence of the frequency involved.

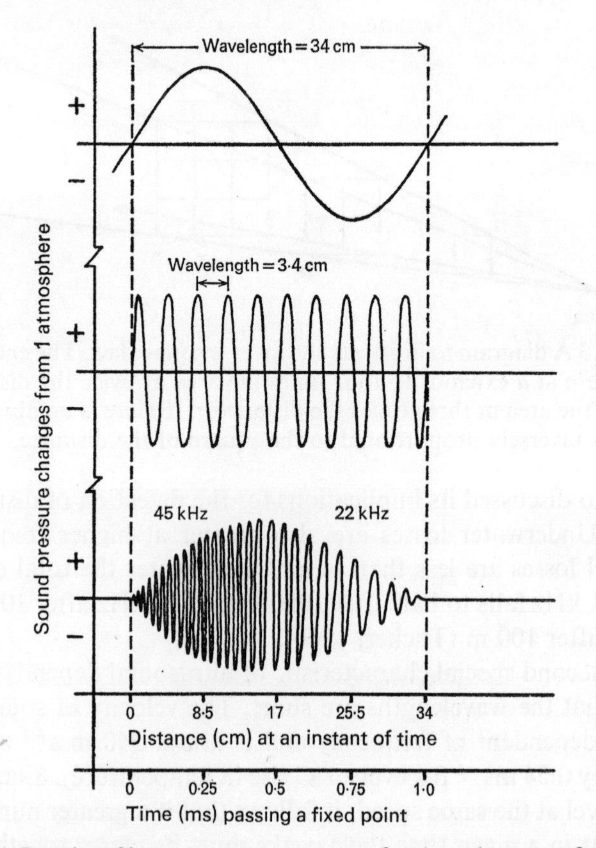

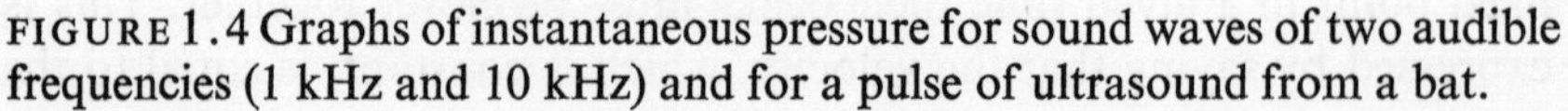

FIGURE 1.4 Graphs of instantaneous pressure for sound waves of two audible frequencies (1 kHz and 10 kHz) and for a pulse of ultrasound from a bat.

Ultrasound is not uncommon in the natural environment, indeed it is present at quite high intensities in human speech. Dental sounds (tt-tt) and especially sybilants (ssss and sh-sh) contain energy up to very high frequencies and it is often said that it is better to speak in undertones rather than whispering when observing small mammals. A brief hiss makes tame rats give an immediate startle reaction and some apparently silent door-hinges have a similar effect. Other common sources of man-made ultrasound are the jingling of keys or coins, some motor-car brakes, tearing of paper, striking matches, clicking the finger-nails and even a slight stuffiness of the nose. In general, however, the world is quieter at ultrasonic frequencies than at audible ones, probably because of atmospheric attenuation. A jet engine presumably produces a great deal of ultrasound but low-flying aircraft do not interfere with ultrasound recordings made in the open. Similarly waterfalls and lightning emit strong sounds of high frequency but they cannot be detected at a short distance even where conditions are ideal for propagation.

Another natural influence on ultrasound is fog which strongly absorbs high frequencies. Small spherical drops of water vibrate by changing their shape at frequencies which depend on their diameters. They should therefore absorb sound of their own 'resonant' frequency and re-radiate it slowly over a long time. Pye (1971) calculated the resonant frequencies for the range of drop sizes measured in fogs by meteorologists and found that they coincide with the spectrum of frequencies used by bats. Thus although fog scatters all wavelengths of visible light and is impenetrably white to us, to a bat it should appear impenetrably dark with perhaps a faint 'afterglow' due to the re-radiated sound. This probably explains why bats seem to avoid flying in fog, for it would be equivalent to 'flying blind'. A similar situation occurs in water with air-bubbles, although these are compressible and can resonate by changes in volume as well as by changes in shape (Richardson, 1952). Screens of bubbles have been used by engineers and physicists to absorb ultrasound underwater, for instance when reflections from the walls of a test-tank interfere with acoustic measurements.

Thus there is nothing unusual or uncommon about ultrasound and its properties are exploited by many different animals for different purposes. The aim of this book is to review these uses of ultrasound and to discuss the special problems of research into the field.

CHAPTER TWO

Methods of Detection and Analysis

The first item in any set of acoustic instruments is a microphone. This acts as a transducer, that is it converts the sound waves impinging on it into electrical waves that can then be examined in a number of ways by electronic instruments. Broadly, three different sets of electronic equipment have been used for studying animal ultrasounds: the oscilloscope, the tape-recorder and the ultrasound detector or 'bat detector'. These different instruments will be reviewed in turn.

ULTRASONIC MICROPHONES

Microphones for low frequency sounds use a wide variety of different principles but few of them are capable of responding at ultrasonic frequencies. The first detection of ultrasound from bats (Pierce and Griffin, 1938; Griffin and Novick, 1955) used microphones based on a piezo-electric crystal. These make splendid hydrophones for use underwater but in air they are very insensitive unless they are specifically manufactured to respond only over a very narrow frequency range.

A tremendous advance was therefore made when in 1954 Kuhl, Schodder and Schröder invented the solid-dielectric capacitance microphone that is very sensitive in air over a wide frequency range and which is described below. These microphones are quite easy to make and have been used for the majority of studies of animal ultrasound up to the present. The only competitor came with the commercial introduction of small air-dielectric microphones by Bruel and Kjaer of Denmark. These have very flat responses, up to 100 kHz in the quarter-inch (6·3 mm) model and to 140 kHz in the eighth inch (3·2 mm) model. The sensitivity is very stable so that they can be used to measure sound intensities and are individually calibrated by the manufacturers. But they are easily damaged, require stabilized power supplies and are less sensitive than the cheaper, rugged, solid-dielectric forms whose only disadvantage is that their sensitivity fluctuates somewhat with time. The choice between these two types then depends on the application: for field work the solid-dielectric type is better but for intensity measurements the air-dielectric type is essential. The operation of these two instruments will be described briefly.

A capacitor is an electrical device consisting of two 'plates' of conducting metal, connected to wire leads and separated from each other by a layer of insulating material, the dielectric. A familiar example from school physics is the Leyden jar in which the plates are tin mugs, one fitting inside and one outside a glass jar. When a capacitor is connected to a steady voltage a current flows but the charges are unable to cross the dielectric gap and so they accumulate on one plate and are depleted on the other. This gives a voltage difference between the plates that opposes the flow of current. When the plates reach the applied voltage, the current is finally reduced to zero and the capacitor is said to be charged to this voltage. The number of charges required to reach this state for a given dielectric material varies inversely with the distance between the plates and directly with the area of the plates and the applied voltage. If, therefore, the plates are moved closer together, a further current flows and a larger charge accumulates until a new equilibrium is reached, or if they are moved apart the charge is excessive and current flows back to the source of the charging voltage. Placing a large resistor in one lead slows down the flow of current so that the voltage between the plates momentarily changes when the plates are moved.

This is the basic action of a capacitance microphone. The diaphragm of the instrument consists of a thin, flexible sheet of metal, generally earthed by being clamped at its periphery to the casing. It forms one plate, the front-plate, of the capacitor and close behind it, but insulated from it and from the casing, is another disc of metal, the back-plate, which is connected to a high positive or negative voltage through a large resistance. Alternating sound pressures bend the diaphragm in and out by a minute amount and so cause a fluctuation in the voltage between the plates.

In the air-dielectric microphone the gap between the diaphragm and the back-plate is a thin layer of air (Fig. 2.1a). The metal diaphragm must be very thin if it is to respond to high frequencies, and as it is supported only at its periphery it is very susceptible to damage and cannot easily be replaced. In the solid-dielectric microphone the diaphragm incorporates the dielectric; it consists of a thin film of plastic (the dielectric) coated with a layer of aluminium (the front-plate) on its outer surface (Fig. 2.1b). This is stretched across the actual surface of the back-plate which therefore gives considerable mechanical support. The diaphragm can even be touched while in use and, although it is remarkably tough, if it should be torn it can be replaced in a few minutes. The

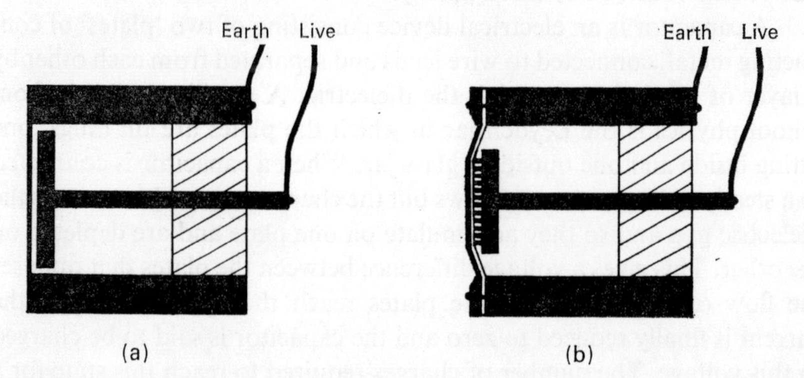

FIGURE 2.1 Diagram of a section through two different types of capacitance microphones. (a) Air-dielectric type: a flat, metal backplate behind a self-supporting metal diaphragm. (b) Solid dielectric type: a grooved, metal backplate pressed against a thin, plastic film with aluminium coated on the outer side. For details of operation see text.

best material, not available to Kuhl, Schodder and Schröder but now used almost universally, is Mylar, or polyethylene terephthalate, known in fibre form as Terylene. It is cheaply available as films of various thicknesses, often with the aluminium layer already on one side. Films 3·5 μm thick can stand a polarizing (charging) voltage of about 120 V and the tougher 6 μm film gives as great a sensitivity if it is charged to about 200 V. These voltages can be obtained from a stack of dry cells or from a very simple, compact generator described by Pye (1968).

A flat back-plate would not allow the diaphragm to move so it is shaped in one of a number of ways. Kuhl, Schodder and Schröder either sand-blasted the surface to leave it irregularly rough or cut a series of circular grooves 0·25 mm wide and 0·5 mm deep on its surface. Matsuzawa (1958) found that the best method was to drill a large number of tiny holes in the back-plate. Pye and Flinn (1964) used a hair-spring from an alarm clock to form a spiral rail; this was cemented to a plastic disc which obviated an insulating liner inside the metal case of the microphone. Another arrangement used a disc of fine copper mesh and several other possibilities suggest themselves for trial.

The only problem with the capacitance microphone lies in the need for a rather special pre-amplifier which must be close to the microphone itself and is usually mounted in the same casing. The actual charge on the capacitor is very small and can do little work without leaking away. Even a few feet of screened cable needs a small amount of current to charge and discharge its own capacitance (two wires separated by insulation) and forms a relatively heavy 'load' for the microphone if connected directly to it. Similarly the pre-amplifier must draw only a minute current from the microphone. Originally this was achieved by a small thermionic valve connected as a cathode follower; the voltage change was not increased but the valve was able to drive the cable and subsequent amplifiers without loss. Valves, however, are wasteful of power and may need careful mounting because they also respond to direct vibration. Junction transistors, being 'current operated' are unsuitable for the job but in the nineteen-sixties the emergence of cheap field-effect transistors (fets) solved the problem admirably. A simple microphone pre-amplifier with a voltage gain of 10 was described by Pye (1968) and has proved entirely satisfactory.

CATHODE-RAY OSCILLOSCOPES

The voltage change produced by a capacitance microphone in response to sound is very small, at most a few millivolts, but the signal from the pre-amplifier can easily be increased by a transistor amplifier with an adequate frequency response. It can then be displayed on a cathode-ray oscilloscope. This instrument consists of a screen, like that of a small television set, on which a small spot of light appears. This spot can be deflected vertically and horizontally by electrical means. In practice the spot is usually moved steadily from left to right and then flicked back again by an internal circuit, called a time-base because it corresponds to a horizontal time-axis on a graph. The signal voltage to be displayed is then made to deflect the spot up and down, thus giving a complete graph of voltage (and therefore sound pressure) against time.

At moderate time-base speeds, say one sweep per 10 ms (ten milliseconds or ten thousandths of a second) the horizontal deflection appears to be a steady line because the eye fuses 100 sweeps per second; the vertical movements overlap due to the diameter of the spot and so a single short sound is seen fleetingly as a broadening of the baseline. Due to persistence of phosphorescence on the screen and to visual persistence in the eye, even very short sounds of high frequency can readily be seen. The individual waves can only be resolved on the screen if faster time-base speeds are used; 100 kHz generally requires a sweep time of 0·5 ms or less. This can give a confusing picture if the sound lasts longer than one sweep because the successive traces appear to overlap in an unco-ordinated way. If the sound is of constant frequency and amplitude, however, it is possible to synchronize the time-base so that the sweep

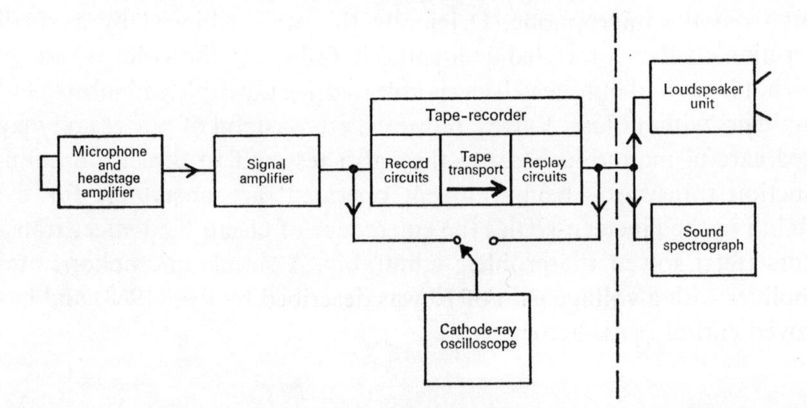

FIGURE 2.2 A block diagram of the apparatus used to record and analyse ultrasonic signals.

lasts for exactly a whole number of cycles and successive traces are exactly superimposed to give a steady picture from which the sound frequency may be measured.

The oscilloscope is useful in a number of ways. It gives an indication that ultrasounds have been received by the microphone and, as it responds to all frequencies present, it has been used in conjunction with a tuned bat detector to make sure that nothing was missed through incorrect tuning (Sales, 1972a). With a Bruel and Kjaer microphone and a calibrated amplifier, the oscilloscope gives a ready indication of sound intensity since the height of the trace can be related directly to sound pressure at the microphone.

An oscilloscope is also essential to the tape-recording of ultrasonic signals (Fig. 2.2), first by indicating when it is worth starting the high-speed tape transport so that tape is not wasted during silent periods, and secondly by acting as a voltmeter to ensure that the recorder receives a signal of suitable amplitude. Most recorders are fitted with a meter to monitor the recording level but many ultrasounds are very short and the needle has insufficient time to reach its correct position; it thus gives an average reading that takes no account of the high peak values that may be present. The action of the oscilloscope (or of a 'magic eye' indicator) is practically instantaneous and serves splendidly as a monitor.

In the early days, before about 1960, the oscilloscope provided the only way of analysing ultrasounds. The 'waveform' on the screen was photographed and the trace could then be measured with a ruler to assess the frequencies and durations of the sounds. The interpretation of complex traces, with multiple harmonics for instance, was often uncertain for the appearance of the waveform is greatly dependent on the relative timing (phase) of different components, while different harmonic patterns can look extremely similar. The method is also technically difficult because of overlapping traces as discussed above. This could be partly overcome by moving the film steadily so that the traces were separated out, but for details of the waveform the sounds had to be sampled by photographing single sweeps with a still (or cine) camera and ignoring the intervening traces. Nevertheless tremendous advances were made in the study of bats by these methods (Griffin and Novick, 1955; Novick, 1958a, b). Griffin even studied fishing bats in the field by mounting the whole apparatus in a dugout canoe at the end of a long power line and has given a vivid account of this exploit (Griffin, 1958).

TAPE-RECORDING AND ANALYSIS

In a tape-recorder the electrical signal from the microphone is converted to a varying current in the winding of an electromagnet (the recording head) with a very narrow gap between its poles. A magnetic tape, drawn through this changing field at a steady speed, retains the signal as a varying state of permanent magnetization along its length. When the tape is again drawn past an electromagnet, a fluctuating current is induced in the winding and after amplification this can be used to drive a loudspeaker to reproduce the original sound. For complicated reasons a faithful recording is only obtained if the recording head is also magnetized by a high frequency current called the bias current.

The frequency of this must be at least twice the highest frequency to be recorded.

Designing a tape-recorder with a frequency response that covers ultrasounds is not easy and domestic recorders cannot be modified. Specially constructed machines did not appear until about 1960 and they are still very expensive because they must be engineered to very high standards. There are several good reasons for this. A large number of waves in a second must either be 'written' very close together on the tape, which needs tape of high quality and an extremely small gap between the poles, or else the tape must be moved at very high speed. The first recorders to reach 100 kHz used tape speeds of 60 in s^{-1} (152 cm s^{-1}) or even 120 in s^{-1}. Gradually the manufacturers have been able to reduce the gap between the magnetic poles to only 1 μm or so, allowing 150 kHz to be recorded at only 30 in s^{-1}, that is five thousand waves per inch of tape or two thousand waves per centimetre. Even at this speed it is difficult to ensure that the tape travels smoothly and steadily, so that the transport mechanism must be made to very high standards. In addition to having a minute pole-gap, the electromagnetic heads must be able to take a very high frequency of bias current, about 500 kHz, and this also makes them expensive.

Consequently an ultrasonic tape-recorder costs as much as a 'superior quality' motor car but it does offer tremendous possibilities for studying ultrasound. Not only is the whole signal preserved for subsequent examination but a whole range of standard audio-frequency techniques can then be applied. If the tape is replayed at a much reduced speed the sound is drawn out in time and all frequencies are reduced in proportion. Thus 100 kHz replayed at 1/100 of the original speed is reduced to 1 kHz and is clearly audible through a loudspeaker. This lower frequency signal is much easier to photograph from an oscilloscope screen and can also be analysed by a sound spectrograph, or sonagraph, which does not by itself cover ultrasonic frequencies.

There are several kinds of analysis that can be produced by a sonagraph but by far the most common in general use is called a sonagram. The sonagraph re-records a short example of sound on the edge of a magnetic drum. It is then replayed about 500 times through a narrowly tuned filter. As the filter is gradually tuned to higher frequencies a pen moves up a sheet of sensitive paper wrapped around the drum and records the output of the filter. When the paper is removed from the drum it thus presents a graph of frequency (vertically) against time (horizontally) with sounds represented by darkening of the paper. Examples of this display are shown in Figs. 3.2–3.5 and 7.1–7.7. In conjunction with a photograph of the waveform from an oscilloscope, the sonagram gives a very clear picture of the acoustic 'structure' of the sounds (Plate III).

Another form of display that will be presented in Chapter 5 is a close approximation to the total spectrum of the recorded sample. For this purpose the sonagraph must be slightly modified as will be described elsewhere (Pye, in preparation). The total sound energy at each frequency is averaged and arranged to draw a line of proportional length. The result is that the ends of the lines form a graph of sound energy against frequency without taking any account of changes in frequency with time. Examples are given in Plate X.

BAT DETECTORS

Although the tape-recorder and its associated instruments can give very detailed analyses, the process is very time-consuming. Even for listening with a loudspeaker the sounds must first be recorded, the tape wound back and then replayed at reduced speed. Quite short recordings need a long listening time and silent intervals can seem interminable. The ultrasound detector or 'bat detector' forms a much cheaper way of listening to ultrasound while it is being emitted and also of determining many of its characteristics.

The bat detector modifies the electrical signal from the microphone so that when an ultrasonic sound is received an audible sound is simultaneously produced by a loudspeaker or earphone. The process of transformation varies in different instruments (see below) but in each case different ultrasounds produce characteristic noises from the loudspeaker and, if the instrument is tuned, the original frequency can be determined. Naturally there are drawbacks: a complete frequency analysis can only be performed by altering the tuning and so rests on the assumption that the ultrasounds are unchanging; this is often, but not always, true. The detector is also unable to resolve temporal relationships so that it cannot distinguish between upsweeps and downsweeps or between a sweep with harmonics and a single deeper sweep. Nevertheless much valuable work can be done with a detector only and it is perhaps the most basic single tool for research on ultrasound. It can be made by anyone who is handy at constructing radios and at least one type is available commercially at a reasonable cost.

Three basically different forms of detector have been developed and they all depend on two simple principles used in radio reception: the envelope detector and the heterodyne detector. A description of each instrument should make these principles clear.

The most straightforward instrument is the broad-band envelope detector developed and described by McCue and Bertolini (1964). Here

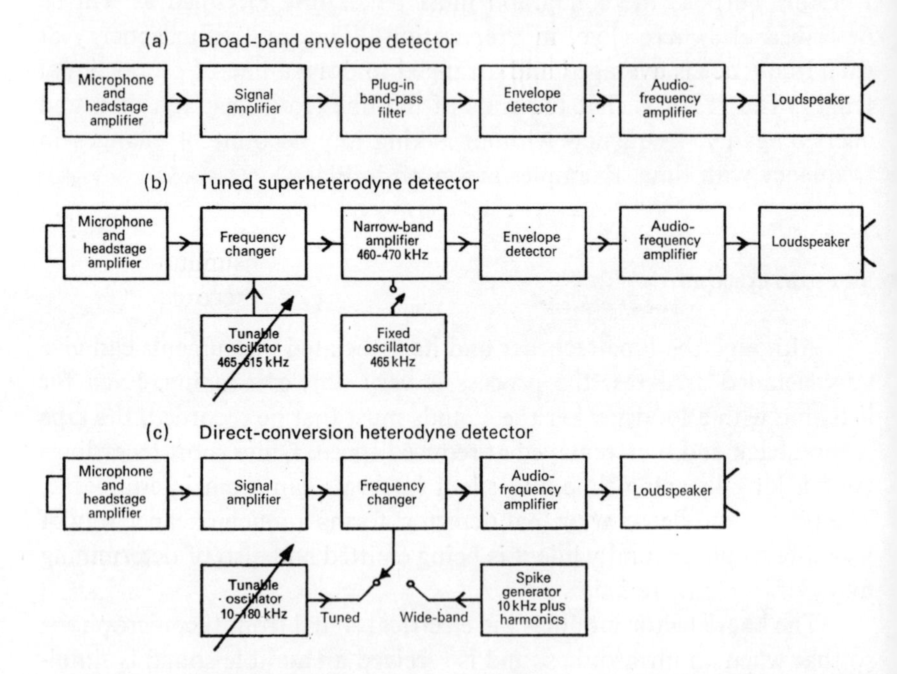

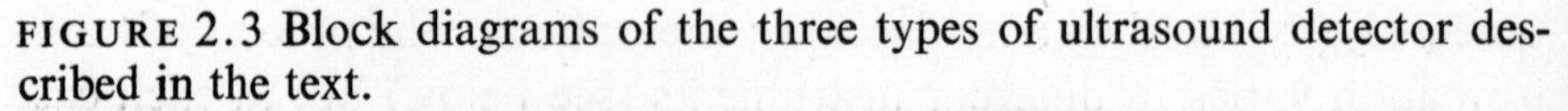

FIGURE 2.3 Block diagrams of the three types of ultrasound detector described in the text.

the microphone signal is amplified by an amplifier which responds to all frequencies although a set of plug-in filters can be used to restrict the frequency range if desired. The signal is then detected by a diode rectifier or envelope detector (Fig. 2.3a). This simple circuit eliminates the actual high frequency waves and instead responds only to their total amplitude, jumping from the peak of one wave to the peak of the next. On an oscilloscope it thus traces the outline of an 'envelope' that could just be fitted to one side of the 'packet' of waves (Fig. 2.4a). This envelope signal is then amplified and fed to a loudspeaker.

Obviously the envelope signal is very much lower in frequency than the original waves and for many ultrasounds it is clearly audible. Short bat pulses of less than 4 ms duration, for which this detector was originally designed, produce a sharp click which can be heard or recorded on a simple tape-recorder. The train of clicks from a flying bat allows the pulse repetition rate to be measured. Flight manoeuvres are clearly dis-

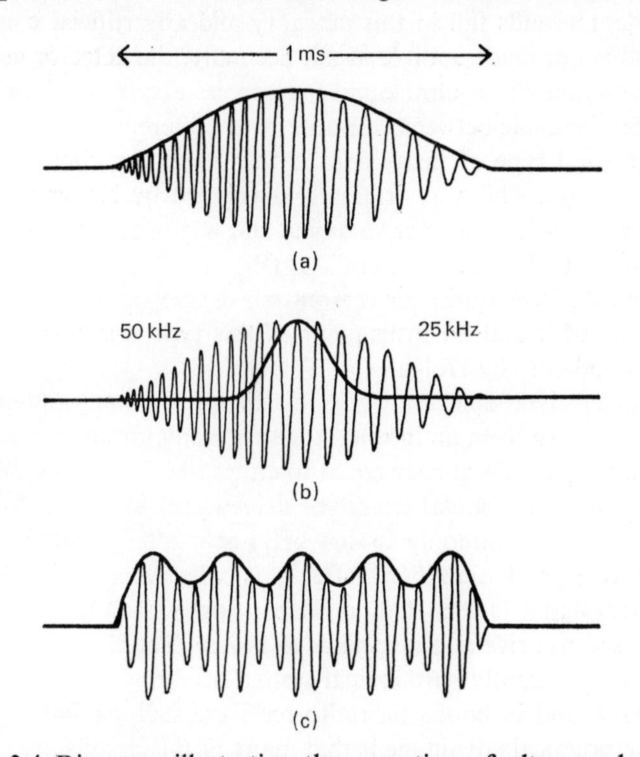

FIGURE 2.4 Diagrams illustrating the operation of ultrasound detectors. (a) Broad-band envelope-detector. The heavy line shows the response to a short bat pulse with a frequency sweep; this response would be clearly audible through a loudspeaker. (b) Heterodyne detector tuned to the centre frequency of the same pulse; tuning to high or lower frequencies would give weaker clicks earlier or later respectively. (c) Heterodyne detector tuned to a longer pulse of constant frequency and amplitude; the addition of a similar frequency from a beat-frequency oscillator produces amplitude fluctuations resulting in a pure-tone response of lower frequency.

tinguished by a rise in the repetition rate and during the terminal phase of prey interception the rate is often over 100 pulses per second. Clicks at this rate are fused by the ear into a brief 'buzz', a term that is now often applied to the original bat signal.

This detector also works very well for most insect signals because the envelope is usually very 'spiky' and produces a clearly audible sound.

This pattern of detector has two drawbacks however. First, the envelopes of very long pulses produce dull 'thumps' from the loudspeaker instead of clicks; if the pulses start and decay gradually, the main energy of the envelope may be too low in frequency to make a very distinct sound. Many rodent sounds fall in this category and an artificial continuous ultrasound is not made audible at all. Secondly, the detector gives very little information about ultrasound frequencies except for the comparison that can be made between responses with different band-pass filters.

The second type of instrument is the tuned superheterodyne detector (Fig. 2.3b). This was originally developed by Noyes and Pierce (1938; Pierce, 1948, gives later versions) and was used for the first detection of ultrasound from bush crickets (Pierce, 1948) and bats (Pierce and Griffin, 1938). A transistor version was developed by Pye and Flinn (1964; Pye, 1968) and an instrument of this type was then marketed quite independently by Holgates of Totton Ltd.

The heterodyne detector works by mixing the microphone signal with another wave from an internal tunable oscillator so that a new signal is produced at a frequency equal to either the sum or the difference of the first two. In the superheterodyne detector the ultrasonic frequency is actually raised (commonly to 464 kHz) before being amplified by a tuned, narrow-band amplifier. After this selective amplification the heterodyned signal is then fed to an envelope detector and the new envelope used to drive a loudspeaker as before. The theoretical advantages of this apparently paradoxical system need not be discussed here and can be found in books on radio receivers such as Witts (1934 to 1961). One practical advantage is that many of the circuits are identical to those of domestic radios and can even be purchased as ready-made 'building bricks', or a complete radio can fairly easily be modified to make a bat detector.

To use this detector it is not necessary to understand its principles, only the way in which it behaves. It responds only to sound within a narrow frequency band, about 10 kHz wide, which can be tuned by a knob (controlling the internal oscillator) to any frequency from 10 kHz to over 150 kHz. By changing the tuning, the frequency spectrum of the original signal can be roughly measured. Thus the short, frequency sweep pulses of a vespertilionid bat may give weak clicks at 80 kHz on the dial, loud clicks at 50 kHz and weak ones again at 35 kHz (Fig. 2.4b).

A further advantage of the superheterodyne detector is that long pulses can be clearly detected. Since only a narrow frequency band is amplified at a fixed frequency, a second continuous signal of similar frequency can be added from another internal oscillator, the beat-frequency oscillator or b.f.o. (Fig. 2.3b). The envelope detector then responds to beats between the two, giving a clearly audible difference tone (equal in frequency to the difference between the amplified signal and the b.f.o.) as long as the ultrasound stays in tune (Fig. 2.4c). Thus constant frequency bat pulses are effectively 'translated down' to any desired audible frequency by adjusting the tuning knob. Slowly drifting frequencies of rodents or some bats produce audible tones of varying pitch. As long as the frequency changes are not too great, the superheterodyne detector also allows one to search for harmonics at simple ratios of any detected frequency. Thus a number of details of frequency patterns can be revealed.

Although the tuned superheterodyne detector gives more information than the wide-band detector and can respond to any ultrasonic signal, it has the great disadvantage that it does not respond at all if it is tuned to the wrong frequency. Important signals can therefore be missed altogether. Comparative tests with both instruments suggested that a combination would be best of all; it is also desirable to be able to heterodyne any constant or slowly changing frequency without having to tune to it. These aims have now been achieved by the direct-conversion heterodyne detector (Pye *et al.*, in preparation) which forms the third type of bat detector.

In the direct-conversion principle the internal, tuned oscillator beats directly with the microphone signal to produce an audible difference frequency which is then selectively amplified before going to a loudspeaker (Fig. 2.3c). A second oscillator and an envelope detector are no longer necessary; constant frequencies produce an audible note directly and short pulses still produce clicks. Although this seems, and is, much simpler in principle than the superheterodyne plan, it is somewhat more difficult to achieve in practice. The problems have now been overcome and it is interesting that the same arrangement is now gaining favour for radio reception too (Hawker, 1972a, b). The two bat detectors are indistinguishable in use but the direct-conversion instrument can very easily be untuned so that it responds to all frequencies at once. A switch simply replaces the tuned oscillator by a 'spike generator' that produces a sharply spiked waveform ten thousand times a second (Fig. 2.3c). This signal represents a fundamental wave of 10 kHz together with all its harmonics at 20 kHz, 30 kHz and so on up to at least 180 kHz. Thus no incoming signal can be more than 5 kHz from one of these harmonics and an audible difference frequency must be produced. In this mode the instrument corresponds to a collection of 15–18 tuned instruments adjusted so that their frequency ranges just overlap and no signal can be missed. During field trials in 1972 it satisfied a long-standing ambition; while listening to short-pulse vespertilionid bats with greatest energy at 40–50 kHz, it detected a passing greater horseshoe bat with a constant frequency of 83 kHz. Using the envelope detector the horseshoe bat might well have been overlooked and a superheterodyne detector tuned at 40–50 kHz would certainly have missed it altogether.

It is hoped that this instrument will shortly be available commercially. A simpler form, operating only in the untuned mode, has also been developed and may form the basis for a really cheap bat detector.

CHAPTER THREE

Bats

There can be little doubt that of all animals using ultrasound the bats come first to mind and are also the best studied. Indeed many of the techniques described in the previous chapter were originally developed for the study of ultrasound in bats, and the availability of these techniques led in turn to the discovery of most of the phenomena to be described in later chapters of this book.

The majority of bats have poor eyesight (but see Suthers, 1970), yet they fly at high speed in the dark, often through complex environments and many live exclusively on small insects intercepted in mid-air. In 1794 Lazzaro Spallanzani first experimented on the guidance mechanisms of of vespertilionid and rhinolophid bats (see Galambos, 1942b; Dijkgraaf, 1960). He found that complete darkness or even blinding made no difference to the bats' ability to detect fine wires in their path but that deafening completely destroyed this ability. The experiments were repeated by others (see Griffin, 1958) and Jurine (unpublished) found that the bats were seriously disorientated if plugs were placed in their ears but not if small, open-ended brass tubes were put there instead. Other scientists were sceptical about these experiments, considering that the observers were mistaken or that excessive sensory damage had been produced by the operations, some of which were undoubtedly very cruel. Rollinat and Trouessart (1900) and Hahn (1908) repeated some of these experiments. Hahn used a regular barrier of wires. He calculated the probability that the wires could be avoided by chance and showed that the bats did much better than this. But it was not until 1920 that Hartridge suggested that bats use echoes from their own, still undetected, ultrasonic sounds and it was another 18 years before technology allowed these sounds to be discovered.

The ultrasonic signals of bats were first detected by Pierce and Griffin in 1938. Griffin with a number of collaborators went on to elucidate the phenomenon of echo-location, and published the work in an important series of papers (e.g. Griffin and Galambos, 1941; Galambos and Griffin, 1942). In 1958 he summarized this work and its historical background in a fascinating monograph and much further information has emerged since then. Before the subject is described in detail, it may help to examine the animals themselves and their ways of life so that the demands made upon their senses can be better appreciated.

THE BIOLOGY OF BATS

It is commonly supposed that bats show little variation beyond a major division into 'fruit bats' and 'insectivorous bats'. It is true that, being the only mammals capable of powered flight and hence highly specialized, they form a very cohesive group, instantly recognizable by their unique wings formed from enormous webbed 'hands' and extending to the hind legs and the tail if present. But so slight is the contact between bats and man that few people realize that bats form a highly complex group with a wide range of habits. There are about 800 different species, grouped into 17 families (Table 3.1), a diversity among mammals that is second only to that of the rodents. They show several unusual phenomena reviewed briefly by Pye (1969) and a major, modern review of some aspects of the 'Biology of Bats' has been published in two volumes by Wimsatt (Editor, 1970).

The affinities of bats with other mammals are difficult to establish because the fossil record is scanty. The oldest known bat and also one of the very few complete fossils was found in the early Eocene strata deposited 50 million years ago, and yet it resembles present-day bats except for a few trivial details (Jepson, 1966, 1970). During the Eocene times other groups of mammals were very different from today, for horses were only represented by the terrier-sized *Eohippus* and man's ancestors were still rather like lemurs. Nevertheless it seems certain that bats evolved from an insectivore stock, probably quite close to that of the primates, and despite appearances bats are therefore closer to man than most other major groups of mammals. The diversity of present-day bats is shown by their external appearance, especially their faces (Plate II), by their roosting sites and by their feeding habits. All these features are related to some extent to their ability to fly and the need for a flight guidance system.

Faces of course bear vocal transmitting and auditory receiving structures which may be highly specialized and will be discussed later. The nose-leaves, however, give an immediate clue to the identity of many bats. As will be seen from Table 3.1, the 16 families of the Microchiroptera are divided into four super-families and this is done mainly by the nature of their shoulder articulation. Two super-families have simple noses (Plate II b, f) and shout their ultrasound through open mouths, while the other two, one entirely from the Old World and the other only from the New World, have various forms of nose-leaf (Plate IIc–e and Fig. 3.6) and hum ultrasonically through their nostrils. There are exceptions to this rule, which will be discussed in a later section, but they are few in number.

The faces of the simple-nosed bats are not so easily distinguished although each family has its own facial characteristics which, together with those of the wings and tail, permit immediate identification. The Pteropidae are distinctly dog-like in appearance (they include the flying foxes) with a long muzzle, large eyes and ears that are tubular at their bases (Plate IIa) while almost all have a claw on each index finger. The Rhinopomatidae are instantly recognizable by the extremely long, whip-like tail extending far beyond the interfemoral membrane. In both the Noctilionidae and the Emballonuridae, however, the tail is short and the tip is free of the flight membrane, but the former are characterized by deep jowls (Plate IIb) and the latter have glandular sacs in the forewing or under the chin.

Tails and interfemoral membranes are also useful in distinguishing between the Old World nose-leaf bats, in addition to variations in the nose-leaf itself. The Nycteridae have no true nose-leaf (Fig. 3.6b), the nostrils lying in a slit bordered by two pairs of fleshy pads. They have long legs and the last vertebra of the tail is T-shaped, apparently to help in supporting the large interfemoral membrane, although the Megadermatidae have just as large an interfemoral membrane and no tail at all. The tail and interfemoral membrane of the Rhinolophidae and Hipposideridae are unremarkable except that at rest they are folded flat against the back. These two families are distinguished by the nose-leaf. This is complex but remarkably constant in the Rhinolophidae, which have a sculptured 'leaf' on the forehead, while in the Hipposideridae it is either simple above the eyes or has a varied structure, often with three pointed lobes forming a 'trident' (Fig. 3.6).

TABLE 3.1 The families of bats.

A. **Sub-order Megachiroptera:**

Family Pteropidae: a single family with about 130 species throughout the warmer parts of the Old World. Most are 'fruit bats' but many eat flowers and pollen and some are specialized for drinking nectar. All fly visually and only *Rousettus* appears to use echo-location.

B. **Sub-order Microchiroptera:** divided into four super-families. All use ultrasonic echo-location.

1. Super-family Emballonuroidea: simple-nosed bats with simple shoulder joints.
   - (i) Family Rhinopomatidae: four species of 'mouse-tailed' or 'long-tailed' desert bats from the Sahara to Sumatra. Insectivorous.
   - (ii) Family Emballonuridae: about 40 species of 'sheath-tailed' bats distributed throughout the tropics. Insectivorous.
   - (iii) Family Noctilionidae: two species of 'bulldog' bats that hunt over water in the neotropics. One catches fish, the other is apparently insectivorous.

2. Super-family Rhinolophoidea: Old World nose-leaf bats. Semi-specialized shoulder joints.
   - (i) Family Nycteridae: about 10 species of 'slit-faced' bats, one from the Far East and the rest from Africa and the Middle East. Insectivorous.
   - (ii) Family Megadermatidae: five species of 'false vampires' from Africa to Australia and the Philippines. Insectivorous and carnivorous, hunting small vertebrates.
   - (iii) Family Rhinolophidae: about 50 species of 'horseshoe' bats throughout tropical and temperate parts of the Old World. Insectivorous.
   - (iv) Family Hipposideridae: about 40 species of 'leaf-nosed' bats from the Old World tropics, closely related to, but more varied than, the Rhinolophidae. Insectivorous.

3. Super-family Phyllostomidea: New World nose-leaf bats. Semi-specialized shoulder joints.
   - (i) Family Phyllostomidae: about 140 species with very diverse habits and diets, including insects, vertebrates, fruit, flowers, pollen and nectar. Mainly tropical.
   - (ii) Family Desmodontidae: three species of vampires. Blood-drinking parasites; one attacks only mammals, the others mainly attack birds. Neotropics.

4. Super-family Vespertilionoidea: simple-nosed bats with specialized shoulder joints.
   - (i) Family Natalidae: about 10 species of 'long-legged' or 'funnel-eared' bats from the neotropics. Insectivorous.

TABLE 3.1 (*contd.*)

(ii)	Family Furipteridae: two species of 'thumbless' bats, otherwise similar to the Natalidae. Neotropical and insectivorous.
(iii)	Family Thyropteridae: two species of 'disc-winged' or 'sucker-footed' bats from the neotropics, otherwise resembling the Natalidae. Probably insectivorous.
(iv)	Family Myzopodidae: one species of rare 'disc-winged' or 'sucker-footed' bat from Madagascar but not closely related to Thyropteridae. Probably insectivorous.
(v)	Family Vespertilionidae: about 280 species of 'common' bats, distributed throughout tropical and temperate parts of the world. Mainly insectivorous but at least two fishing bats and some may be partly carnivorous.
(vi)	Family Mystacinidae: one species of 'short-tailed' bat from New Zealand. Insectivorous.
(vii)	Family Molossidae: about 80 species of 'mastiff' bats or 'free-tailed' bats found throughout the tropics and warmer temperate regions of the world. Insectivorous.

The Phyllostomidae have very variable tails and interfemoral membranes, from no trace of either to a large membrane with the tail partly free above it. The nose-leaf is also variable from a small, simple 'spear' (Plate IId) to a large and elaborate, fleshy leaf. The Desmodontidae have a small interfemoral membrane, no tail and a fleshy, pad-like nose-leaf (Plate IIe); they are also characterized by very large, chisel-like incisor teeth whereas most other bats have small incisors.

The families of the Vespertilionoidea lack a nose-leaf and are recognizable by a variety of features, although the Vespertilionidae themselves are difficult to characterize. Like the Natalidae, they have a tail running to the edge of the interfemoral membrane but generally lack the funnel-shaped lips and ears and the enormous expanse of the wing, supported by very long legs and tail. The Molossidae have heavy-looking ears and jowls (Plate IIf and Fig. 3.7) and a tail that projects beyond the interfemoral membrane when at rest, although the membrane can slide up to the tip of the tail during flight. The other four minor families have restricted distributions and peculiar features as suggested by their common names. Two of these have suckers at the wrists and feet which are also found in some vespertilionids and in all these cases they are used for clinging to smooth surfaces.

The actual wings of bats have been little mentioned so far but although the arrangement of the hand is essentially the same in all bats there is considerable variation in its proportions. This is clearly related to the manner of flight and is of little taxonomic use. Slow, flapping flight is accompanied by large, rounded wings (Pteropidae, Rhinolophidae) while high speeds are achieved by bats with long, pointed wings (Emballonuridae, Molossidae). Many bats can hover, for example the nectar-feeders (Pteropidae or Phyllostomidae) as well as some insectivorous bats that take insects from solid surfaces (Nycteridae, and *Plecotus* of the Vespertilionidae). *Natalus* flies skilfully and quite rapidly on a very large area of membrane whereas *Mimetillus* (Vespertilionidae) whizzes about on tiny wings with a very rapid beat.

The feeding habits of bats are closely related to the demands they make on their echo-location systems. The variety of diets is roughly indicated in Table 3.1 and includes all imaginable possibilities for a purely aerial group of animals (Pye, 1969). Within each kind of diet there is a great deal of specialization at the species level. Insectivorous bats hunt for insects of different kinds in different ways and at different heights, from non-flying insects on the ground or on water surfaces to insects flying as much as 3000 m above the ground. Carnivorous bats take different kinds of prey and one phyllostomid (*Trachops*) specializes in catching geckos, while the three vampires select different hosts. Even the fruit bats fly at different levels and seek out different kinds of fruit and flowers (Jones, 1972).

Finally the nature of the roost site is to some extent determined by the method of flight guidance, for a bat must be able to 'find its way home'. Thus the safety of deep, completely dark caves would seem to demand acoustic or some other form of non-visual guidance. The oil-bird, *Steatornis*, and the swiftlets, *Collocalia* (described in Chapter 8), are able to roost in such caves only because they have developed a form of echo-location and species of *Collocalia* without this ability build their nests on open cliffs (Medway, 1967). A similar position is seen in the Pteropidae. Most species fly visually and must roost on trees in the open; but *Rousettus* which roosts only in caves is able to echo-locate and, like the birds, when flying in darkness uses a continuous train of clicks audible to man. A number of other pteropids also roost in caves but they stay near the entrance, although *Notopteris* has been said to penetrate completely dark regions without producing any detectable sound or ultrasound.

Although all the Microchiroptera examined so far (nearly a third of the species from all but two families) are able to fly in complete darkness by ultrasonic echo-location, not all of them roost in caves. In most parts of the world caves are few and far between, whereas in all warm regions bats are extremely abundant. Thus few species are habitual cave-dwellers and others find alternative roosts in foliage, under bark, in clefts between rocks and similar situations.

THE ECHO-LOCATION SIGNALS OF BATS

It should now be clear that in discussing the echo-location of bats one is not dealing with a small, obscure, homogeneous group of animals but with a large assemblage of species with very diverse habits and varied navigational requirements. Bats seek their food in many different ways, mostly using sound because they are active only at night (some reasons why bats are only nocturnal have been discussed by Pye, 1969). It should not be surprising therefore to find that echo-location behaviour is also diverse. Human radar systems have now been developed to work in many different ways for different purposes (e.g. Skolnik, 1962) and the same is true of echo-location in bats; indeed the parallels between the two are often very close. Radar engineers have found that different types of emitted signal are best for obtaining different kinds of information and just the same kinds of signal, using ultrasound instead of radio-waves, can be found in different bats. Something is now known about the signals of 180 of the 800-odd species of bats and these represent 15 of the 17 families (all but the Furipteridae and Myzopodidae), while over 100 species of 14 families have been recorded on magnetic tape by Pye (in preparation). These signals fall into three main categories: short clicks, frequency-sweep pulses and constant frequency pulses, with mixtures of the last two. The next four sections will discuss each of these in turn.

Short-pulse echo-location in Megachiroptera

As pointed out above, the majority of the Megachiroptera-Pteropidae fly without any known form of echo-location. Although a fairly representative number of species have been examined, only the genus *Rousettus* has been found to produce orientation sounds. Indeed *Lissonycteris*, which was formerly regarded as a rather doubtful sub-genus of *Rousettus*, has been removed to form a separate genus by Lawrence and Novick (1963), largely because careful tests showed that it is completely incapable of echo-location. By contrast another sub-genus,

Stenonycteris, has been said to produce sounds in flight like those of *Rousettus* (Kingdon, 1974).

Rousettus shares the superb night vision of other Pteropidae and this sense will be described briefly. It has been reviewed for the genus *Pteropus* by Neuweiler (1967) and for the family as a whole by Suthers (1970). The eyes of all Megachiroptera are large so that the iris can take in as much light as possible; as this depends on the absolute dimensions, the eyes of small Megachiroptera (and some are quite small despite the name) are proportionally larger than those in larger forms. The eye also has a lens of short focal length, giving a smaller but brighter image, and the retina shows several special adaptations. There are no cone cells but the rods are packed very densely and the surface area of the retina is increased by being folded over a series of projecting hillocks. A large number of rods are connected to a single optic nerve fibre, giving greater sensitivity while degrading the discrimination of fine details. But behavioural experiments have shown that this poor visual acuity is maintained even at very low light intensities whereas the acuity of man and other diurnal animals is better in a good light and much worse in dim light. Far from being 'as blind as a bat', the Megachiroptera have night vision that is probably not surpassed by any other animals.

No eye can function in absolute darkness, however sensitive it is, and in pitch darkness among strange surroundings most fruit bats can only blunder about, literally blindly. But *Rousettus* can fly skilfully under such conditions. Since Möhres and Kulzer (1956a) first established that in *Rousettus aegyptiacus* this is achieved by acoustic means, the sonar of these bats has been studied in some detail and two more species have been examined by Novick (1958b). When flying, or even when excited while at roost, *Rousettus* emit a series of high-pitched, but clearly audible clicks, rather like small pebbles being rattled together. In this they resemble the birds *Steatornis* and *Collocalia*, described in Chapter 8, and differ markedly from all microchiropteran bats which produce pulses containing a pure wave or a harmonic series, nearly always without any major audible components.

The way in which *Rousettus aegyptiacus* produces these clicks was examined by Kulzer (1958; 1960). The muscular lips of *Rousettus* extend far back on the jaws (Plate IIa) and can be parted in a peculiarly inane 'grin' through which the clicks apparently emerge. Kulzer showed that the larynx is not involved and that the sounds are produced in the mouth by clicking the tongue, first on one side and then on the other. The clicks are therefore produced in pairs. The intervals between pairs of pulses ranged from 138 ms to 435 ms, giving a maximum of about seven pairs of pulses per second. Kulzer also found that the interval between clicks in each pair was variable from 20 ms to 44 ms with a regular mean at 30 ms. Pye (unpublished) has remeasured this interval from an ultrasonic recording of several hundred pulses produced by the same species in flight. The inter-pulse interval was analysed as an interval-histogram by playing the tape into a Biomac 1000 computer and the results showed a narrow peak at 20 ms with the spread effectively only from 18 ms to 22 ms. These bats were recorded in Uganda whereas Kulzer obtained his bats from Egypt, so there may be a geographical variation here.

The waveform of each click has been analysed by Möhres and Kulzer (1956a), Novick (1958b) and Pye (1967a and unpublished). In shape it is roughly saw-toothed, starting at high amplitude and decaying rapidly and approximately exponentially over several cycles until it disappears into the noise level (Fig. 3.1a). Louder pulses thus show a greater number of waves and it is not possible to give a definite figure for pulse duration; however waves are seldom detectable more than 1–2 ms after the start. Novick described a second 'packet' of waves closely following the first, but this may have been due to echoes.

Because each sound has a very sudden onset, the spectrum shows a wide spread of frequencies (see also Chapter 5) extending from about 10 kHz to about 50–60 kHz in both *Rousettus aegyptiacus* and *Rousettus amplexicaudatus* (Fig. 3.1). Novick (1958b) showed a similar lower limit in *Rousettus seminudus* but the upper frequencies were attenuated by the microphone then used. Wide-band tape-recordings of the first two species now show that the sounds are rather complex, with no single 'carrier' frequency but a number of major components that vary from pulse to pulse (Fig. 3.1). Pulses of the two species are almost indistinguishable.

Pye (1967a) suggested that the waveform represents the decay of a shock-excited resonator, perhaps due to the mouth cavity resonating when the tongue clicks. This is in accord with Kulzer's conclusions and with what happens when the human tongue is clicked; varying the shape of the mouth cavity gives clicks of different pitch and some people can play tunes by tapping their teeth with a pencil and varying their mouth in this way. Such a mechanism in *Rousettus* would account for the 'triangular' waveform decaying exponentially; for the complexity of components, since the mouth is not a simple cavity; and possibly for the variability between consecutive pulses if the mouth changes its configuration slightly.

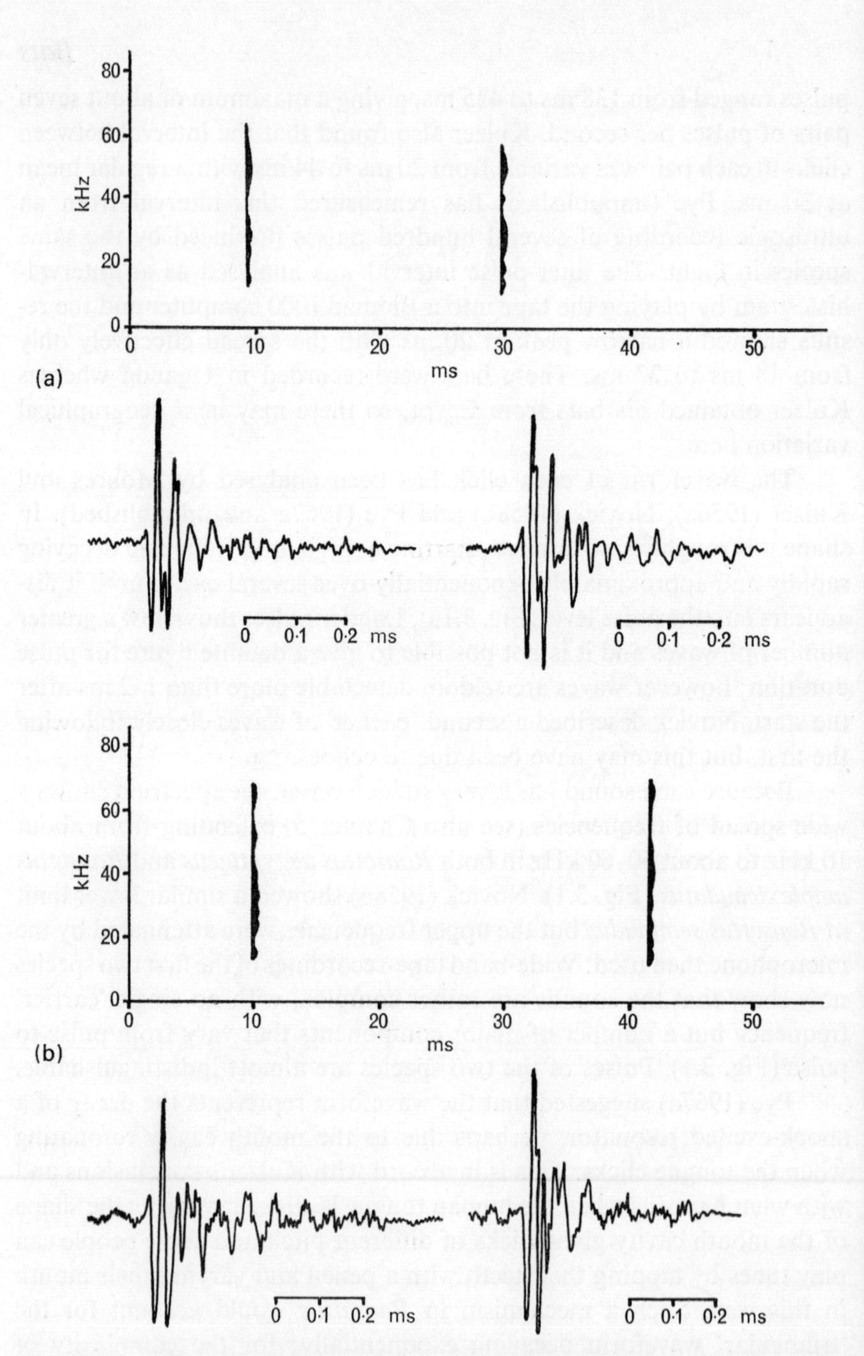

FIGURE 3.1 The echo-location clicks of *Rousettus*. (a) *Rousettus aegyptiacus*. (b) *Rousettus amplexicaudatus*. The upper figure in each case shows a sonagram of a pulse-pair, the lower figure shows the expanded waveforms of the same pulses.

An experimental approach to this theory has now been made by Roberts (1973b) who recorded *Rousettus amplexicaudatus* both in air and in diver's gas in which the nitrogen and other heavy, non-respiratory gases of air are replaced by helium. In light gases, such as helium or hydrogen, the velocity of sound is greater and since the wavelength of a resonant cavity is determined by its dimensions and is therefore fixed, the resonant frequency rises. In a respiratory mixture of 20% oxygen and 80% helium the resonant frequency of any cavity should increase by a factor of 1·86, or nearly an octave. Roberts analysed statistically the major components of a large number of clicks recorded in each medium and found that the frequencies increased as predicted.

The ability of *Rousettus aegyptiacus* to navigate by acoustic means in complete darkness was examined by Griffin, Novick and Kornfield (1958). They made a single specimen fly in a large darkened room across which was hung a barrier of vertical rods spaced 53 cm apart, or about two-thirds of the wingspan. The accuracy of the bat in flying through this barrier was then measured for different diameters, from wire of 0·28 mm to cardboard tubes of 25 mm. The bat was highly successful with 3-mm wires and its performance declined with smaller diameter wires although it remained better than chance even with wires of 0·46 mm diameter. By contrast, the vespertilionid bat *Myotis lucifugus* in a similar situation could avoid wires of only 0·12 mm (Curtis, 1952, cited in Griffin, 1958). The performance of *Rousettus* was completely degraded by the presence of filtered white (wide-band) noise of high intensity whether this extended upwards from 15 kHz or downwards from 25 kHz. Altogether these results suggest that *Rousettus* uses a wide range of frequencies but that lower frequencies than those of *Myotis* are especially important.

The hearing of *Rousettus* has been examined in several ways (explained in a later section) and the results are curiously paradoxical. In 1968 Pye (Brown and Pye, 1974) measured the cochlear microphonic potentials at the round window of three *Rousettus aegyptiacus* and found that the response showed a sharp peak of sensitivity at 12 kHz. This result was examined more carefully and confirmed by Brown (1973c), using two independent methods for stimulating the cochlea and analysing its responses. From the peak at 11–12 kHz the response fell steeply on both sides, reaching the noise level 30 dB below at 5 kHz and at 40 kHz. In one case a small second peak was indicated at 30–35 kHz, but otherwise the slopes were smooth.

This agrees well with some experiments on acoustic trauma in the same species reported by A. Pye (1971). After exposure to intense pure tones of different frequencies, the cochleae of guinea-pigs and *Rousettus aegyptiacus* were micro-dissected and examined for damage to the sensory hair-cells in the organ of Corti. In guinea-pigs suitable exposure times produced discrete damage at various points along the cochlea which depended on the frequency used. This was as expected, but in *Rousettus* the results were quite different. Despite prolonged exposure to intense sounds at 20 kHz, no cochlear damage was produced and of two specimens exposed to 4 kHz only one showed partial damage in the second turn. But with tones of 8 kHz to 15 kHz extensive damage occurred and at 10 kHz almost complete destruction could be produced by exposure times as short as 3 min. It thus looked from both these studies as if *Rousettus aegyptiacus* has a cochlea that is especially sensitive to 10–12 kHz, the bottom end of its emitted spectrum, with very little response at higher frequencies.

A very different picture has been given by Grinnell and Hagiwara (1972b) who examined the evoked potentials at the inferior colliculus of the mid-brain in *Rousettus amplexicaudatus* from New Guinea. They found a broad response from 10 kHz to 100 kHz with sensitivity increasing gradually up to 50 kHz and then decreasing rather rapidly. This response was exactly comparable with that of five other genera of small, non-echo-locating fruit bats from the same area of New Guinea, and all six showed comparable abilities to discriminate tone pulse stimuli in the presence of continuous masking background tones. *Rousettus* did differ from these other forms in showing a much more rapid recovery following a response to the first of paired stimuli, as might be expected in an echo-locating species. As the clicks emitted by *Rousettus aegyptiacus* and *Rousettus amplexicaudatus* are almost indistinguishable, the contrast between a highly specialized, low frequency cochlear response of one and a broad, unremarkable collicular response peaking two octaves higher in the other is startling and suggests that further work on the hearing of these bats would be of great interest.

This discrepancy is also functionally important for understanding the mechanism of echo-location in these bats. Radar theory shows (see Appendix) that the ability to resolve two targets side by side at the same distance, what we may call angular resolution, depends on the wavelength of the transmitted signal. Since higher frequencies have shorter wavelengths, they are capable of better angular resolution and also give stronger echoes from small targets. The strength of the echo from a cylindrical wire thus depends on its diameter and the frequency of the sound used. The finding by Griffin, Novick and Kornfield (1958), that *Rousettus aegyptiacus* can only avoid wires about four times the diameter of those detected by *Myotis lucifugus*, suggests that *Rousettus* is using about four times the wavelength. Since *Myotis* has its main energy at about 50 kHz (see next section), a figure of 12 kHz for the frequency actually used by *Rousettus* would seem entirely reasonable. This rather over-simplified argument is only offered tentatively, however, and a firm conclusion must await further investigations.

Rousettus is therefore able to echo-locate quite well and uses this ability to roost in the safety of dark caves. It is not known to what extent it uses acoustic guidance outside the caves or in feeding.

Frequency sweep pulses in Microchiroptera

The most obvious way to define the position of an object is to establish its distance in a known direction: its range and bearing. This is the way in which the majority of man-made radars and sonars work although the following section will describe an alternative mode of operation. The accuracy with which distance can be measured is called the range resolution and it is important not only for location but also for the extent to which a cluster of two or more small objects at slightly different distances can be separately distinguished instead of being 'seen' as a single, deeper object.

Radar theory shows (e.g. Skolnik, 1962; and Appendix) that range resolution depends on the bandwidth of the emitted signal and in the early days of radar this was achieved by making very short pulses of the transmitted carrier wave. The connection between range resolution, pulse length and bandwidth is not obvious but it can be explained in two stages. First, it is clear that a long pulse reflected by two objects at slightly different ranges will give two echoes that overlap at the receiver and cause confusion; the echoes can only be separated clearly, and so resolved, if the physical length of the pulse as it travels is less than twice the difference in range between the two targets, so that there is a gap between the echoes when they are received. Thus short pulses give better range resolution. Secondly, since a pure tone or single frequency is continuous and of infinite duration, a short pulse must be 'less pure' than a longer one and therefore contains a wider range of frequencies. Without going into the details, it is the bandwidth that is important in range discrimination and there are other ways of achieving a wide bandwidth. The actual duration of the pulse only matters if the argument is limited to square pulses of constant carrier frequency, as is assumed in many simpler books about radar. The pulses of *Rousettus* achieve a wide bandwidth with only a sudden onset and the tail of the pulse decays gradually; this pattern will clearly give good range discrimination even if the echoes do overlap to some extent.

All short pulses have a drawback, however. The maximum range or sensitivity of an echo-location system depends upon the total energy in the pulse and a shorter pulse must therefore have a greater amplitude to maintain the total energy. Since the maximum output amplitude of any transmitter is limited, engineers have searched for a method of increasing the bandwidth without shortening the pulse. In 1960, Klauder, Price, Darlington and Albersheim investigated the theoretical advantages of a relatively long pulse of moderate amplitude but high total energy in which the frequency of the carrier wave is swept upwards (or downwards) from start to finish, a pattern that they called 'chirp radar'. Two overlapping echoes in such a system can be distinguished because the earlier one has risen (or fallen) somewhat in frequency by the time the second one arrives. The frequency sweep pulse, used with suitable receivers, is capable of very good range resolution and great sensitivity for a given maximum amplitude at the transmitter.

In view of these advantages, it should not be surprising to find that the majority of microchiropteran bats also use a frequency sweep system (Plate III). The first bats investigated by Pierce and Griffin in 1938 were members of the family Vespertilionidae (*Myotis lucifugus* and *Eptesicus fuscus*) and the frequency pattern of their signals could not be assessed accurately with Pierce's tuned detector. In a series of later papers Griffin and his various collaborators (Griffin, 1950 and reviewed in 1958), soon showed that each pulse was frequency modulated, starting at a high frequency and falling about an octave by the end. The pulse duration was 2–4 ms and the maximum amplitude occurred about halfway through the pulse. In most cases the sound wave within the pulse was fairly 'pure' although traces of second harmonic could be detected towards the end when the frequency reached the bottom of its sweep (Plate IIIa, b). In *Myotis lucifugus* a typical pulse started at 78 kHz and swept down to 39 kHz in 2·3 ms, reaching its maximum amplitude at about 50 kHz (Fig. 3.2), while *Eptesicus fuscus* swept from 50 kHz to 25 kHz in 2·7 ms with its maximum amplitude at about 40 kHz. In both cases the sweep covered just one octave, a factor of two in frequency. Application of the formulae given by Klauder, Price, Darlington and Albersheim (1960) shows that such pulses are respectively capable of 91 times and 68 times the range resolution of pulses with the same duration but with a constant frequency and a 'square' shape (Pye, 1963).

When Griffin examined the sounds of a *Myotis* or an *Eptesicus* as it was about to land or to negotiate a barrier of wires in flight, he found that the pulse repetition rate rose dramatically and the nature of the pulses changed. A 'cruising' bat emits pulses of the kind described above at a rate of 8–15 pulses per second but whenever a tricky manoeuvre is required, the repetition rate rises as high as 150–200 pulses per second, over about 0·5 s. At first these rapid pulses are grouped, perhaps four, then six, then eight, but the final group may contain an unbroken train of thirty pulses within about 160 ms at a steadily accelerating rate (Fig. 3.2b, f). With an ultrasound detector the cruising pulses are heard as a rapid train of discrete clicks, but the much more rapid train of clicks during a manoeuvre cannot be separated by the ear and form a brief 'buzz'. This sequence is also produced during the interception of prey, as discussed below, and can readily be elicited by tossing a pebble in the flight path of a hunting bat while listening to it with an ultrasound detector. Through familiarity with this effect, the name 'interception buzz' has come to be applied to the rapid sequence of pulses although strictly it should only be used for the response of the detector as interpreted by the human ear.

Even at these very high pulse rates, each pulse is a discrete and beautifully formed frequency modulated sweep but the nature of the pulse changes somewhat. As the rate accelerates, the pulse duration becomes shorter and its amplitude decreases. This gives a marked reduction in pulse energy but presumably the bat is temporarily more concerned with the accurate location of one nearby target than with the detection of more distant ones. The pulse also sweeps over a smaller range at lower frequencies, often with the appearance of weak second and even third harmonics that were not previously appreciable. At the end of an interception buzz, the pulses of *Myotis* may last only 0·25 ms and sweep from 30 kHz to 20 kHz (Fig. 3.2f). As is shown in the Appendix, these pulses give a more accurate range resolution despite their apparently narrower bandwidth. This is gained at the expense of maximum range (and also of velocity information; explained later, and see Cahlander, 1967; Altes and Titlebaum, 1970) but under the circumstances this presumably does not matter.

In 1955 Griffin and Novick described the pulses emitted by a variety of neotropical bats from four other families. Later Novick visited several

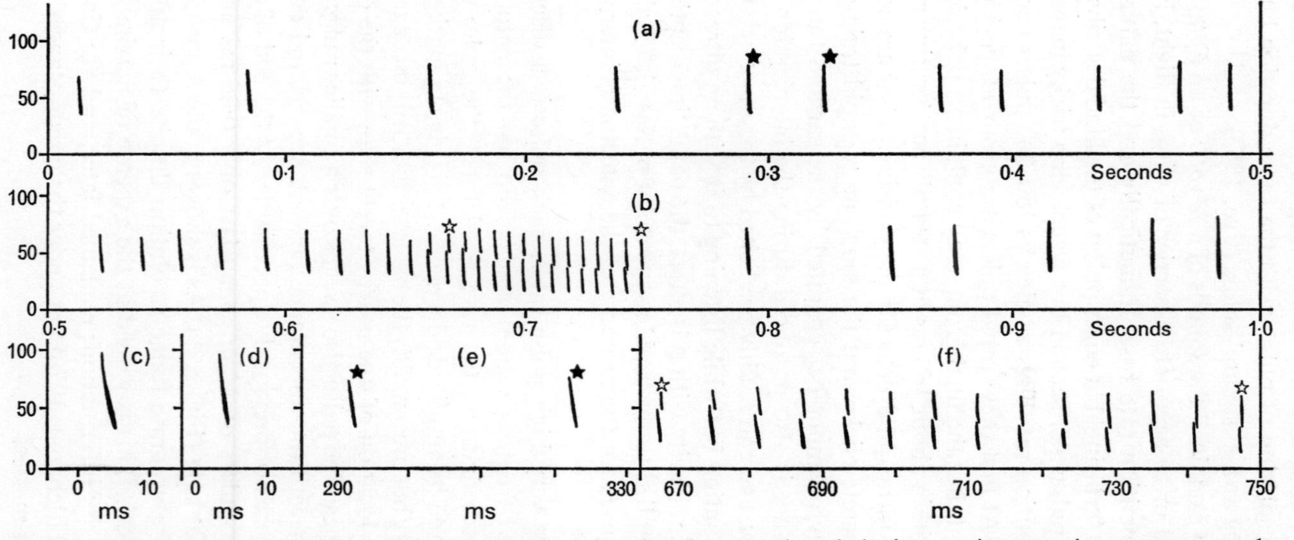

FIGURE 3.2 Sonagram analysis of frequency-modulated pulses produced during an interception manoeuvre by *Myotis lucifugus*. (a–b) Continuous analysis of one second of recording. (c and d) Two very loud pulses to show full extent of sweep. (e) Two approach-pulses (★ in (a)) on an expanded time scale. (f) Terminal part of the buzz sequence (☆ in (b)) on an expanded time scale.

places in Asia and Africa and described the pulses of a number of species from another five families from the palaeotropics (1958a), some of which had been examined in less detail by Möhres and Kulzer (1956b). Since then further comparative work on these groups has been done, especially by Novick (1962, 1963a), Möhres (1967a; Neuweiler and Möhres 1967) and Pye (1967a, but largely unpublished). The great majority of these bats use frequency-modulated pulses, at least under certain conditions, although in many cases the pulses are more complex than those of the Vespertilionidae. The following description of complex pulses with frequency modulation applies generally to some Emballonuridae, the Rhinopomatidae, the Nycteridae, the Megadermatidae, most Phyllostomidae, the Desmodontidae and the Mystacinidae, although the details differ somewhat in many cases. The Natalidae and some Vespertilionidae, such as *Plecotus*, emit a pulse intermediate between these and the simpler pulses of *Myotis* (Plate IIIc, d).

The complex pulses are rather short with a rich harmonic content extending up to quite high frequencies (Plate IIIe, f). Four families emit their complex pulses through a nose-leaf and in all these cases the fundamental-frequency component is weak, a feature that was apparently first noticed by Roberts (1973a, b). In the others, the fundamental is stronger but it is not the predominant component of the pulse since much or most of the energy is at higher frequencies. A typical pulse may last for 0·5 ms, starting with a fundamental of 30 kHz and harmonics at 60 kHz, 90 kHz and 120 kHz. The main energy at this time may lie in the second harmonic. As the frequency sweeps downwards, all the components remain in simple arithmetic proportions by frequency so that at the end of the pulse the fundamental might be 20 kHz with harmonics at 40 kHz, 60 kHz and 80 kHz. But the frequency of principal energy remains the same so that the second harmonic becomes weaker, the third becomes the main component and the fourth becomes somewhat stronger. Towards the end of such a pulse even the fifth or sixth harmonics may appear, while the fundamental, if present at all, gets weaker and may disappear.

In some cases the sweeps of the fundamental and harmonics are apparently the same but a narrower range of frequencies is actually emitted, so that there may be no more than two or three components present at a time. When this happens, the transfer of energy from one harmonic to another often results in a pulse with two amplitude peaks at the same frequency but separated in time to give a ∞-shaped envelope (Plate IIIf). Such pulses are difficult to interpret on the oscilloscope screen and it is especially difficult to tell which harmonics are present, say the second and third or the third and fourth. The harmonic structure can be clearly established by analysing tape-recordings with a sound spectrograph (sonagraph), but it may vary between different pulses of the same bat although the actual frequency of the main energy is remarkably constant.

Another complexity in pulses of this type is that on the oscilloscope screen the different components of the waveform are seen to change their relative timing or phase, so that they are strictly not quite harmonics of a common fundamental frequency. This may be an artefact introduced by the phase response of the microphone or of other instruments such as the amplifier or tape-recorder, but this does not seem to be the whole explanation and leads to a consideration of how these pulses are produced, which will be discussed in the next section of this chapter.

Complex pulses are nearly always of much lower intensity than those of vespertilionids. In the first case examined, the phyllostomid *Carollia perspicillata*, Griffin and Novick (1955) found some difficulty in detecting these sounds with the equipment then available. Many of the pulses reached only a level of 1·3 μbar r.m.s. (76 dB) a few cm from the bat, roughly equal to the conversational level of the human voice in quiet surroundings, and led to the designation 'whispering bats'. The simple pulses of *Myotis* may reach an amplitude of 36 μbar r.m.s. (105 dB), roughly equal to the sound level close to an unsilenced road-drill and 1000 times the intensity of *Carollia*. Some other vespertilionids are even louder: Novick (1958a) recorded the highest values measured so far as over 1000 μbar peak-to-peak (350 μbar r.m.s. or 125 dB) at 10 cm in some species. The loud vespertilionid *Nyctalus noctula*, with longer, slower sweeps can be heard with a detector at 100 m when hunting and some molossids can probably be heard at greater distances, but even now *Carollia* can only be detected clearly at a few metres and then only if the bat is facing the microphone.

As in vespertilionids, changes also occur in the sounds of complex pulse bats during interception and pursuit manoeuvres but here they are much less marked. The repetition rate rises to similar very high levels but although the pulses shorten even further in some cases, the frequency of the main energy does not change appreciably.

The discovery by Curtis (cited by Griffin, 1958) that *Myotis* can detect a barrier of wires down to a diameter of 0·12 mm has already been mentioned in a previous section. In 1958, Grinnell and Griffin showed that *Myotis* could detect a barrier of 3-mm wires from at least 2 m and wires of only 0·18 mm were often detected at 1 m. Griffin and Grinnell (1958; Griffin *et al.*, 1963) showed that similar abilities in *Plecotus* were adversely affected by the presence of extraneous white noise covering all relevant ultrasonic frequencies. Grumman and Novick (1963) also performed wire-avoidance tests on the phyllostomid *Macrotus mexicanus*. This bat, which emits multiple-harmonic sweeps through a nose-leaf, could easily detect wires of 0·27 mm and performed better than chance with wires of 0·19 mm. Möhres and Neuweiler (1966) found that *Megaderma lyra* could detect plastic threads as small as 0·08 mm, again using short, multiple-harmonic pulses emitted through a nose-leaf.

In order to see how the natural behaviour of bats is influenced by their echo-location, one must go back to Spallanzani in 1794 (see Galambos, 1942b; Dijkgraaf, 1960), for one of his experiments was crucial to Griffin's development of the subject. Spallanzani had blinded some bats by surgery and released them again in the open. Some time later he recovered some of these specimens from their normal roost and found that they were well fed. Later, when Griffin's experiments had shown beyond all reasonable doubt that microchiropteran bats navigate and avoid obstacles by echo-location, it was natural to wonder whether they also find their prey by this means. So in 1951 the study was extended into the field, with some difficulty since only mains-powered equipment was then available (Griffin, 1958). The results were again clear. When *Myotis* or *Eptesicus* swerved or dived to catch an insect or to inspect a pebble tossed up by an observer, the bat produced a full interception buzz. It seemed most unlikely that the bat would produce such a barrage of sound if it were in fact locating its food by listening passively to the wing-beat of the prey. Observations of many insectivorous species of different families over the last 20 years suggest that echo-location is not dispensed with and probably forms the major means of detecting aerial prey and intercepting it.

Later, some *Myotis* and other species were persuaded to hunt for insects indoors so that the actual manoeuvres could be photographed by multiple-exposure techniques (Griffin *et al.*, 1960; McCue, 1961; Webster and Griffin, 1962). This subsequently led to the production of cine-films slowed down by 64 times and accurately synchronized to an ultrasonic tape-recording of the orientation pulses. The results were intriguing. The bats appeared to predict the course of the prey and adopted a true interception course rather than a mere pursuit course. It is not always possible for the bat to capture the insect in its mouth and more commonly it is 'scooped up' in the tail membrane before being transferred to the mouth. In many cases where the prey is finally off to one side, the bat nets it in one wing (Plate IVa) and transfers it first to the interfemoral membrane and finally to the mouth. The bat thus uses the flight membrane to increase its catchment area and to allow for some inaccuracy in interception. But the bat's task is not always an easy one and the story of the discovery that many nocturnal insects listen for bats at ultrasonic frequencies and take avoiding action will be told in the next chapter of this book.

Webster and others photographed bats flying indoors in order to examine the discriminating abilities of the bat's system with moving targets (Webster, 1963, 1966; Webster and Brazier, 1965). The bat learned to fly regularly over an apparatus that could toss a mealworm into its path in front of a multiple-exposure camera. If a cluster of 8–15 mealworms was tossed up at once, the bat was well able to pick one out without being confused, and could catch a mealworm tossed up with some small nylon balls. Observations were also made of the bat's ability to catch insects close to large obstacles or spiky foliage.

Simmons (1968, 1970, 1971; Simmons and Vernon, 1971) has used a completely different technique to obtain very detailed information on the echo-location abilities of blinded bats. He trained individual bats of two species, *Eptesicus fuscus* and *Phyllostomus hastatus*, to sit on a platform and to distinguish between two targets before flying to one of them for a reward; flight to the other target was unrewarded. The two targets differed in various ways (range, size or shape) in different experiments and their positions were interchanged randomly between left and right so that the bat was forced to examine them carefully before choosing which to fly to. Perhaps the most revealing experiment investigated the bat's ability to choose the nearer target of the two. The further target was set at 30 cm, 60 cm or 120 cm from the bat but in each case the other target was up to 10 cm nearer. *Eptesicus* gave almost 100% correct responses down to 3 cm difference, falling to 75% correct at 1·5 cm and 50% correct (random choice) at 0 cm. *Phyllostomus* scored almost 100% at 2 cm difference but about 78% at both 1·0 cm and 0·5 cm before the score fell further to 50%. The bats' ultrasonic signals were recorded on tape and processed by a special computer (an autocorrelator) to derive an indication of their maximum possible range resolution by radar theory. Theoretical curves obtained in this way for the two species exactly fitted their observed performance, even to the ambiguity at 0·5–1·0 cm in *Phyllostomus* which is probably due to the complex waveform of this bat's pulses. Thus the bats are able to measure target range as accurately as their pulses can allow, and their auditory echo-processing system must conform in some way with an ideal (or correlation) receiver matched to the pulse structures used.

Constant frequencies in Microchiroptera

Although the polar co-ordinates of direction and range may seem the most obvious ones to use in any radar or echo-location system, an alternative scheme that produces much useful information measures instead the relative velocity of targets. This works by a phenomenon known as the Doppler shift. Whenever an observer moves towards a sound source the frequency received is higher than that transmitted and the faster the movement the greater the change. The same effect occurs if the transmitter is moving towards a stationary observer or if both move towards each other; it is the relative movement that matters. If the two move further apart, then the observed frequency is less than that transmitted. In the first case the later sound waves have less distance to travel and so 'catch up' with the earlier ones, giving a shorter wavelength and so a higher frequency, while in the second case they have further to go and so lag behind somewhat, increasing the wavelength.

If the 'observer' merely reflects the received sound back to the transmitter, as happens when a relatively moving target reflects an echo to a radar or echo-location system, the Doppler shift occurs twice and the effect is doubled. A comparison of the transmitted frequency and the frequency of the echo thus gives a measure of the velocity of the target, or in other words of the first derivative (rate of change) of range. Such a system is more useful than might appear at first sight and finds many applications in man-made systems, not only for catching the speeding motorist but for parking giant tankers gently or for distinguishing flying aircraft from stationary ground clutter and rain clouds. Passive observation of the Doppler shift from a satellite's radio transmitter gives sufficient information to plot its orbit since both velocity and the nearest range can be measured. Without going into the details, a range-measuring radar can estimate relative velocity from successive measurements of range, and conversely an integration of velocity can, under certain circumstances, give the remaining range (Pye, 1967b). Furthermore the Doppler system does not require a wide bandwidth, only a long pulse, preferably of constant frequency so that as much noise as possible can be filtered off (Appendix). Narrow filtering in turn leads to high sensitivity so that a given transmitted power can be used to detect

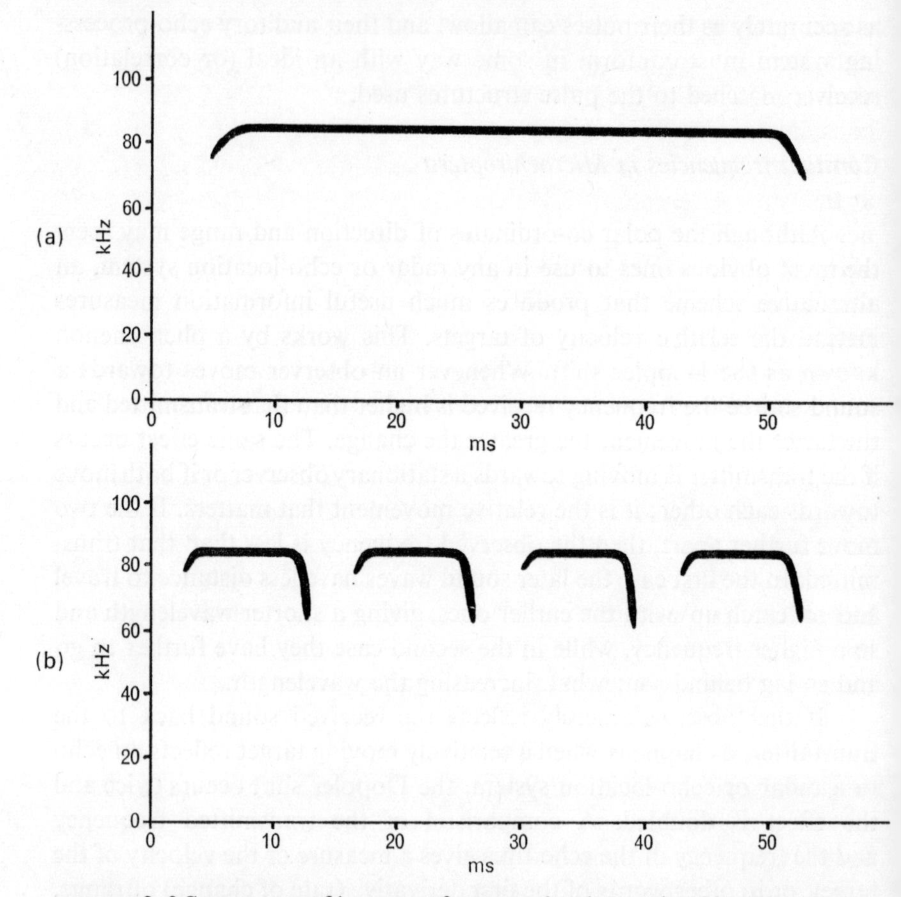

FIGURE 3.3 Sonagrams of 'constant frequency' pulses emitted by *Rhinolophus ferrumequinum*. (a) A cruising pulse. (b) Part of a rapid train of shorter pulses. The fundamental components are too weak to show and the lines represent the predominant second harmonic.

targets at longer distances. It is not surprising, therefore, to find that some bats operate highly successfully in the Doppler mode

The first, and still the best, example known of the use of Doppler in bats is that of the Rhinolophidae or horseshoe bats. The echo-location of these bats was first investigated by Möhres in the late nineteen-forties and described by him in a very important paper in 1953. The pulses emitted by rhinolophids have since been analysed by several other workers and are quite unlike any of those previously known or discussed so far. In an unexcited, cruising or resting animal they last up to 80 ms and have a fairly constant, very high amplitude except for rather gentle slopes in the first few and last few milliseconds. The note is almost pure although there are weak components at 0·5 and 1·5 times the frequency, showing that it is really the second harmonic. The frequency is very constant throughout most of the pulse but it sweeps down by about 15 kHz in the last 2–3 ms and in many pulses it also sweeps up somewhat at the very beginning of the pulse (Fig. 3.3a). When the bat is excited, negotiating an obstacle or landing, the pulse repetition rate rises from about 6–10 s^{-1} to as high as 60–80 s^{-1}. The pulses shorten at this time, sometimes to less than 10 ms (Fig. 3.3b) but the pattern of constant frequency followed by a terminal sweep is maintained.

Pye (1967a) and Schnitzler (1967) showed that either the maximum amplitude may occur during the constant frequency, with the terminal sweep falling rapidly in amplitude (sweep-decay pulses), or the constant frequency may have a relatively low amplitude, and the sweep may rise to a high amplitude peak before decaying (peak-sweep pulses). Schnitzler showed that the sweep-decay pulses are typical of a bat at rest while the peak-sweep pattern is typical of the short pulses produced at higher repetition rates as when a bat is about to land. It appears that the bat is then relying more on range information and less on Doppler information.

That rhinolophids change their frequency somewhat from pulse to pulse is immediately obvious when listening to a bat with a heterodyne detector. This instrument effectively lowers the constant frequency to an audible level but transposes changes in frequency intact. The note produced by a flying horseshoe bat clearly shows changes due to the Doppler shift as it flies to and from the observer but it also fluctuates independently of this. Schnitzler (1967) measured the velocity of bats flying down a long room to a landing platform and observed the pulses they emitted. In 1968 he showed that the constant frequency is varied by the bat so as to compensate exactly for the Doppler shifts of the echo received from the landing platform. As the bat accelerated at the start of its flight, the frequency actually emitted by the bat fell about 2·5% below its value at rest and remained low until the bat started to slow down again for landing. Schnitzler also recorded the pulses of a resting *Rhinolophus* while a pendulum swung slowly in front of it; the constant frequency varied from pulse to pulse by as much as ±0·4% of its resting value to compensate very closely for the varying Doppler shifts of echoes from the pendulum.

Despite this last observation, Schnitzler still cautiously considered that the flying bat might achieve Doppler compensation not by a real measurement of the Doppler shift, but alternatively by simply measuring its own air speed, perhaps by the whiskers on its face, and applying the appropriate frequency shift. By two crucial experiments Schnitzler (1973) has now shown that real frequency measurement is indeed involved. In the first experiment, a bat was made to fly in a wind-tunnel in both directions at various wind-speeds up to 8 ms^{-1}. Although the air speed and ground speed of the bat differed considerably, the emitted frequency always varied so as to compensate for the Doppler shifts of echoes from the landing platform produced by the bat's actual approach velocity. In the second experiment, the bat was made to fly in a large plastic tent filled with helium-oxygen mixture. The increase in sound velocity by 1·68 times under these conditions would have three effects: first, it would change the energy in each harmonic as discussed in a later section, secondly, it would make echoes return much sooner from a given range and thirdly, it would reduce the Doppler shifts produced by a given relative velocity. Understandably the bat was at first reluctant to fly in the light gas mixture and had difficulty in orientating, but after a time it flew skilfully and again varied its frequency in accordance with the actual Doppler shifts experienced under these strange conditions. It seems certain, therefore, that *Rhinolophus* can not only measure very small Doppler shifts down to a fraction of $\pm 1\%$ deviation but can also control its emitted frequency to a comparable degree.

Another remarkable phenomenon shown by horseshoe bats concerns their ear movements. Many frequency-modulated bats have essentially fixed external ears; in the vespertilionid, *Plecotus*, Griffin (1958) showed that even a slight distortion of the pinnae leads to severe disorientation. However, Möhres (1953) found that *Rhinolophus* not only turns each ear independently through wide angles but also shows rapid alternating movements of the two pinnae. Schneider and Möhres (1960) analysed the complex musculature associated with the ears and showed that preventing the movements by denervation of the muscles caused a complete loss of orientation ability.

In 1962 two groups of workers (Griffin *et al.* and Pye *et al.*) showed by independent methods that a forward movement of one ear and a backward movement of the other is associated with the emission of each ultrasonic pulse up to the highest pulse rates of 60–80 s^{-1}. The movements are not precisely synchronized to the pulses but there is an almost exact one-to-one ratio in their occurrence. The function of the movements is unknown but they may be used to scan the surroundings (Möhres, 1953) or to impose further Doppler shifts on the echo that could increase the inherent directionality of the ear (Pye, 1960, 1963).

Möhres (1953) thought that *Rhinolophus* with one ear plugged was not disorientated (whereas Griffin (1958) found that plugging one ear of a vespertilionid upset its echo-location completely). However, Flieger and Schnitzler (1973) have now shown that a *complete* block of one ear caused very severe disorientation and, surprisingly, so does lightly plugging one ear, although lightly plugging both ears restores normal behaviour.

In many ways, therefore, *Rhinolophus* are remarkable animals but they are not unique, for several other bats are now known to be convergent on them in many respects. The family Rhinolophidae contains only one genus with great conformity between the 50 or so species. In the 11 species in which echo-location has been examined, the main variation is of frequency, from 40 kHz in the larger forms to 120 kHz in the smaller ones. The closely related family Hipposideridae contains about 40 species but they are more diverse in form and habits. Nevertheless the 13 species examined so far show considerable convergence with the Rhinolophidae. Their emitted signals follow exactly the same pattern (Novick, 1958a; Pye, J. D., 1972) although they are generally shorter in duration, rarely exceeding 25 ms, and the frequencies tend to be higher for the same body size, ranging from 58 kHz in giant *Hipposideros commersoni gigas* to 160 kHz in the tiny *H. caffer*. A special discussion of the pulses of *Asellia tridens* will be reserved for the next section but Pye and Roberts (1970) showed that ear movements are synchronized with pulse emissions at all rates up to 70 pulses/s and this seems likely to apply to other hipposiderids. The muscles of the ear of *Asellia* have been described by Schneider (1961) and compared with those of *Myotis*.

Most constant frequency bats show very little frequency variation between individuals and Kay and Pickvance (1963) found that 60 *Rhinolophus hipposideros* all produced constant frequencies within a band of $\pm 1{\cdot}3\%$ that might be largely accounted for by individual Doppler compensation. But J. D. Pye (1972) reported that a large, crowded colony of *Hipposideros commersoni* in Kenya showed a bimodal frequency spectrum with constant frequencies at 58 kHz and 66 kHz and a silent band 8 kHz wide in between. Forearm lengths in a sample from each frequency group showed no overlap and it is possible that two subspecies (*H. c. gigas* and *H. c. marungensis*) were present. However, the frequency distributions within each group were curiously skewed away from the silent band and it is possible that this interval is kept free to prevent masking of Doppler-shifted echoes when flying in large numbers.

Now 8 kHz is just about the shift expected from the moving wing-tips of flying bats. This possibility was strengthened by evidence that *H. caffer* and *Triaenops afer* both showed a bimodal frequency distribution when in the presence of other bats of the same species, but when alone some bats emitted pulses of intermediate frequency.

Another bat that is very convergent on the Rhinolophidae is the phyllostomid *Chilonycteris rubiginosa* (=*C. parnellii*), first investigated by Griffin and Novick (1955), Novick (1963a) and later in greater detail by Schnitzler (1970a, b). The frequency emitted by this species rises 1–2 kHz at the start of a pulse to a constant level of 57 kHz, and falls rapidly by about 8 kHz at the end. Most of the energy is again in the second harmonic although the fundamental at 29 kHz and the third harmonic at 87 kHz, are stronger than in *Rhinolophus*. The maximum duration is about 20 ms and such pulses are emitted in pairs during the up-beat of the wings which move at about 10–11 beats per second. Pulse repetition rates higher than 100 s^{-1} can be produced when landing and the duration then falls to about 7 ms of which the first 5 ms or so are at constant frequency (Schnitzler, 1971). Schnitzler (1970a) showed that these bats compensate for Doppler shifts in their echoes and also produce ear movements correlated with pulse emission. Thus, except for the absence of a nose-leaf, *Chilonycteris* appears to converge on *Rhinolophus* in every way, including its auditory responses which will be discussed in a later section.

Yet another example of the same pulse type is shown by two emballonurid genera *Rhynchonycteris* (=*Rhynchiscus*) and *Saccopteryx*. These very delicate little neotropical bats were first examined by Griffin and Novick (1955) and have now been recorded on tape (Pye, J. D., 1973). *Rhynchonycteris* emits rather short pulses up to 5–7 ms in duration with a fairly pure second harmonic held constant at just over 100 kHz before a rapid terminal sweep of up to 20 kHz. During interception buzzes, the constant frequency duration is reduced and may disappear altogether in the final stages to leave short sweeps at lower frequencies and at rates up to 200 s^{-1}. *Saccopteryx* produces pulses of a similar pattern but is remarkable because the two species that have been examined both emit two constant frequencies about 2 kHz apart in alternate pulses. Again the second harmonic predominates and *S. bilineata* from Panama emits a predominant second harmonic of 46–50 kHz while animals from Trinidad use 38–42 kHz; *S. leptura* is smaller and its pulses are about 8–10 kHz higher at both locations. By measuring the frequency changes of bats flying in the open air, J. D. Pye (1973) has been able to estimate the flight velocity. The results show that the difference between the frequencies of alternate pulses is the same as the Doppler shift of echoes produced by the bat's own speed. It was suggested, by analogy with Schnitzler's work on *Rhinolophus*, that the upper notes of *Saccopteryx* constitute a 'navigational radar' for echoes say from the ground with no Doppler shift and so returning at the preferred frequency, while the lower notes are pre-compensated for Doppler shifts due to flight speed and form a forward-looking, 'hunting radar'. This idea is supported by the fact that the interception buzz uses only the lower note, the upper one being temporarily omitted, while a resting bat emits a train of pulses all at the upper frequency. Further investigation of these delicate experimental subjects is clearly desirable. They move their ears rapidly but it is not known whether this is correlated with pulse production.

There is less information about the performance of constant frequency bats than there is for frequency-modulated bats but some tests have been carried out. Möhres (1953) stated that *Rhinolophus ferrumequinum* could detect whether or not the 20-cm-square door of their cage was open up to a distance of 6·4 m. Schnitzler (1968) showed that *Rhinolophus ferrumequinum* with a frequency of 83 kHz could detect wires down to 0·08–0·05 mm diameter while *Rhinolophus euryale* at 104 kHz could detect 0·05-mm wires. The maximum distance of detection of these fine wires was under 1 m and even with 3-mm wires the bat actually showed that it had detected them only at 1·4 m. Novick and Vaisnys (1964) showed that *Chilonycteris* can detect fruit flies up to 3·8 m away, just the distance at which the start of the echo returned in time to overlap with the end of the transmitted pulse.

Mixed signals in Microchiroptera

In general terms three types of echo-location signal have been described so far. Two are broad-band, the clicks of *Rousettus* and the frequency sweeps of many Microchiroptera, while the third is essentially narrow-band although it generally terminates in a sweep. The different bandwidths provide different kinds of information and their performance is mutually exclusive (Appendix). But the information requirements of animals must almost certainly change with their circumstances and there is increasing evidence (reviewed by Pye, J. D., 1973) that many bats are more flexible in their use of echo-location than the descriptions so far might have suggested. The signals of some bats are intermediate in

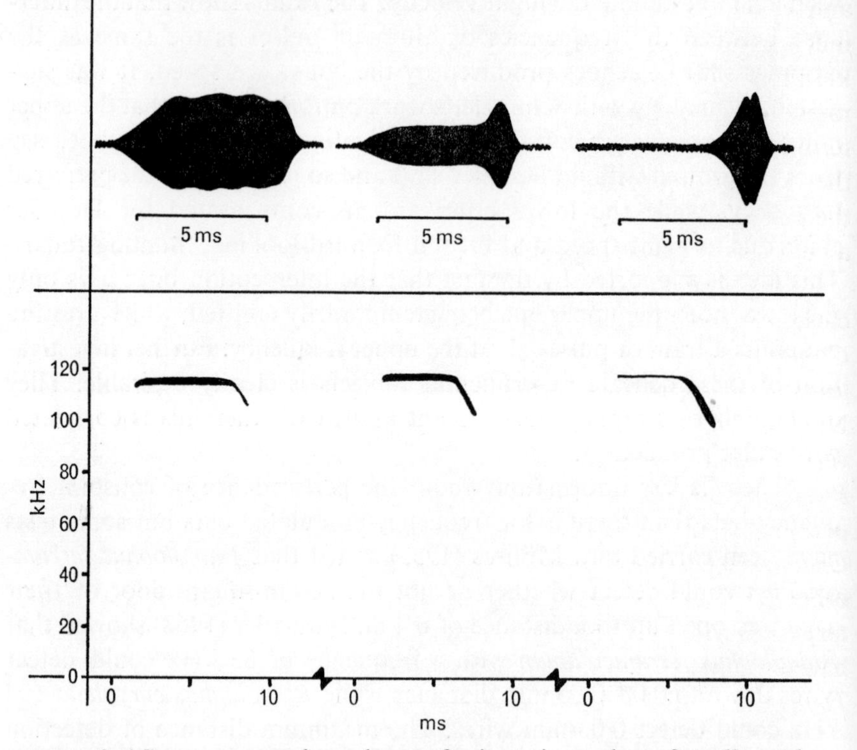

FIGURE 3.4 Sonagrams and envelopes of orientation pulses of *Asellia tridens*. Greater amplitude may occur during the constant frequency or during the frequency-sweep; the constant frequency part is retained but may fall below the noise-level on the oscilloscope screen.

form and combine features of both frequency-sweep pulses and constant frequency pulses so that either characteristic can be accentuated when required, and some other bats appear to be able to switch from one pulse structure to a completely different one on different occasions.

It has already been shown that *Rhinolophus* and *Chilonycteris* when landing can reduce the amplitude and duration of the constant frequency part of their pulses and accentuate the sweeps into a peak. A more extreme form of this was found by Möhres and Kulzer (1955, 1956b) in the hipposiderid *Asellia tridens*. Here the initial, rather short constant frequency part may disappear altogether on an oscilloscope screen, leaving only sharply peaked sweeps resembling the pulses of vespertilionids. Pye and Roberts (1970) and J. D. Pye (1972) showed from an analysis of tape-recordings that the duration of constant frequency is not reduced to zero, as it is during the buzz of *Rhynchonycteris* for example, but its amplitude may fall so low that it may no longer have much significance for echo-location (Fig. 3.4). The duration of the 120 kHz constant frequency varies from about 7 ms in longer pulses at low rates, to about 3 ms at rates up to 70 s^{-1} and all pulses have a deep terminal sweep lasting 1–2 ms. The echo-location of *Asellia* well deserves the description 'kombinierter-typ' used by Möhres and Kulzer, since the pulses may effectively be short sweeps as in vespertilionids or include a period of constant frequency at high amplitude as in rhinolophids.

Pulses with much shorter constant frequency have been found in close relatives of *Chilonycteris rubiginosa*; even within the same genus *C. personata* has only a very brief constant frequency portion lasting less than 2 ms before a rather slow and pronounced sweep (Griffin and Novick, 1955; Novick, 1963a, 1965). Similar pulses, sometimes with very brief or no constant frequency, have been found in two species of *Pteronotus* and in *Mormoops* (Novick, 1963a, b; Pye, J. D., 1973). This is perhaps surprising in view of the convergence in several respects of *Chilonycteris rubiginosa* with the highly specialized rhinolophids. Another taxonomic anomaly appears in the Emballonuridae, where *Peropteryx* strongly resembles and is closely related to *Saccopteryx* in form and habits but emits multiple harmonic sweeps like those of the less similar *Taphozous* (Pye, J. D., 1973).

A different pattern of combined pulse has been discovered in the British vespertilionids *Nyctalus noctula* and *Pipistrellus pipistrellus* and in the African *Scotophilus nigrita* (Pye, 1967a; Pye, J. D., 1973). Indoors all these bats emit short sweeps similar to those of *Myotis* but with a more pronounced second harmonic (Plate IIIc). When hunting in the open, however, all three species produce a constant frequency portion of quite long duration at the bottom end of the sweep. Thus *Pipistrellus* sweeps rapidly down from 85 kHz to a constant level of 45–50 kHz for 8–10 ms, while *Scotophilus* does much the same at 35 kHz (Fig. 3.5a–c) and *Nyctalus* may lose its sweep altogether to give an almost constant note of 20–25 kHz for 25–30 ms. During interception manoeuvres all these species lose the constant frequency and emit a very rapid buzz of short, frequency-sweeps (Fig. 3.5d). Hooper (1969) has confirmed these findings for the two British species and has also reported constant frequencies of 20–25 kHz in *Nyctalus leisleri* and of 40–50 kHz in *Eptesicus serotinus* flying out of doors. The long pulses of hunting *Eptesicus fuscus* observed in America by Griffin (1958) may also be of this type and Cahlander (1967) published a single pulse of this kind from a cruising *Lasiurus borealis*.

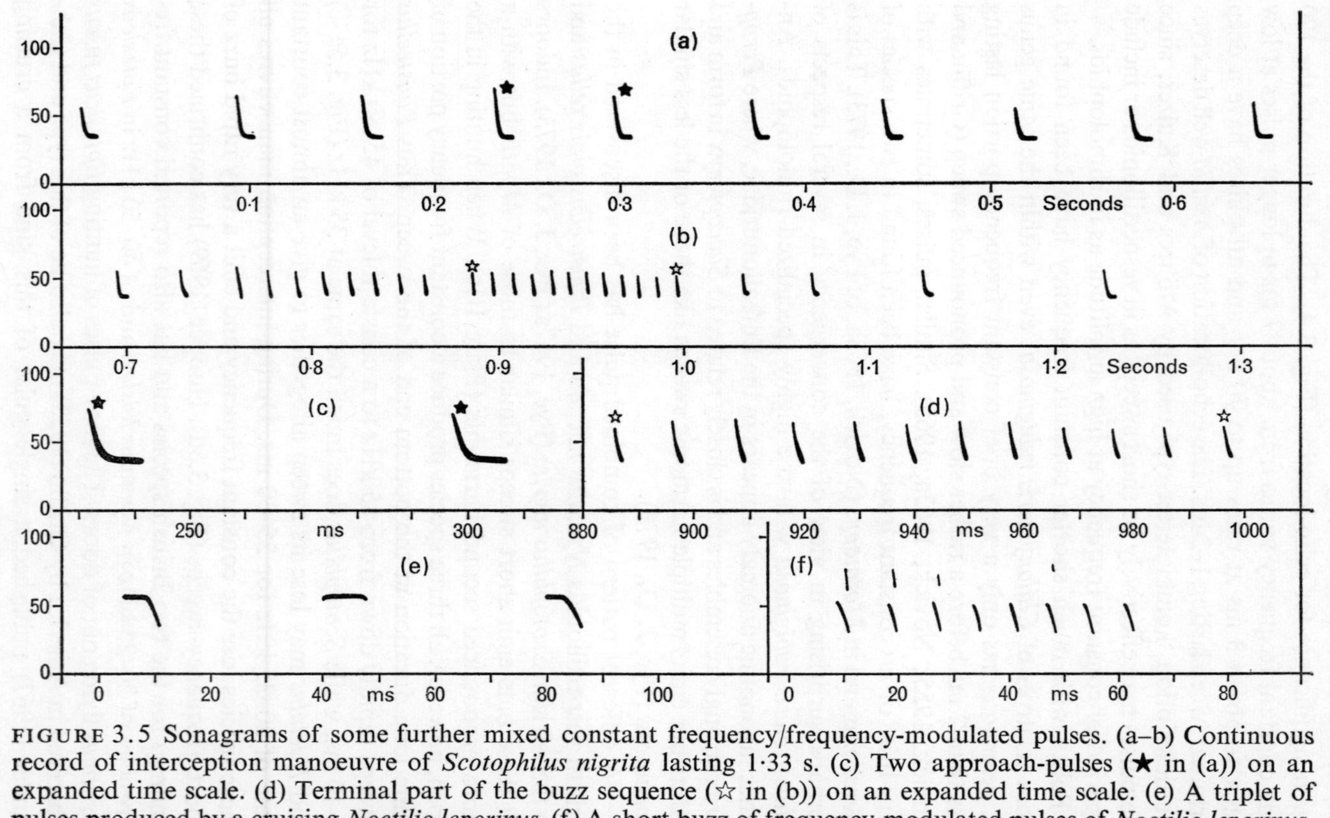

FIGURE 3.5 Sonagrams of some further mixed constant frequency/frequency-modulated pulses. (a–b) Continuous record of interception manoeuvre of *Scotophilus nigrita* lasting 1·33 s. (c) Two approach-pulses (★ in (a)) on an expanded time scale. (d) Terminal part of the buzz sequence (☆ in (b)) on an expanded time scale. (e) A triplet of pulses produced by a cruising *Noctilio leporinus*. (f) A short buzz of frequency-modulated pulses of *Noctilio leporinus*.

Very similar pulses are also emitted by molossid bats which hunt high in the air, often at great speed. Perhaps because of this habit, they are reluctant to fly indoors and only do so with great difficulty, often landing on the first wall they come to. Novick (1958a) examined three species of *Tadarida* indoors and found that in each case the pulses were rather variable. They were fairly long, 3–30 ms in different species, and showed a rather pure sweep starting at low frequencies, 22–47 kHz and descending slowly over about an octave throughout. Vernon and Peterson (1965) studied pulses emitted by *Tadarida* suspended from a trolley moving down a wire and found frequency sweeps from 45 kHz down to 25 kHz in pulses 1·5–4 ms long. Pye has recorded 11 species including *Tadarida, Otomops, Cheiromeles* and *Molossus* indoors and obtained similar results (reported briefly in 1973). Outdoors the species cannot be identified and often cannot even be seen, but signals that almost certainly come from visible or high-flying molossids have been recorded in Africa and the neotropics. These pulses varied greatly from long durations with slow, shallow sweeps throughout or almost constant low frequency, to rapid downsweeps terminating in a short period of constant frequency. During manoeuvres the pulse rate rose to a buzz in which the pulses shortened and the frequency swept rapidly as in the Vespertilionidae. It seems that molossids are generally very flexible in their echo-location behaviour and if some way could be found to work with them the results might be of great interest.

Yet another case of flexibility is seen in the Noctilionidae, first examined by Griffin and Novick (1955) and later by Suthers (1965, 1967) and Suthers and Fattu (1973). *Noctilio leporinus* (Plate IIb) catches fish; Suthers and Fattu (1973) have now shown that *N. labialis* (=*Dirias albiventer*) takes insects from the water surface, and have reported that orientation cries are similar in the two species. The pulses of *N. leporinus* vary from a constant frequency, at about 60 kHz with a duration of more than 11 ms and a barely perceptible terminal sweep, to deep sweeps more than 40 kHz in extent and lasting only 1–2·5 ms. An intermediate form of pulse consists of several milliseconds of constant frequency followed by a deep sweep and such pulses are often emitted before and after a constant frequency pulse to form a triplet. Interception manoeuvres involve a rapid buzz of short sweeps at up to 200 s^{-1}. These descriptions can be confirmed from recordings made by Pye (Fig. 3.5e, f). Suthers and Fattu (1973) found that in *N. labialis* the constant frequency is a little higher at 70 kHz and the terminal sweep is seldom as reduced as it sometimes becomes in *N. leporinus*. Unlike most other constant frequency bats with terminal sweeps, *Noctilio* emit an almost pure fundamental component.

The final story in this section concerns some typical phyllostomids which usually produce very brief sweeps with multiple harmonics. Bradbury (1970) first showed that the large, carnivorous *Vampyrum spectrum* flying indoors sometimes emitted much longer pulses, up to 15 ms in duration with a slow, shallow sweep and multiple harmonics. In tests where the bats were trained to discriminate in flight between a plastic sphere and a prolate spheroid, the pulse structure was varied along the flight path and the two individuals behaved in different ways, suggesting that each chose to rely on a different kind of information. Another example has been given, somewhat tentatively, by J. D. Pye (1973) who recorded pulses exactly like those of *Chilonycteris rubiginosa* (predominant second harmonic constant at about 60 kHz, followed by a brief sweep) in the corridors of a cave in Panama, but attributed them to a nectarivorous bat, *Lonchophylla robusta*. In the main chamber of the cave, similar pulses also appeared to be produced by *Carollia* and *Glossophaga* (Plate IId) since the sound levels were very high and these were by far the most abundant species present. But captured individuals of all three species produced only short, multiple harmonic sweeps when flying in the laboratory. It was suggested that constant frequencies may have considerable advantages when flying in large numbers, as long as all bats use the same frequency. Possibly the bats in the laboratory showed different acoustic behaviour because they were stressed and nervous of the strange surroundings. It is well known that most phyllostomids have very mobile ears and that they are moved very rapidly in response to extraneous sounds. The movements do not appear to be correlated with the production of short sweep pulses but if long pulses are produced they might then show a correlation.

The significance of the examples described in this section is that bats appear to be much more flexible in their echo-location behaviour than was suspected a few years ago. There are some bats, such as *Myotis*, that appear to produce only frequency-modulated, short pulses. Others, such as *Rhinolophus*, the Hipposideridae, and *Chilonycteris rubiginosa*, use only long, constant frequencies and show sharp tuning of the second harmonic, compensation for Doppler shift, correlated ear movements and very sharply tuned auditory responses (described later). Yet these remarkable adaptations do not always appear even in close relatives, for *Chilonycteris personata* seems to lack them. There are several examples of close relatives that use different types of pulse; there are constant frequency bats in each of the four super-families of the Microchiroptera and there are several examples of flexibility in single individuals. Perhaps one should conclude that the pulses recorded so far are not necessarily 'typical' of each species but only represent the kind of pulse that each prefers to use in the conditions under which it has been observed. A comprehensive statement about the nature of microchiropteran echo-location signals cannot be made without a great deal more investigation in a wide variety of situations.

SOUND PRODUCTION AND EMISSION IN MICROCHIROPTERA

The voiced sounds of most mammals are produced in the larynx by bringing together a pair of lateral folds, usually called the vocal cords, so that the respiratory passage is blocked. If air pressure is built up by the action of the thorax and diaphragm, the cords part momentarily to allow a puff of air through and then close again, the process being repeated cyclically and passively as in the action of the reeds in some woodwind instruments or the human lips with brass instruments. The frequency at which puffs of air are produced depends on the length, mass and tension of the vocal cords and constitutes the fundamental frequency of the voice, but since the train of puffs is not sinusoidal there is also a rich sequence of harmonics.

The structure of the larynx of bats was examined in detail by Robin (1881) and Elias (1907) who showed that the normally cartilaginous skeleton was ossified and fused for greater strength, that the intrinsic muscles, especially the cricothyroids responsible for tensioning the vocal membranes, were well-developed and that the vocal membranes themselves were very thin and light. At that time all these features suggested adaptations for producing the well-known high-pitched squeaks of bats but later they also suggested that the larynx was the source of the newly discovered ultrasonic pulses.

This was conclusively demonstrated by Novick and Griffin (1961) who experimented upon the complex and delicate larynges of bats from five different families. They surgically sectioned the three pairs of nerves that serve the larynx in order to determine the effect upon the sounds produced, and they also recorded action potentials from the cricothyroid muscle during ultrasound production. They showed that the cricothyroid muscles, innervated by the superior laryngeal nerve (a branch of the vagus nerve), are mainly responsible for the high frequency of sound and for its downward sweep but that the inferior constrictor muscle, innervated by the pharyngeal plexus, is also involved to some extent. The emission of sound in discrete pulses appeared to result from a complex interplay of several factors and could not be satisfactorily explained.

In the human voice the larynx is responsible for the initiation, during exhalation, of voiced sounds but the sound that emerges through the mouth and nose is much modified by the acoustic influence of the mouth, lips, teeth and the nasal cavity (Fant, 1960; Flanagan, 1972). Even in vowel sounds, where the lips and teeth play no active part, the original series of harmonics is greatly influenced by acoustic resonances in the throat, mouth and nose, the so-called vocal tract. Instead of a fundamental with a steadily declining series of harmonics, as produced by the larynx, certain harmonics are accentuated in four groups called formants, and altering the shape of the mouth and position of the tongue varies the lower three formants to produce all the various sustainable vowel sounds. Pye (1967a) suggested that a similar mechanism, using only a single formant resonance, could account for the shaping of all bats' pulses and for the complexity of the waveform that is sometimes seen. A single resonant cavity would not only explain why certain frequencies are accentuated but would also produce the complex phase-shifts that are seen in the waveform. A simple electronic model, called *Pseudochiroptera electronica*, has been constructed and can produce pulses imitating those of any bat in every detail (Pye, 1967a, 1968a).

Frequency-sweep bats

Actual correlation of pulse production with respiration has been achieved in a number of frequency-modulated bats. Schnitzler (1968) used a hot-wire anemometer to observe the breathing of *Myotis myotis* and found that either single pulses or bursts of pulses were produced during single exhalations. Roberts (1972b) used a similar method on *Eptesicus serotinus* and *Plecotus auritus* but found that although most pulses occurred during exhalation, some pulses were also emitted at any other part of the respiratory cycle. Schnitzler (1971) correlated pulse production with the wing-beat in flying *Myotis lucifugus* and in the phyllostomid *Carollia perspicillata*; pulses were only produced on the up-beat in *Carollia* but *Myotis* produced a few pulses at other points anywhere in the beat. Suthers, Thomas and Suthers (1972) mounted a thermistor anemometer and radio-transmitter to the head of *Phyllostomus hastatus* and were able to record its respiratory cycle during free flight indoors. By simultaneous photography they found that respiration is exactly co-ordinated with the wing-stroke at about 10 s^{-1} (inhalation occurring on the down-beat and exhalation on the up-beat) and ultrasonic recording showed that pulses are generally but not always produced during exhalation. Pulses produced during inhalation were associated with a slight interruption of the air-flow (also seen by Roberts in *Eptesicus*) and it was suggested that special momentary exhalations for the production of each pulse were superimposed on the slower respiratory cycle.

Roberts (1973a, b) has confirmed that vocal-tract resonance plays a part in shaping the pulses of the nose-leaf bats but not of the open-mouth bats. He recorded various species in diver's gas (80% helium, 20% oxygen) and also in air. In bats which emit sound through their noses he confirmed the presence of a vocal tract resonance that shifted the accentuated formant frequency upwards in light gas by a predictable amount without affecting the frequencies of the harmonics themselves. For example in *Desmodus*, the vampire, the main energy of the pulse could be shifted from the second harmonic to the third harmonic. At the same time the fundamental, which is normally suppressed and which should decline even further in light gases, appeared at fairly high intensity. Roberts suggested that all nose-leaf bats have a second resonance, in series with the vocal tract which suppresses the fundamental, and also that retuning the cavities with light gases allows the fundamental to appear (as proposed for Rhinolophids by Pye, 1968a). Even in *Plecotus*, a vespertilionid that normally emits its pulses through its nostrils instead of its mouth, light gas mixtures accentuated the normally weak fundamental and also the third harmonic at the expense of the second. But in other vespertilionids no such shift occurred and Roberts concluded that there the purity of the fundamental waveform and the rise and fall in amplitude were generated at the larynx and were not a property of resonance in the vocal tract.

Constant frequency bats

Schnitzler (1968) and Roberts (1972b) have shown that the pulses of constant frequency bats are also synchronized with the respiratory cycle, one long pulse or as many as nine short pulses being emitted during each exhalation. Schnitzler (1971) also showed that pulse production is synchronized with the wing-beat, with one or a group of pulses during each wing-beat. Wing movements are rather slow in these bats and the respiration rates range from about 2 s^{-1} at rest to 8–12 s^{-1} in flight. Apparently no sound is produced during inhalation but this is very rapid and so sound can be emitted for more than 75% of the time.

It was first pointed out by Novick (1958a) that, in some African and Asian rhinolophids, the emitted note is in fact an almost pure second harmonic since there was a faint trace of a component at half this frequency. Pye (1967a) suggested that this second harmonic might be selected by a vocal tract resonance and the relative amplitudes of constant frequency and frequency sweep parts could then be due to slight changes of emitted frequency. If a vocal tract resonance is involved, then light gases should transfer the main energy to the third or even higher harmonics. Pye (1968a) observed *Rhinolophus hipposideros* in an uncontrolled concentration of hydrogen and found that the normally very weak third harmonic was much enhanced. But a trace of the light gas made the fundamental appear first, whereas with a simple single resonance it would be suppressed even further. It was suggested that *Rhinolophus* has a series-resonance at the fundamental frequency that normally suppresses this component and slight mistuning with light gas caused it to appear. Schnitzler (1970b, 1973) has repeated this test on *Rhinolophus ferrumequinum* and Roberts (1973b) has examined the responses of *Rhinolophus luctus*. Both authors used a known mixture of helium and oxygen and confirmed Pye's conclusions although the actual resonators have not yet been identified. The larynx and vocal tract of *Rhinolophus* contain several specialized cavities that could act in the way required. In comparing *Chilonycteris* with *Rhinolophus*, Schnitzler (1970b) also reported that an oxygen–helium mixture suppresses the second harmonic of *Chilonycteris* and produces prominent fundamental and third harmonic components. Roberts (1972a) has also carefully analysed the frequencies of sweep-decay and peak-sweep pulses in several constant frequency bats and shown that the two types are apparently produced by a change in the vocal tract resonance acting on the second harmonic and not by changes in the emitted frequency as Pye has originally suggested.

Nose-leaves

It has long been recognized that a nose-leaf around the nostrils is typical of two super-families, the Rhinolophoidea and the Phyllostomoidea (Table 3.1) as the roots of these names show. More recently it was realized that bats with nose-leaves emit ultrasound for echo-location

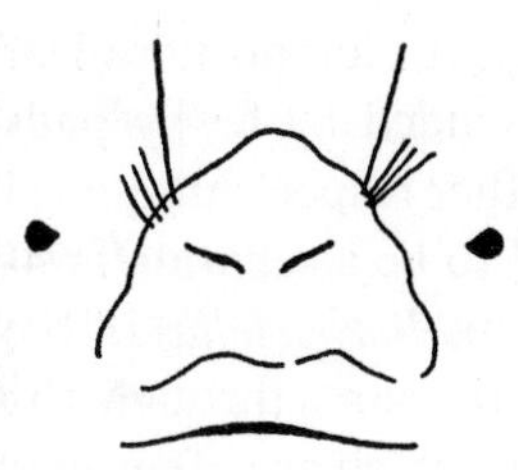
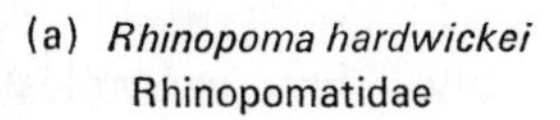

(a) *Rhinopoma hardwickei*
Rhinopomatidae

(b) *Nycteris arge*
Nycteridae

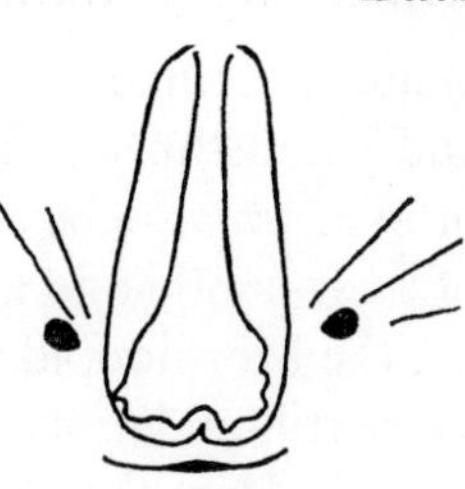

(c) *Lavia frons*
Megadermatidae

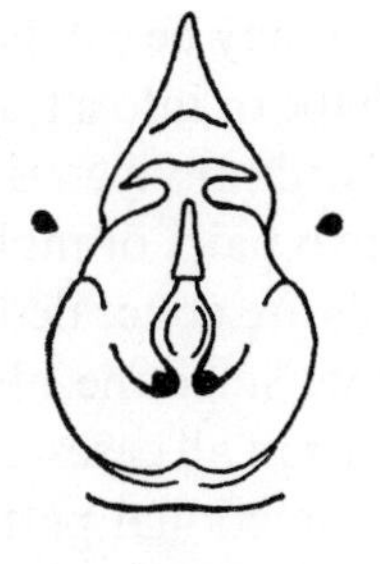

(d) *Rhinolophus landeri*
Rhinolophidae

(e) *Hipposideros commersoni*
Hipposideridae

(f) *Triaenops afer*
Hipposideridae

(g) *Phyllostomus hastatus*
Phyllostomidae

(h) *Artibeus jamaicensis*
Phyllostomidae

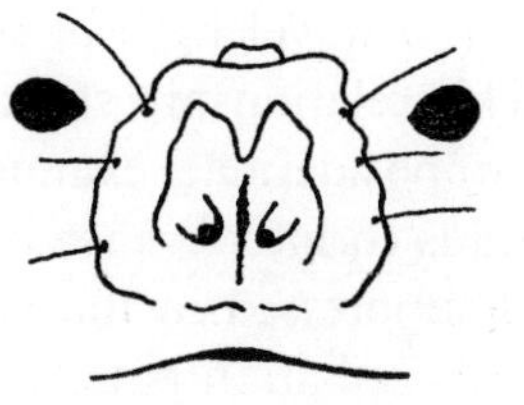

(i) *Diaemus youngi*
Desmodontidae

FIGURE 3.6 Outlines of the nose-leaves of some microchiropteran bats.

through the nostrils (Möhres, 1950) and those without it shout through their open mouths. There are only a few exceptions to this rule, *Rhinopoma*, of the Emballanuroidea, has a simple nose-leaf-like structure (Fig. 3.6a) but this is not used for sound emission and appears to be for closing the nostrils which open only momentarily in each respiratory cycle. Perhaps a desert bat needs to keep dust out of its nose. The

Chilonycterinae of the Phyllostomidae also have little or no nose-leaf and shout through their mouths which are surrounded by fleshy pads or in *Mormoops* by a chin-leaf. These bats are in other respects untypical of their family and are now sometimes considered to be a separate family, the Mormoopidae. Finally *Plecotus* (and perhaps *Barbastella*) of the Vespertilionidae fly with the mouth shut and emit sound through the nostrils. Here the nostrils are backed by fleshy pads which are often very reminiscent of some simple nose-leaves.

The structure of nose-leaves varies considerably. Three examples can be seen in Plate II and more are sketched in Fig. 3.6. The leaf may consist of a simple fleshy pad surrounding the nostrils or be sculptured into grooves or be adorned by complex folds of skin. It may be extended upwards into a pointed lancet of simple or complex shape or into a three-pronged structure as in many Hipposideridae. The nostrils of Nycteridae lie in a deep vertical groove on the face, flanked by two pairs of mobile fleshy lobes, while in the Megadermatidae the nostrils are concealed by lateral extensions from a central fold of the nose-leaf. Since the ultrasounds are always emitted through the nose-leaf, it must in all cases have an acoustic function or at least an influence on the directional pattern of sounds emitted. But the complexity and variety both of nose-leaves and of the sounds emitted make this influence difficult to interpret.

The first advance came when Möhres (1953) pointed out that the exposed nostrils in two species of *Rhinolophus* are spaced apart by exactly half a wavelength of their constant frequency at 83 kHz and 110 kHz. This relationship still appears to hold good for all Rhinolophidae and Hipposideridae examined, from frequencies of 40 kHz–160 kHz. It leads to interference between the two sound sources so that the sound is suppressed in the sideways direction but is accentuated forwards. This radiation pattern has been measured by Schnitzler (1968).

This simple explanation cannot apply, however, to say the Phyllostomidae where multiple harmonic sweeps are emitted. Pye (1967a) pointed out that the interference pattern in this case would be different for every harmonic and each would 'fan out' as the frequency falls; this action would strongly influence the nature of the echo from a given direction and its harmonic content may allow the bat to make an unambiguous judgment of target direction. But even this treatment does not explain the shape of other nose-leaves, and even the complex dorsal lancet of *Rhinolophus* itself is so far mysterious. A more complete explanation must await a detailed investigation of the acoustic properties of various nose-leaves. Möhres and Neuweiler (1966) measured the directionality of sound emission of *Megaderma lyra* and the results are very similar to those of Simmons (1969) for *Eptesicus* and *Chilonycteris*. These bats have no nose-leaves and obtain directionality from their funnel-shaped mouths, with the constant frequency *Chilonycteris* being slightly more directional.

HEARING IN THE MICROCHIROPTERA

The slender evidence on the hearing abilities of *Rousettus* was discussed in an earlier section but the hearing of Microchiroptera has been saved for a special section because of the very large literature on the subject. When Pye reviewed the field in 1968(b) he listed 11 principal papers

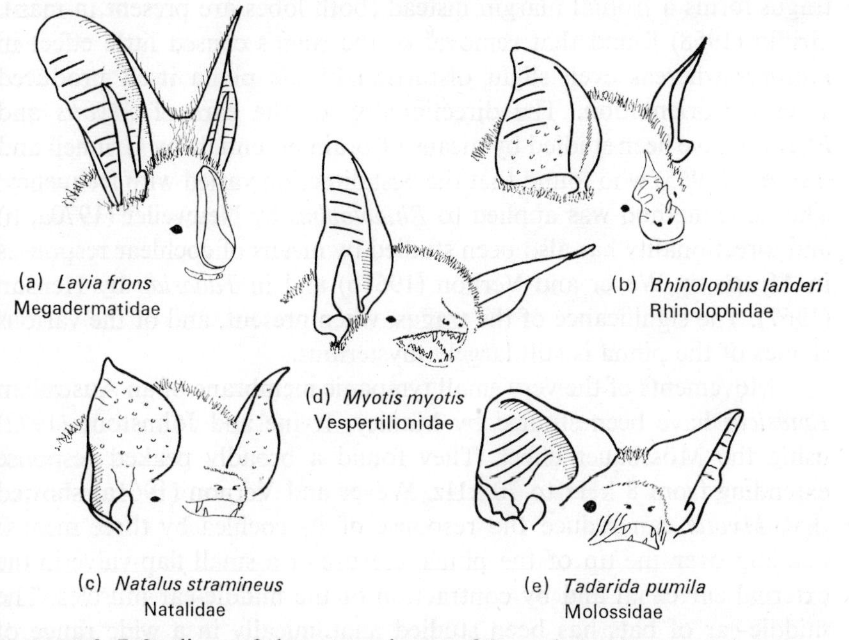

FIGURE 3.7 Outlines of the external ears of five microchiropteran bats.

on neurophysiology of the auditory nervous system of bats and the number has more than doubled since then. It will be impossible here to do justice to the many elegant and painstaking studies involved, so the processes of hearing will be followed 'from outside in', listing the references for further reading if the reader wishes. More detailed reviews have been given by Henson (1970) who lists some recent Russian literature and by Brown and Pye (1974).

The external ears of Microchiroptera are generally large but very variable in both development and form (see Plate II and Fig. 3.7). They may be simple curved flaps or deeply folded in a complex way, they may stand free or be joined together in the mid-line, the free ears may be fixed or highly mobile, they may curl up when the bat is at rest and in *Plecotus* the ears are carefully folded beneath the wings for protection. Finally the ears may be strengthened by transverse ribs or occasionally (in the Nycteridae and the Natalidae) by a 'pimply' structure. The complex musculature of the pinna has been studied in *Rhinolophus* by Schneider and Möhres (1960) and in *Asellia* and *Myotis* by Schneider (1961).

Just in front of the ear is a lobe called the tragus that may form a striking 'secondary ear' or be quite insignificant. In the Rhinolophidae and Hipposideridae it is completely absent and another lobe, the antitragus forms a frontal margin instead (both lobes are present in man). Griffin (1958) found that removal of the tragus caused little effect in *Plecotus* whereas even slight distortion of the pinna itself produced severe disorientation. The directionality of the ears of *Myotis* and *Plecotus* have been studied by means of brain potentials by Grinnell and Grinnell (1965) who found that the best direction varied with frequency. The same method was applied to *Rhinolophus* by Neuweiler (1970a, b) and directionality has also been studied by means of cochlear responses in *Myotis* by Wever and Vernon (1961b) and in *Tadarida* by Henson (1967). The significance of the tragus, when present, and of the various shapes of the pinna is still largely mysterious.

Movements of the very small tympanic membrane of an Australian *Eptesicus* have been studied by Manley, Irvine and Johnstone (1972) using the Mössbauer effect. They found a broadly peaked response extending from 8 kHz to 70 kHz. Wever and Vernon (1961a) showed that *Myotis* can reduce the response of its cochlea by three means: bending over the tip of the pinna, closure of a small flap-valve in the external ear canal and by contraction of the middle-ear muscles. The middle-ear of bats has been studied anatomically in a wide range of species and physiologically in *Tadarida* and *Chilonycteris* by Henson (1961, 1965, 1967a; Henson and Henson, 1972). The ossicular chain is small, light and tightly coupled, with remarkably large stapedius and tensor tympani muscles attached (Fig. 3.8). By recording stapedius contraction and cochlear responses in active, and even flying bats, Henson showed that the muscle contracts just before each pulse is emitted and relaxes to restore sensitivity as later echoes are received from greater distances. Just such an action was originally suggested by Hartridge

(1945). Further unusual features of the middle ears are found in a variety of bats (Wassif, 1946; Pye, A. and Hinchcliffe, 1968; Hinchcliffe and Pye, A., 1969).

The cochlea of bats has been extensively studied by A. Pye (1966–1972) and by Hinchcliffe and A. Pye (1968). Certain unusual features, such as thickenings at the basal end of the basilar membrane, appeared to be related to the habitual use of constant frequency and higher frequency bats tended to have narrower basilar membranes. The response of the cochlea to high frequency sounds was first studied by Galambos

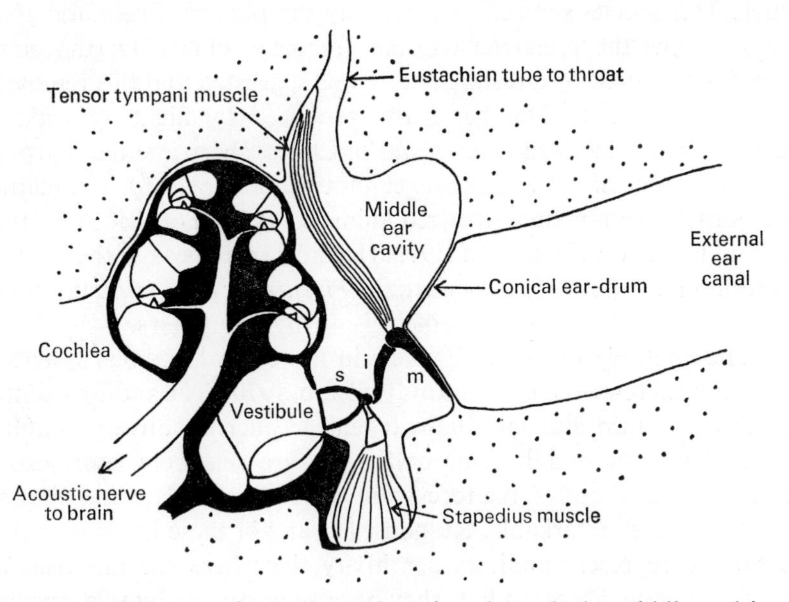

FIGURE 3.8 Diagram of a horizontal section through the middle and inner ear of *Pipistrellus pipistrellus*. (Based on a section cut by A. Pye.) *m*, malleus; *i*, incus; *s*, stapes.

(1941, 1942a) who recorded cochlear microphonic potentials up to 98 kHz, strongest at 25–45 kHz, in a number of vespertilionids. As the name implies this electrical response, recorded from the cochlea of anaesthetized animals, follows the waveform of acoustic stimuli; it is often assumed to play an intermediate part in the excitation of auditory nerve fibres and therefore to imply sensitivity of the ear as a whole, although this has been disputed.

Cochlear microphonic potentials of *Myotis* have also been studied by Wever and Vernon (1961b), Vernon, Dalland and Wever (1966) and by Harrison (1965) who confirmed and elaborated Galambos' results.

McCue (1969) also studied *Myotis* to test whether the cochlea might compress frequency sweep pulses, as some radars do, but found that it did not. Dalland, Vernon and Peterson (1967) studied the cochlear response curve of *Eptesicus fuscus* and Vernon and Peterson (1966) examined the vampire, *Desmodus rotundus*. In both cases the response matched the spectrum of their echo-location sounds rather well. Henson has recorded from active *Tadarida* and *Chilonycteris* as mentioned above. Recently Pollak, Henson and Novick (1972) obtained remarkable results from constant frequency *Chilonycteris* which were awake and active. This species showed an extremely sharply tuned response about 1·5 kHz above the 'preferred' constant frequency of 60 kHz, (the second harmonic) emitted by a resting bat. It was suggested that this represents a pre-adaptation to Doppler shifts produced by flight velocity. In anaesthetized animals the tuning was much less sharp and the sharpness of the response of active animals cannot yet be explained. The method of recording from unanaesthetized animals has been described by Henson and Pollak (1972). Brown (1973c) found a small peak in the cochlear microphonic response of anaesthetized *Pipistrellus pipistrellus* just above the constant frequency emitted in cruising flight outdoors.

The anatomy of auditory tracts in the central nervous system of bats has been reviewed by Henson (1967a, b, 1970). The auditory centres of the hind-brain and mid-brain (cochlear nucleus, olivary complex, lateral lemniscus and inferior colliculus) are relatively enormous in most bats but those of the fore-brain (medial geniculate and auditory cortex) are less remarkable. Responses at any of these levels may truly be said to represent auditory sensitivity since here the messages are carried by nerve fibres. In bats they have been studied by two principal methods. First, fine wire electrodes were applied to the surface of the brain or inserted into its tissue; because such an electrode records the responses of large numbers of neurones in response to sound (registering a larger signal from those nearest to its tip) the recordings made this way are called gross evoked responses. A large response at one sound frequency may either represent great sensitivity to that frequency or a large number of neurones which respond at that frequency. In the second method a micro-electrode was inserted so that it penetrated a single neurone without damaging it and then responded only to the excitation of that one cell, called single unit responses. This gives essential information about the capabilities of single cells, for example in responding to different frequencies or to faint echoes just after a loud sound, but the procedure must be repeated for a very large number of cells in order to build up a composite picture of the response of the whole system. Both techniques have been applied to bats, notably by Grinnell and by Suga but also by several other workers (e.g. Frischkopf, 1964; Henson, 1967a). The favourite recording site has been the inferior colliculus because it is easily accessible and is the highest hypertrophied centre, but other parts of the brain have also been investigated.

Grinnell (1963a) was the first to record gross responses from the colliculus of bats. Responses were obtained up to 150 kHz in *Myotis* with the best responses at 40 kHz, while *Plecotus* responded up to 110 kHz with greatest sensitivity at 15–35 kHz and again at 55–65 kHz. These results closely match the spectrum of the ultrasonic pulses emitted by each species, including the fundamental and prominent second harmonic of *Plecotus*. In a series of later papers (Grinnell, 1963b, c, d, 1967) he showed that single units were sometimes sharply tuned to one frequency, that some gave maximum responses for a sound intensity only slightly above their threshold of sensitivity, that some showed an upper intensity threshold above which they were inhibited and that many showed extremely rapid recovery, being able to respond separately to two stimuli only a few milliseconds apart. Some units were very sensitive to the direction of the sound source and Grinnell found that each side of the colliculus responded to signals from both ears. Binaural interactions as a clue to sound direction were studied by Griffin, McCue and Grinnell (1963) and, as already mentioned, Grinnell and Grinnell (1965) used the gross response to map the directionality of one ear (the other being plugged). Grinnell and McCue (1963) studied the gross responses and single unit responses to sounds that swept rapidly up or down in frequency. They found that some units responded better to constant frequencies while others showed a greater sensitivity to a sweep through their best frequency than to this frequency alone. In some cases the response to one frequency alone was increased by preceding it with a frequency sweep.

All these responses not only showed that the bats involved were sensitive to ultrasonic frequencies but also were relevant to the problem of hearing faint echoes of their frequency-swept cries. The work was soon taken up by other physiologists who concentrated on various different aspects. Harrison (1965) showed that collicular gross potentials of *Myotis*, especially at high sound frequencies, are very sensitive to body temperature whereas cochlear microphonic potentials are much less affected; this is of great physiological interest and is very significant in an animal that allows its body temperature to fall when it is inactive.

Friend, Suga and Suthers (1966) examined the response of single units of *Myotis* to paired frequency-sweep pulses when the second was less intense (simulating a cry and its echo) and found that the responses of a few units to a faint pulse were actually increased by preceding it with a more intense one.

The greatest single body of work has been that of Suga (1964–1972; Suga and Schlegel, 1972) who has performed elegant and comprehensive studies of single units in the cochlear nucleus and auditory cortex as well as in the colliculus of *Myotis*. He concentrated largely on responses to frequency sweeps and to the second (echo) pulse of a pair in relation to echo-location ability. Briefly, he found that ascending the auditory tract from cochlear nucleus to cerebral cortex the neurones showed a greater degree of specialization in response, attributed to the progressive processing of auditory information; the units become more specialized to respond to sweeps, to pulses of various intensities or to fainter second pulses. Some units in the colliculus (but not in the cochlear nucleus) responded to downward frequency sweeps but not to upward sweeps over the same frequency range and this was explained in terms of excitatory and inhibitory interactions between 'lower' units tuned to different frequencies. Some neurones in the colliculus showed responses that were extremely stable in time after the stimulus, despite wide fluctuations in stimulus intensity (an unusual property in sensory systems) and these were designated specialized 'echo-ranging neurones' (Suga, 1970). In the lateral lemniscus of unanaesthetized bats Suga and Schlegel (1972) have shown that when a bat is emitting its own pulse the auditory responses are greatly attenuated before passing to the colliculus, but this does not happen for ultrasounds from external sources. This shows one more way in which echoes are selected and suggests that there is a common centre in the brain that initiates the production of a pulse by the larynx, contracts the middle-ear muscles in advance and momentarily suppresses the auditory responses of the brain itself.

Finally Suga (1969b, c) studied the ability of *Myotis* to avoid wire obstacles in flight after various parts of the auditory nervous system had been removed by surgery. Removal of the ventral half of the colliculus of both sides had no effect. Removal of the auditory cortex on both sides had variable effects, some bats were unaffected but some others were severely affected and collided even with large obstacles. Suga concluded that the auditory cortex is probably not important for echo-location and that surgery might have affected other cortical areas in the two disturbed bats.

Comparative work on the hearing of other species has been done in recent years. Grinnell (1970) and Grinnell and Hagiwara (1972a) studied gross and single unit responses in 12 species from four families (Emballonuridae, Phyllostomidae, Hipposideridae and Vespertilionidae) from Panama and New Guinea. The responses obtained are of great interest in relation to the different echo-location signals employed by these species. Species with wide-band signals showed broad responses, well suited to their acoustic spectrum, while constant frequency bats (*Chilonycteris rubiginosa* and all species of *Hipposideros*) showed very

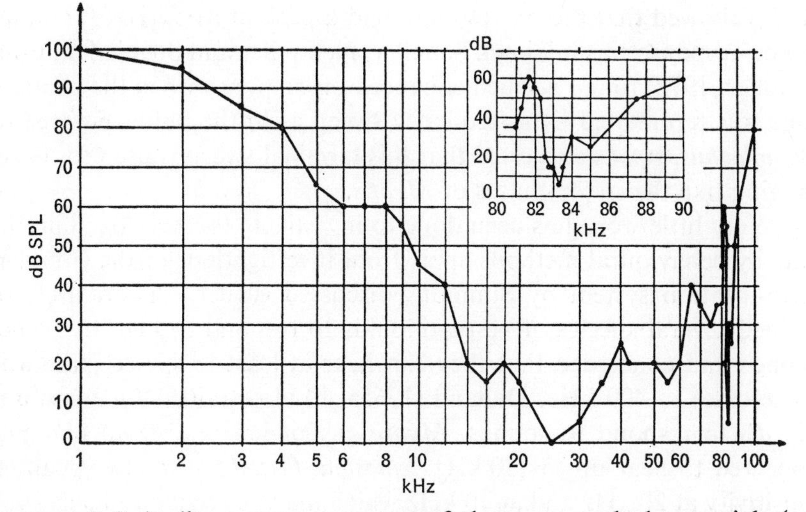

FIGURE 3.9 Auditory response curve of the gross evoked potentials (on-responses) at the inferior colliculus of *Rhinolophus ferrumequinum*. Inset: the sharp peak at 83 kHz on an expanded frequency scale. (After Neuweiler *et al.*, 1971)

sharply tuned responses at their emitted frequency. The latter bats also showed pronounced 'off'-responses to the end of a stimulus pulse and in some cases the 'on'-response to the start of the stimulus was tuned to a slightly higher frequency. The relationship between these two responses has recently been examined in some detail by Grinnell (1973) in *Chilonycteris*. This work promises to have important implications for the physiology of hearing and in particular for echo-location by constant frequencies.

Neural responses in *Rhinolophus ferrumequinum* have been studied in Germany by Neuweiler and his colleagues. Neuweiler (1970a, b) found that the collicular gross on-response was very sharply peaked at 83·3 kHz (Fig. 3.9), the frequency found by Schnitzler to be the preferred echo-frequency for animals from the same population of this species. The response at this frequency was so sharply tuned that a cry emitted in flight at a slightly lower frequency, in order to compensate for Doppler shifts, would generally fall in a region of minimum response (81·5 kHz). This then provides some explanation of the Doppler compensation discovered by Schnitzler; *Rhinolophus* appears to devote a great part of its auditory system to one narrow band of frequencies in order to cope with small Doppler shifts but must then ensure that any echo of interest returns at just that frequency. Later, Neuweiler, Schuller and Schnitzler (1971) showed that the off-response had a peak at 81·5 kHz, for which the on-response was minimal. Schuller, Neuweiler and Schnitzler (1971; Schuller, 1972) found a much enhanced off-response when the stimulus tone was terminated by a frequency sweep as in the pulses emitted by *Rhinolophus*. It was suggested that this terminal sweep is used for echo-ranging like the sweep-pulses of *Myotis*.

Very little work has been done to investigate the hearing ability of bats by behavioural methods apart from investigations of the complete echo-location system by Simmons discussed earlier. Dijkgraaf (1957) trained several species of vespertilionids to respond to high frequency sounds and concluded that they can hear at least to 175 kHz, and in some cases to 400 kHz. Dalland (1965a, b) also trained *Eptesicus* and *Myotis* to respond to sounds. *Myotis* responded best at 40 kHz and appeared to hear up to 120 kHz. *Eptesicus fuscus* showed two peaks of sensitivity at 20 kHz and at 60 kHz which may be related to its use of a prominent second harmonic as in *Plecotus*.

In conclusion, a great deal of detailed, painstaking work mainly in the last 10 years has shown conclusively that bats can hear well, and indeed are at their best, at sound frequencies well above the upper human limit. It has also shown a wonderful specialization for the detection and analysis of echoes, to which man, even at lower frequencies, is very insensitive. But it is not yet possible to present a complete picture of the bats' 'radar receiver' and much more work will be necessary to obtain a plan of the matched or correlation system by which bats obtain close to the maximum possible amount of echo-information.

Plates II–VIII

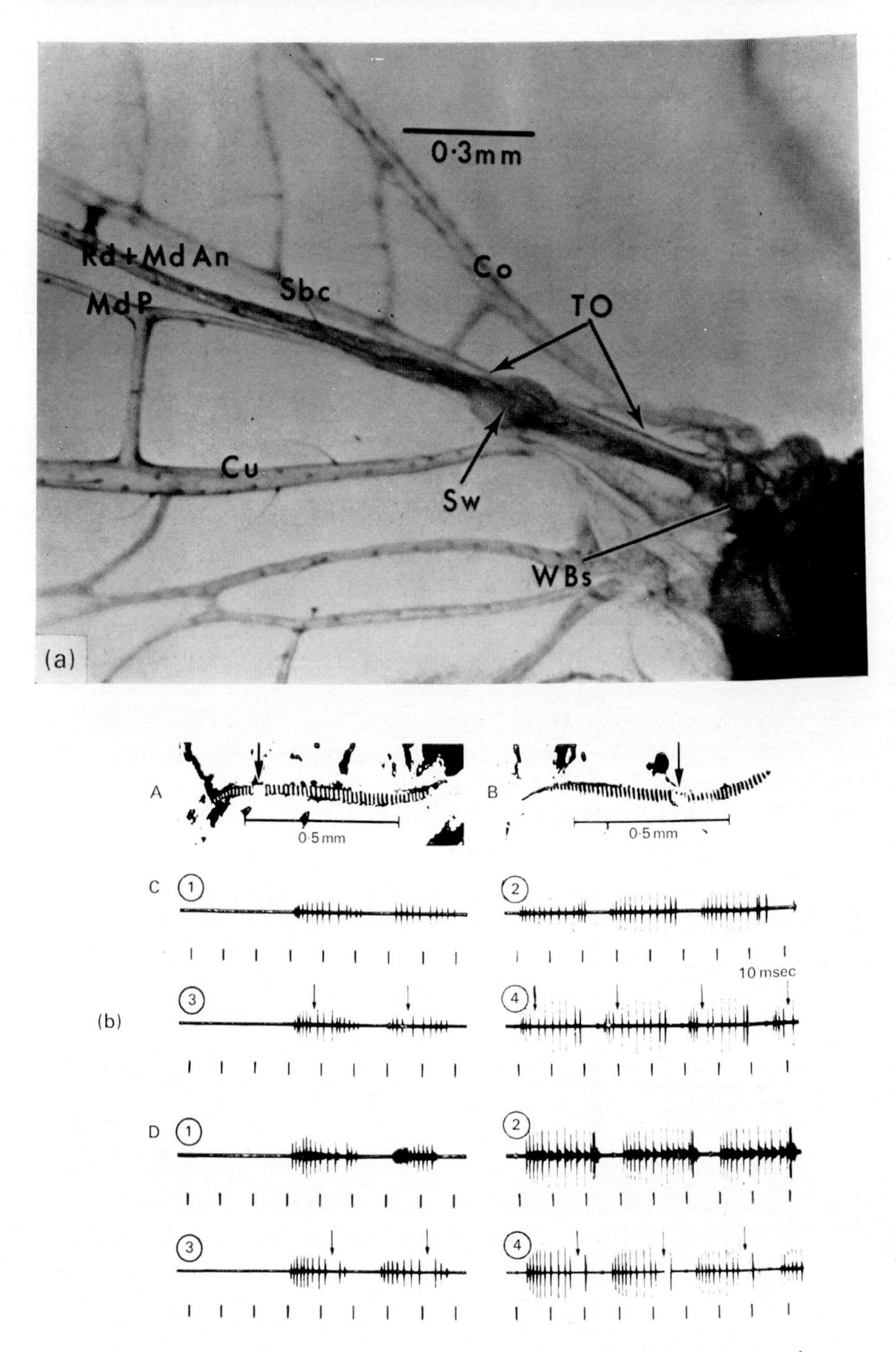

PLATE VIII (a) The tympanic organ (TO) in the forewing of a lacewing, *Chrysopa carnea*. About 25 sensory cells lie within the swelling (Sw) which is situated near to the base of the wing (WBs). The other letters indicate various wing veins. (From Miller, 1970.) (b) The relationship between the teeth of the file and sound production in a tettigoniid, *Phlugis sp*. A and B show two files from which a few teeth have been removed at the points indicated by arrows. C shows oscillograph traces of sounds produced before (1 and 2) and after (3 and 4) the operation on insect A. The arrows in 3 and 4 show where spikes are missing from the normal sequence of the syllable. D shows the results obtained before (1 and 2) and after (3 and 4) the operation on insect B. (From Suga, 1966.)

OTHER SENSES AND SOCIAL USE OF ULTRASOUND IN MICROCHIROPTERA

The echo-location of bats has been much studied because of its intrinsic fascination as a sensory feat and because of the wide range of beautiful and unusual adaptations that different bats possess. But it would be wrong to assume that bats use only echo-location for flight guidance or that ultrasound is used by bats only for echo-location. Some other aspects of bat behaviour will be discussed here to complete the consideration of the group.

Other senses of bats have been rather little studied although Whitaker (1906), Hahn (1908) and Dijkgraaf (1946, 1957) for instance considered the complete sensory equipment. The inertial sense of the inner ear and the bodily muscular senses must surely be of great importance to any flying animals and it has been shown that bats have a very good kinaesthetic memory for spatial relationships, quite apart from their echo-location. Möhres (1953) found that *Rhinolophus* would collide with large, novel obstacles in otherwise familiar surroundings, although they were still emitting ultrasound and must have detected the barrier easily. It seems that bats may often ignore unsuspected echoes in familiar places and this may explain why so many bats can be caught in mist-nets strung across their usual flight-paths. A remarkable case was shown by Möhres and Neuweiler (1966) and Neuweiler and Möhres (1967) in several *Megaderma lyra* which were trained to fly through a doorway from one room to another. The opening was strung with a grid of fine wires leaving apertures 14 cm square through which a bat with a wingspan of 40 cm could only fly by momentarily folding its wings. After some time, when the bats were expert at negotiating this barrier, the wire grid was removed and replaced by fine light-beams in the same positions. The bats still folded their wings when passing through the doorway and photoelectric cells showed that the beams of light were not interrupted. The bats followed exactly their accustomed path, perhaps guided by echo-location of the doorway, and avoided touching wires that were no longer there.

The vision of Microchiroptera has been studied by Suthers and Chase (summarized by Suthers, 1970) who found that visual acuity varies considerably between species, as does the size of the eyes, and that many species are by no means as 'blind' as is commonly supposed. Nevertheless no microchiropteran appears to approach the visual ability of the Megachiroptera and it is significant that even the frugivorous species still use echo-location. Microchiroptera are also supposed to have little sense of smell but it is likely that this applies only to insectivorous species and that the sense is well developed in frugivorous species and in vampires (again reviewed by Suthers, 1970).

The use of ultrasound for social purposes by bats has been much neglected. The familiar audible squeaks of bats extend well up into the ultrasonic frequencies but have not been adequately analysed. It may well be that excited squeaking attracts other bats to a good feeding site and also serves in aggression and mating. Gould (1970, 1971) has examined the use of ultrasound and other sensory signals in mother–young relationships of bats; especially *Eptesicus* and *Myotis*. Both the mothers and the young emit ultrasonic calls which appear to allow individual recognition, so that a mother returning to a communal crèche after hunting can find her own baby for nursing. The cries of young *Rhinolophus*, which may be developing echo-location cries or social signals or both, have been studied by Kay and Pickvance (1963). Möhres (1967) showed that echo-location cries themselves can have a communicatory function in *Rhinolophus* and other bats.

The behaviour of bats therefore seems to be remarkable for its variety and its achievement. Much remains to be learned but ultrasound is clearly very important, and in echo-location the bats appear to be very capable and experienced physicists.

CHAPTER FOUR

Countermeasures by Insects

The last chapter was concerned with mechanisms employed by bats to locate and capture their prey – mainly nocturnal flying insects. These mechanisms are so successful that bats provide an important selection pressure which acts directly on the survival of the insects. It is not surprising therefore that some of these insects have evolved means of protecting themselves from attack. This chapter will examine the countermeasures that have been adopted by moths and some other nocturnal insects in their battle for survival.

As long ago as 1877, White suggested that moths might be able to avoid bats by listening to their 'shrill squeaking'. He was referring to their audible cries, but nevertheless his suggestion has recently been substantiated, in a sense, for the ultrasonic cries of bats. A large number of nocturnal moths possess 'ears', or tympanic organs, and Schaller and Timm (1949, 1950) and Treat (1955) have shown that in many species these are sensitive to artificial ultrasounds. The detection of ultrasounds by moths and their subsequent evasive behaviour have been best studied in the Noctuidae, or owlet moths, and were reviewed in detail by Roeder (1963, revised 1967). Less detailed investigations have been made on other families of moths and also on an entirely different group of nocturnal insects, the Neuroptera or lacewings.

NOCTUIDAE

Noctuid moths are common throughout the world and the larvae of many species such as *Agrotis*, the cutworm, and *Heliothis*, the bollworm, are serious agricultural pests. The adults are heavily built, mostly dull-coloured insects which have a pair of ultrasonic ears on the third thoracic segment, just in front of the waist, behind and below the second pair of wings.

The evasive behaviour of noctuid moths

Noctuid moths are known to be captured on the wing and eaten by bats but they are not taken easily. When approached by hunting bats they often show a variety of flight patterns, loops, spirals and changes in both speed and direction (Webster, 1963; Agee, 1967; Roeder, 1963). Webster photographed the tracks of moths in the presence of hunting bats (Plate IV b, c). He reported that although many of the moths escaped, others were not able to out-manoeuvre the bats. The survival value of these evasive reactions has been studied by Roeder and Treat (1962) who observed 402 encounters between moths and feeding bats. They found that approximately half of the non-reacting moths were captured by bats but only 7% of the reacting ones were caught. Agee (1967) found that in 56 cases, 8% of the reacting moths were caught by the bats as well as 39% of the non-reacting ones.

The first observations of the reactions of moths to artificial ultrasounds were apparently made by Schaller and Timm in 1949 and 1950. These authors found that captured moths responded to signals of 10–200 kHz by taking flight if they were sitting or walking, or if they were already in flight, by changing their flight pattern. The moths were most sensitive to signals of 40–80 kHz but the reactions were abolished when the tympanic organs were destroyed.

Roeder (1962) and later Agee (1967, 1969a) studied the reactions of moths to artificial ultrasonic pulses relayed from a loudspeaker out of doors. Roeder set up an ultrasonic loudspeaker on a tall mast in his garden. After dusk pulses similar to those of bats were relayed at a rate of 30 s^{-1}. These pulses were 5 ms long and were generally at 50 kHz to 70 kHz, but they did not contain the frequency sweeps that occur in many bat pulses. The area around the mast was illuminated by a floodlight and when a moth flew into the area a camera shutter was opened and the source of ultrasonic pulses was switched on.

About a thousand tracks of moths across the floodlit area were photographed and many other visual observations were made with lower illumination. Some of the moths did not change their flight path when they encountered the ultrasounds, others within 3 m or so of the loudspeaker reacted with a variety of loops and dives which did not appear to be in any definite direction but which generally carried the moths downwards, often to the ground. Insects some distance from the mast gave directional responses to the sound source; they turned and flew away. Similar responses were seen nearer to the mast when the intensity of the signals was reduced; insects below the level of the loudspeaker generally flew downwards, those above the loudspeaker flew upwards and out of sight.

The insects giving these responses could not always be identified and many that fell to the ground could not be found. Most of the reacting insects, however, could be recognized as noctuid or geometrid moths or unrelated insects called lacewings. Many of the non-reacting moths were identified as belonging to the families Saturniidae and Sphingidae, which lack tympanic organs and which, except for some Sphingidae, are probably deaf.

Agee (1967, 1969a) made similar observations on the reactions of various species of nocturnal moths to pulsed ultrasound. He set up an ultrasonic loudspeaker and a spotlight near a cotton field in North Carolina and observed the responses of 1878 moths to a variety of acoustic stimuli. Most of the moths were bollworms, *Heliothis zea*, (Noctuidae), European corn borers, *Ostrinia nubilalis* (Pyralididae) or cabbage loopers, *Trichoplusia ni* (Noctuidae). Over 200 of them, mostly *Heliothis*, were captured and identified. Agee studied the effect of both pulse rate and intensity on the behaviour of the moths. He showed that if the pulse rates were high, 4–15 s^{-1}, and the sound pressure levels greater than 80 dB, the moths responded by looping and/or diving to the ground. Sometimes flight was resumed within 2–4 s but at other times the moths remained inactive for 5–10 min. When the sound pressure level was low, the moths made a directional turn away from the sound source. At pulse repetition rates of less than 2 s^{-1} they showed no directional responses but usually the flight pattern showed 'bounces' synchronous with each pulse. Looping and spiralling were occasionally seen in response to these lower pulse rates but usually only if the moths were within 6 m of the loudspeaker.

Agee (1969b) also investigated the use of ultrasound as a possible means of pest control. He studied the effect of artificial signals on various non-flight behaviour patterns, such as feeding and mating, in caged noctuid moths. A 20 kHz signal, 8 ms in duration and produced at a rate of 14·3 s^{-1} was relayed to the moths and their reactions were observed by dim red light. Although mating and egg-laying were sometimes interrupted by the signals, they were generally resumed when the sounds stopped.

Similar results were obtained when bats were released inside the cage. At first the moths that were courting or egg-laying ceased but after 3–5 min these activities were resumed. One bat came within 12–15 cm of a moth that continued courtship behaviour undeterred. A silent period of at least 10 min was necessary before the moths again became responsive to ultrasonic signals.

Agee's observations showed that the evasive reactions of moths to ultrasound occurred most readily when the moths were flying or, when not in flight, after a silent period. It therefore seems unlikely that a system of continually pulsed ultrasound would interrupt the normal reproductive behaviour of these moths and so such a system would apparently be of no value in controlling agricultural pests like the boll-worm. Belton and Kempster (1962), however, have claimed some measure of success in using a similar system to control the infestation of corn crops by the European corn borer, *Ostrinia nubilalis* (see section on Pyralididae).

Acoustic sensitivity of the tympanic organ

The structure of the auditory or tympanic organ of Noctuidae has been described in detail by Eggers (1919), Treat (1955) and by Roeder and Treat (1957). More recently, Ghiradella (1971) has studied the fine structure of the ears of the noctuid moth *Feltia subgothica* using both electron and light microscopy. Each ear lies within a deep, scale-free recess on the posterior wall of the third thoracic segment. A transparent ear-drum or tympanic membrane faces obliquely backwards and outwards into this cavity. Anterior to the tympanic membrane is an air sac, an expanded part of the moth's repiratory system, across which the sensory cells are suspended (Fig. 4.1). There are two acoustic sense cells called the *A* cells which are modified bipolar neurones called chordotonal sensilla or scolopales. They lie close together, suspended in a strand of tissue which runs from the centre of the tympanic membrane towards a skeletal support projecting into the air sac. The proximal nerve fibres from each *A* cell run in the tissue strand towards the skeletal support and are joined by a third nerve fibre from a large, non-acoustic *B* cell which lies near the support. All three fibres then continue together as the tympanic nerve and join the central nervous system of the moth at the large pterothoracic ganglion, formed by the fusion of the last two thoracic ganglia. The *A* cells are sensitive to vibrations received from the tympanic membrane and generate nerve impulses that are sent along the tympanic nerve to the central nervous system.

Between 1957 and 1967 Roeder and Treat investigated many aspects of the response of the noctuid tympanic organ to different ultrasonic stimuli, both natural and artificial. They were interested in the way in

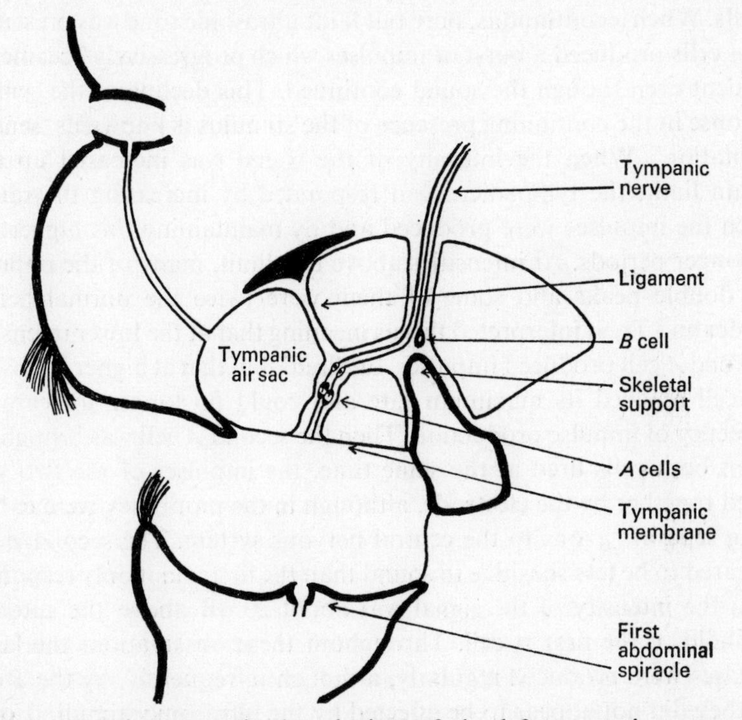

FIGURE 4.1 Diagrammatic dorsal view of the left tympanic organ of a noctuid moth. (Redrawn after Roeder and Treat, 1957 and Treat and Roeder, 1959)

which moths detect the presence and direction of their predators, how they code the information in the form of nerve impulses and also how they make use of this information to avoid capture. They used a variety of species from several genera including *Heliothis*, *Catocala*, *Prodenia*, *Feltia* and *Caenurgina*. For this study, moths were anaesthetized and restrained dorsal side upwards in plasticine. The tympanic nerve was exposed and impulses were recorded by very fine wire electrodes. These nerve impulses were monitored visually on an oscilloscope screen on which they appeared as spikes and audibly through a loudspeaker from which they could be heard as clicks. In the early part of the study only one tympanic organ was studied at a time; the other was deliberately destroyed first.

Roeder and Treat found that even in the absence of sound the *A* cells produced a 'resting discharge' of impulses which appeared to be random. The *B* cell, however, produced impulses regularly 10 to 20 times a second. These were larger impulses than those produced by the *A* cells. When a continuous, pure but faint ultrasonic tone was presented, the *A* cells produced a burst of impulses which progressively became less frequent even though the sound continued. This decline of the sensory response in the continuing presence of the stimulus is known as 'sensory adaptation'. When the intensity of the signal was increased up to a certain limit, the tympanic organ responded by increasing the rate at which the impulses were produced and by maintaining this higher rate for longer periods. At intensities above this limit, many of the impulses had double peaks and some of them were twice the normal height. Roeder and Treat interpreted this as meaning that at the lower intensities only one *A* cell produced impulses, or 'fired', but that at higher intensities this cell reached its maximum rate and could no longer increase the frequency of impulse production. Then the second *A* cell was brought in. When both cells fired at the same time, the impulses of the two were added together by the electrode, although in the moth they were carried along separate axons to the central nervous system. The second *A* cell appeared to be less sensitive to sound than the first and it only responded when the intensity of the signal was about 20 dB above the intensity threshold of the first *A* cell. Throughout these observations the larger impulses were produced regularly, although infrequently, by the *B* cell, but they did not appear to be affected by the ultrasonic stimuli.

Short, artificial ultrasonic pulses, similar to the cries of some bats, resulted in separate bursts of impulses. The interval (the latency) between the presentation of the ultrasonic signal and the onset of the tympanic response decreased with increasing sound intensity and the impulses were produced at a greater rate (Plate Va). They were also produced for longer periods after the end of each sound pulse, as an 'after-discharge' and, over a range of about 40 dB, the total number of impulses produced by both cells together in response to a single artificial pulse was proportional to the intensity of the pulse. At higher intensities, both of the *A* cells apparently 'saturated' and they were unable to increase the length of the after-discharge any further. These results showed that the response of the tympanic organ to short, high intensity sounds was similar to the response to a longer sound of lower intensity and so the moth is probably unable to distinguish between them.

The frequency response of the tympanic organ of noctuid moths has been studied by several authors. Roeder and Treat found that it responded similarly to all pure tones within the frequency range 3 kHz to 150 kHz. There was no evidence that the moth could discriminate between tones of different pitch but in *Caenurgina erechtea* the sensitivity was greatest to frequencies within the range 25 kHz to 50 kHz (Roeder, 1966b, 1971) (Fig. 4.2). Suga (1961) found that in three Japanese species

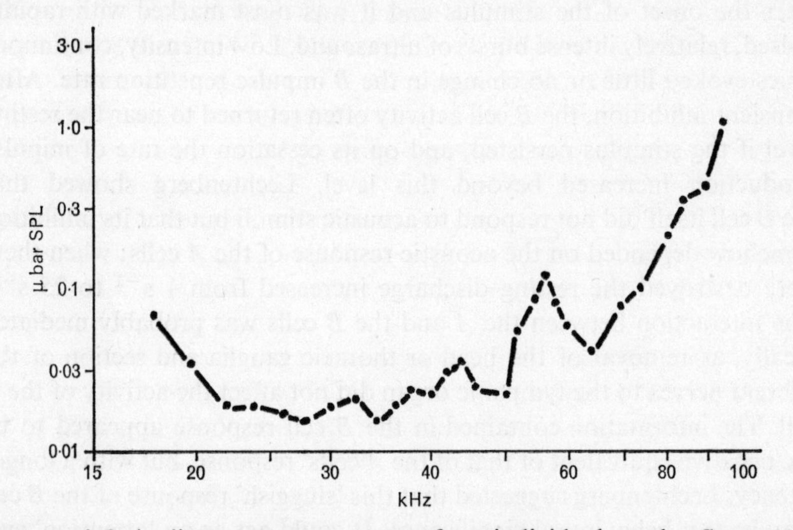

FIGURE 4.2 Auditory response curve of an individual moth, *Caenurgina erechtea*, Noctuidae. (Redrawn after Roeder, 1966b)

of noctuid moths, *Adris tyranus*, *Lagoptera juno* and *Oraesia excavata*, the more sensitive *A* cell was more slowly adapting and responded to a wider range of frequencies than the less sensitive cell. The most effective frequencies for eliciting responses in both types of neurones was between 10 kHz and 20 kHz. Agee (1967) reported a similar range of optimum frequencies, 18–25 kHz, for eliciting acoustic responses from the noctuid moths *Heliothis zea* and *H. virescens*.

Roeder and Treat found that the acoustic response of the tympanic organ was completely abolished if the *A* cells were damaged, interfered with mechanically or were detached from the tympanic membrane, but a severe tear in the membrane did not appear to affect the response. Obviously it is the *A* cells and their connection to the tympanic membrane that are important in the acoustic response. But Roeder and Treat were also interested in the large *B* cell that lay so close to the *A* cells in the tympanic organ. It did not appear to be affected by acoustic stimuli and seemed to be some sort of proprioceptor, indicating mechanical stresses in the body, possibly during wing movements, but its function was a mystery.

Recently (1971) Lechtenberg, a student of Roeder's, made a detailed study of the responses of the *B* cell, and he found that its activity was inhibited by acoustic stimuli within the frequency range of the *A* cell response. This inhibition occurred up to several hundred milliseconds after the onset of the stimulus and it was most marked with rapidly pulsed, relatively intense bursts of ultrasound. Low intensity, continuous tones evoked little or no change in the *B* impulse repetition rate. After transient inhibition, the *B* cell activity often returned to near the resting level if the stimulus persisted, and on its cessation the rate of impulse production increased beyond this level. Lechtenberg showed that the *B* cell itself did not respond to acoustic stimuli but that its inhibition somehow depended on the acoustic response of the *A* cells; when these were destroyed the resting discharge increased from 1 s^{-1} to 25 s^{-1}. The interaction between the *A* and the *B* cells was probably mediated locally, as removal of the head or thoracic ganglia and section of the efferent nerves to the tympanic organ did not affect the activity of the *B* cell. The information contained in the *B* cell response appeared to be the negative equivalent of that of the *A* cells' response, but with a longer latency. Lechtenberg suggested that this 'sluggish' response of the *B* cell may have a behavioural significance. It could act as an 'attention' system which discriminates between rapidly pulsed ultrasounds of relatively high intensity, as in a bat's buzz, and other transient noises. It might also evoke and sustain a particular behaviour pattern such as power diving. Thus the *B* cell may act as a danger warning system and so may only change its rate of impulse production to any great degree in 'dire circumstances' (Lechtenberg, 1971).

Having shown that the tympanic organ of noctuid moths responds to artificial ultrasounds, almost by chance Roeder and Treat (1957) verified that it was highly sensitive to the ultrasonic cries of bats. A bat that had been kept in the laboratory escaped and flew around the room in which a tympanic nerve preparation was set up. Although the bat appeared to be 'silent' to the experimenters, the tympanic nerve produced bursts of impulses. The moth responded continuously as the bat flew around the room but an ultrasonic microphone, also connected to the oscilloscope, detected pulses only when the bat flew close to it.

Roeder and Treat (1961a, b) then decided to observe the tympanic nerve response to natural ultrasounds in the field. On a July evening, they took about 300 lb wt of electronic equipment to a hillside in Massachusetts where bats were known to feed. They captured a moth at a light trap and dissected out the tympanic-nerve region as usual. The light faded before all was ready and the first indication that bats were about was regular bursts of clicks at about 10 s^{-1} from the loudspeaker. The approach of a bat towards the preparation was accompanied by an increase in the number and rate of *A* impulses in first one and then both *A* fibres. The response waned as the bat flew away. A continuous train of *A* impulses appeared to signify the buzz of a bat as it closed in on an insect. On this occasion many of the bats 'detected' by the tympanic nerve preparation were out of range of the floodlight that was set up near the preparation, but on another occasion the *A* cells detected bats which could easily be seen to be 30 m away and 6 m above the ground.

A year later Roeder and Treat had mastered the technique of recording the tympanic nerve response from both ears simultaneously and once again they set up a tympanic nerve preparation out of doors in the presence of flying bats (Roeder, 1963). The approach of a bat was again signified by bursts of regular *A* impulses, first from one ear then, as the bat came closer, from both ears. The responses of the two ears (the latency of the response, the number of impulses produced and their rate) differed most when the bat was some distance away from the moth so that the ultrasonic cries were faint. As the cries became more intense, however, the responses of the two ears became almost identical (Plate Vb).

Directionality

If the moth could compare the different responses of the two ears it would have a means of determining the direction of the sound. Sound coming from one side of the body reaches the opposite ear at a lower intensity due to shielding, and the intensity at each ear could be compared directly by the number of impulses and their rates. Latencies cannot be measured by the moth but the difference in the latencies at each ear could be measured and may provide additional information on the direction of the sound. When the intensity of the signal is 40 dB above the threshold, the tympanic nerve response saturates and if this happens any indication of the direction of the signal is then lost. The moth therefore may be able to tell the direction of a bat that is far away, but not of one that is close by.

A moth then has the necessary information to determine whether a bat is to the right or to the left, but can it tell whether the bat is above or below? The movements of the wings may play a part here. Apart from the changes in the attitude of the moth itself during flight, the wings beat at about 20 to 40 times a second. The tympanic organs lie just below the base of the wings and so the acoustic fields round the moth are probably altered as the wings alternately shield and expose the ear in certain directions. It was impossible to study the tympanic organ response during free flight or even during tethered flight, but Payne, Roeder and Wallman (1966; Roeder and Payne, 1966) plotted the directional sensitivity of the ear of anaesthetized moths when the wings were held in various positions used during flight. With the wings fixed in a certain position, the intensity of 60 kHz pulses needed to produce an arbitrary response in the tympanic nerve was recorded for a large number of loudspeaker positions around the moth. The position of the wings was then altered and the process was repeated.

The results for 12 moths showed that when the wings were at the top of their stroke, each ear was most sensitive to sounds coming at right-angles to the body on the same side and was least sensitive to sounds coming from the opposite side. Near the bottom of their downstroke, the wings shielded the ears and the area of minimum sensitivity was directly above the moth. The intensity necessary to produce the arbitrary tympanic response varied up to 40 dB at any one position of the sound source relative to the body, depending on the position of the wings. The changes in the intensity of a signal reaching the tympanic organ would probably be greatest when the sound source is above the moth's line of flight and least when it is below the line of flight. The moth could therefore possibly determine the vertical position of an acoustic signal from the degree of intensity modulation if the position of the wings were also monitored at least once per wing cycle. This information could come from the motor impulses in the pterothoracic ganglion which activates the wing muscles, or from proprioceptors at the base of the wings. At one time Payne, Roeder and Wallman thought that the *B* cells might play a part in monitoring the position of the wings by responding each time the thorax was deformed to a certain extent during every wing-beat. But Lechtenberg's studies on the *B* cell response show that such a function is unlikely. If the pursuing bat produced echolocation pulses at a rate comparable to that of the movement of the moth's wings, the arrival of the pulses would always coincide with one particular wing position and any sense of vertical direction is then probably lost. This is unlikely to happen for long, however, because a bat intercepting a moth produces several pulses for every wing-beat of its prey.

Central co-ordination

Roeder next (1966c, 1967b) made a closer study of the turning reaction of noctuid moths to ultrasonic pulses of differing intensities. The flight of an insect is determined by the velocity and direction of the airstream that is forced downwards and backwards by the moving wings. A change in the position of this airstream relative to the body axis would indicate an attempt by the insect to change its direction of flight. Roeder designed and built a small electronic device sensitive to differences in the speed of air flowing over two sensors. This was placed horizontally behind a moth in stationary flight (held by the thorax) and used to register the turning attempts made by the moth in response to artificial ultrasonic pulses, which were presented at different angles in the horizontal plane.

Roeder found that noctuid moths attempted to turn away from a lateral source of 40 kHz pulses at intensities from the threshold of the tympanic organ to 40 dB above it. There was no change in either the reaction time or the magnitude of the response with this increasing stimulus intensity and so the moths do not seem to be able to react faster to bats close by. Complex and multi-phasic turning tendencies, towards and away from the sound source, were occasionally seen when still higher intensity signals were presented from one side of the moth. They were also very common when the sound source was directly in front of the moth, and when presumably the response of both tympanic organs was identical. The moths maintained their attempts to turn throughout a train of pulses at 10 s^{-1} but only a very brief response, if any at all, occurred when a single pulse, or a continuous tone was presented. In some cases, when a series of pulses at 1 s^{-1} was presented, the moth failed to respond to the first pulse but did respond to later pulses.

The simple monophasic attempts to turn away from the sound source depended on a comparison of the signals from both ears reaching the central nervous system. This was shown by destroying one ear. The moths then always tended to turn towards the deaf side wherever the sound source was placed, and even in the absence of a stimulus signal.

By photographing the stationary flight of these moths while they were attempting to turn to one side or the other, Roeder (1967b) showed that the chief action producing the turning effect was partial folding of the wings on the side of the moth away from the sound. The wings on the near side beat with extra amplitude causing the moth to yaw away from the sound source. High intensity stimulation was sometimes followed by cessation of flight, either immediately, or after an attempted turn accompanied by a great increase in thoracic vibration. This could have been the equivalent of the sharp turns and dives that were shown by moths in free flight in the field.

The next step in the study of the link between the detection of ultrasonic pulses by the tympanic organ and the evasive behaviour of moths was to look for acoustic responses in the central nervous system of the moth where they could be used to excite the motor organs, in this case the wing muscles, and so cause a change in behaviour. Roeder attempted to record the responses of single neurones, first in the pterothoracic ganglion (1966d) and later in the brain (1969a, b, 1970). Anaesthetized moths were restrained ventral side upwards and the lower surface of the pterothoracic ganglion was exposed by dissecting away the overlying cuticle and muscle. Using a micromanipulator, a very fine stainless steel or tungsten electrode was gently lowered into the ganglion until a clear response from a single neurone was obtained to regularly repeated artificial signals. The acoustic response of this neurone was then studied and its position in the ganglion noted. Many thousands of electrode passes through the ganglion were made at different sites and information about many neurones was gradually collected. The picture was not complete but several types of response could be recognized.

Various neurones were found which responded to different aspects of the *A* cell response. In some cases both the *A* fibre response and the ganglionic neurone response were picked up at the same time by the one electrode but they could be distinguished on the oscilloscope trace. The neurones most frequently encountered were named 'repeater' neurones. These appeared to relay the *A* fibre response almost unchanged from one ear to the opposite side of the ganglion and then to the brain of the moth. Another type of neurone was encountered which produced only one impulse for each ultrasonic pulse, however long its duration. These 'pulse marker' neurones fired only after a long and variable time delay and they apparently required three or four *A* impulses with sufficiently short interpulse intervals for their activation. This 'temporal summation' that was needed to fire the pulse marker neurone was reminiscent of the summation of several closely spaced ultrasonic pulses needed to elicit the turning response in intact moths and it is possible that these particular neurones are involved in that response. The pulse marker neurone also needed a few milliseconds with no *A* activity between bursts of impulses if it was to be reset to fire again. In this it was again similar to the turning away response which was not maintained when a continuous tone was presented.

'Train marker' neurones were only occasionally encountered. These registered the duration of a train of ultrasonic pulses by firing at their own rate, 100 to 200 impulses per second, throughout the length of the pulse train, and ceasing when it ended. On one occasion a neurone was found that fired regularly at about 90 impulses per second in the absence of *A* impulses, but when a train of *A* impulses was relayed by a repeater neurone, its activity was inhibited. Near the midline of the ganglion neurones were sometimes encountered which responded to *A* impulses from either ear, and when both ears were stimulated simultaneously its response was greater than when either was stimulated alone. It therefore appeared to sum signals arriving together from both ears, especially when the stimulus to each ear was of low intensity. These last two types of neurone may form a part of the mechanism that determines the direction of a sound source and allows the moth to steer.

So far it is not known in which order these neurones are activated, or if or how they affect the movement of the wings. But it is clear that they could form part of a system enabling the moth to react optimally only to those signals within a particular frequency range and having certain intensities, durations and pulse repetition rates.

Roeder (1969a, b) next looked for acoustic interneurones in the brain of noctuid moths. The suboesophageal and protocerebral ganglia of the brain were probed with a micro-electrode, in a similar manner to the pterothoracic ganglion, until a neural response to an acoustic stimulus was encountered. Roeder found that when an ultrasonic stimulus was delivered to one ear alone, the other having been destroyed, some 'phasic' neurones on both sides of the brain responded only at the beginning of a 30 ms sound pulse. Other 'tonic' neurones fired throughout the length of the pulse and two types of such units were distinguished; one whose impulses were seen as small spikes on the oscilloscope screen, the 'small tonic unit', and another which produced larger spikes, the 'large tonic unit'. When the intensity was raised above threshold, the 'large' unit increased its response rate at first, but at about the level where the second *A* cell of the tympanic organ was excited, its response gradually decreased again. The 'small' unit, however, continued to increase its response over the whole range of intensity signalled by the tympanic nerve. Both units sometimes increased their responses to test pulses when a brief burst of extra pulses was added. Roeder suggested that this 'facilitation' may be important in the moth's ability to distinguish the pulse sequence (about 10 s^{-1}) of an approaching bat from other, short, random sounds produced at night, for example by other insects.

Roeder has put forward a hypothetical mechanism by which these brain interneurones could modify flight by affecting the steering and driving mechanisms. He suggested that the 'large' tonic units might modify the thoracic flight driving mechanism; at low sound intensities they could cause more rapid flight away from the predator, but at high sound intensities, when their activity is suppressed, they may in turn suppress the flight drive mechanism, causing passive falling and erratic movements. Similarly the 'small' tonic units might influence the steering mechanism at low intensities, but at high intensities the moths might over-react to give the spiralling and looping observed in free-flying moths. It must be emphasized that, as yet, these means of control are merely speculative, but they do show how the acoustic information, initially coded by the *A* cells, might affect the flight of the moth.

Finally, Roeder (1970) discovered that when some interneurones in the brain were studied for several hours, they showed lapses in acoustic sensitivity, which either occurred spontaneously or could be induced by visual stimuli. During some of these lapses, impulses were produced at the wing-beat rate, even if the moth was not flying. These neurones may be part of an 'attention' system that can be directed either towards monitoring flight movements or, possibly by the action of the *B* cell (Lechtenberg, 1971), towards detecting ultrasound.

This story of how noctuid moths detect and evade bats is still not complete. So far experiments have shown that the cries of bats are detected and coded by the tympanic organ and they have indicated the way in which this information is processed by the central nervous system and how it could produce the behavioural responses. This single system has been investigated from many different angles in both the laboratory and the field. As a result, our understanding of the physiological action of this system can be related to the behaviour of the moth in its environment and to the adaptive significance of this behaviour.

ARCTIIDAE, NOTODONTIDAE AND CTENUCHIDAE

At about the same time as Roeder and Treat demonstrated that the ears of noctuid moths respond to ultrasounds, Haskell and Belton (1956) investigated acoustic responses in two other families of moths, the Arctiidae and the Notodontidae. Both of these families are generally found throughout the world and are particularly abundant in the tropics. They are mainly nocturnal in habit and, like their relatives the Noctuidae, they possess tympanic organs on the third thoracic segment.

Haskell and Belton recorded the responses of the tympanic nerves of these moths to various acoustic stimuli. They found that the tympanic organ did not produce a resting discharge in the absence of sound but it did produce nerve impulses in response to pure tones of 3 kHz to 20 kHz, the highest frequency tested. With continuous tones of more than 6 s duration the rate of impulse production fell gradually to zero, but with short pulses a burst of impulses with a pronounced after-discharge was produced for each pulse. Haskell and Belton were not able to test responses to pure ultrasounds but they did find that these tympanic organs were more sensitive than the human ear to frequencies of 16–20 kHz and were also very sensitive to the high frequency squeaks of a glass bottle-top. The authors suggested that the sensitivity of the tympanic organs of arctiid and notodontid moths extended well into the ultrasonic range and that they would respond to short high frequency pulses at rates of up to 45 s^{-1}. It therefore seems probable that these moths could detect the ultrasonic cries of bats, but so far this does not appear to have been investigated any further.

As well as being able to detect sounds, some species of both the Arctiidae and of a related neotropical group, the Ctenuchidae, have the means of producing them. The sounds produced by a Trinidadian arctiid called *Melese* have been analysed by Blest, Collett and Pye (1963) who also described how they were produced. Many arctiid and ctenuchid moths possess paired 'tymbal organs'. These are modified sclerites, or sections of the hard exoskeleton, found one on each side of the third thoracic segment just below and in front of the tympanic organs. In *Melese* the greater part of this sclerite is swollen into a blister, the outer surface of which is thin and translucent. Running dorsoventrally down the blister is a row of 'dimples' or 'striae', which in some other moths are simple grooves or shallow rectangular troughs (Plate VIa).

The tymbal organs are apparently deformed by contraction of some of the thoracic muscles concerned with flight. Sounds are produced when the blister buckles inwards and then returns to its relaxed state. The buckling process begins at the dorsal end of the striated band, but it does not proceed smoothly; each stria in turn is strained to a point when it suddenly buckles, so producing a single pulse of sound or a click. When the tymbal organ returns to its resting state, each stria is reset producing a further click. There are between 15 and 20 striae in the tymbal organ of *Melese* (Plate VIb) and each muscle contraction produces between 10 and 20 clicks, all starting in the same phase (Plate VIc); a similar series, but all starting in the opposite phase, is produced during relaxation. Each complete cycle of buckling and relaxation is called a 'modulation cycle' and bursts of sound consist of a whole number of these cycles.

The moths studied by Blest, Collett and Pye were handled in order to elicit sound production and the pulses were produced at rates of up to 1200 s^{-1}, in 40 modulation cycles per second. This was later found to be approximately equal to the wing-beat rate in flight. Most of the energy of the sound was between 30 kHz and 90 kHz but there was also a weak audible component. As the tymbal organ buckled inwards, the frequency of successive pulses decreased through about one octave but when the organ returned to its original state, the frequency increased with successive pulses (Plate VId). The sounds were within the frequency band used for echo-location by bats so that the addition of these signals to the echoes from a moth's body could be used to 'conceal' the moth from an approaching bat. Unfortunately there was no evidence that these signals are emitted by flying moths and the animals did not produce the sounds when flying freely in small containers or when tethered. Alternatively these sounds could be examples of a nocturnal, non-visual aposematic display which warns predators of the genuine distastefulness of the animals, or of a pseudo-aposematic display, by which palatable animals mimic distasteful ones. Such a mimetic association is called Batesian mimicry. Blest and his colleagues left the function of these sounds as an open question.

The story was then taken up by Dunning and Roeder (1965) who showed that several other arctiid moths produced trains of high-pitched clicks with a dominant frequency at about 60 kHz when exposed to a series of artificial ultrasonic pulses while in stationary flight. They found that if these clicks were recorded and replayed when trained bats were catching mealworms tossed into the air, the bats turned away from most of the targets. But the bats were not affected when similarly exposed to the recorded echo-location cries of another bat. Obviously the sounds of arctiid moths adversely affect the feeding behaviour of the bats and so afford some measure of protection to the moths.

Later Dunning (1968) presented captive bats living in small cages with a variety of live moths including arctiids and ctenuchids which were known to emit clicks as well as with a control group, mainly sphingids, geometrids and notodontids, which were believed not to produce sounds. Not all of the experimental arctiids and ctenuchids produced clicks in the presence of bats, but those that did were rejected more strongly by the bats than the 'silent' ones. Dunning concluded that the clicks produced by arctiid and ctenuchid moths are generally aposematic and operate as warning signals to protect the moths from their predators, and that *Pyrrharctia*, the one palatable species that clicked freely, may represent a case of Batesian mimicry. Using 58 species of arctiids and ctenuchids, Blest (1964) found a negative correlation between readiness to click when handled and distastefulness to monkeys, marmosets and chickens (but not bats). This makes the aposematic explanation seem unlikely but possibly most species were Batesian mimics.

Among the Arctiidae, the sounds of the African genus *Rhodogastria* may be truly aposematic. In 1938 Carpenter described 'sizzling' sounds which were produced when various species of *Rhodogastria* exuded an unpleasant smelling and distasteful froth from vesicles on the prothorax. In an appendix to Carpenter's report, Eltringham described organs in one of these moths that he interpreted as a stridulatory file which produced sounds by friction against the next segment. Later Darwin and Pye (unpublished) found that in *Rhodogastria* and in other African arctiids this organ is a typical tymbal. As in *Melese*, it buckles and relaxes and sound can be elicited in response to artificial pulses or to the cries of a bat held in the hand. The sounds of a large number of arctiids from Africa and the neotropics have now been recorded on tape (Pye, unpublished).

The exact function of these mainly ultrasonic signals is still not clear. Besides 'jamming' or aposematic displays, it is possible that they could be used for communication between the moths themselves and Kay (1969) has even suggested that they would be adequate for echo-location by the moths. The primary function of these sounds, however, is probably to deter predatory bats and so afford some protection to the moths producing them.

Not all arctiids have tymbal organs but some without can produce single or paired clicks when in flight (Webster, 1963; Darwin and Pye, unpublished). Noctuids apparently also lack tymbal organs but Agee (1971) found that at least one species, *Heliothis zea*, could produce ultrasonic clicks by clapping the wings above its back. A mid-ventral organ of tymbal-like appearance was shown not to be implicated.

PYRALIDIDAE

Another group of moths in which tympanic organs are found, this time on the abdomen, is the family Pyralididae. This is a mainly tropical group of moths, some of which are agricultural pests in temperate regions. The adult of the European corn borer, *Ostrinia*, for example, lays its eggs on young maize plants. The larvae feed on the plant and in North America they cause a great deal of damage to the crops.

Belton (1962a), who was looking for a means of controlling this corn pest, studied the behaviour of tethered adult pyralids, including *Ostrinia*, to artificial ultrasonic stimuli. The moths showed a variety of responses: some began to fly, others which were already in flight stopped flying, some moths at rest moved their antennae while others showed no overt response. Belton then investigated the response of the tympanic organ to various sounds by recording from the tympanic nerve. He found that the acoustic sense cells produced an irregular resting discharge in the absence of sound but when the tympanic organ was exposed to a 22 kHz signal from a Galton whistle, or to the sound of two coins being clicked together, a rapid burst of impulses was produced. The tympanic organ adapted to a continuous tone of 32 kHz and generally did not respond to tones below 18 kHz. There was no evidence that these moths could distinguish between different sound frequencies.

Anatomical studies have shown that the paired tympanic organs of pyralids each contains four sensory cells and the oscilloscope traces of the tympanic response of *Ostrinia* indicated that three of these cells are sensitive to sound. Belton found that the sensitivity of the tympanic organs of pyralids was similar to that of noctuid moths and he suggested that their function was also similar: to detect the ultrasonic cries of bats.

Belton suspected that a continuous train of bat-like sounds might modify the breeding behaviour of pyralids and even repel them from their egg-laying sites, and so together with Kempster he studied the effect of broadcasting bat-like pulses over threatened maize crops (Belton,

1962b; Belton and Kempster, 1962). Pulses of 50 kHz were broadcast from rotating transducers over two experimental plots of maize at the repetition rate and intensity of echo-locating bats. The sounds were relayed between 5 p.m. and 9 a.m. from mid-June almost continuously until the corn matured. Belton and Kempster found that the number of larvae in the experimental plots was about one third of that in adjacent control plots and 14% of the ears of corn in the experimental plots were damaged compared with 23% in the control plots. Although this was only a preliminary experiment and the authors themselves pointed out many improvements that could be made, they suggested that it might indicate a possible means of controlling this pest without using pesticides. Belton and Kempster emphasized, however, that the adult moths are only repelled by the sounds and probably move to other areas. But they pointed out that the possibility of the moths becoming resistant to the sounds is small and if it did happen, the moths would lose their protection from predatory bats. However, Agee (1969a, b) was less optimistic that broadcast ultrasounds would be effective for controlling noctuid pests (see section on Noctuidae).

SPHINGIDAE

Moths belonging to the family Sphingidae, or hawk moths, are found throughout the world. The adults are generally large and around dusk they can often be seen hovering near flowering plants, drinking nectar through a long proboscis. The family includes the well-known death's head hawk moth, *Acherontia*, which emits a shrill audible sound when irritated.

In 1967 Roeder was watching about a dozen sphingids, *Celerio lineata*, feeding from a jasmine plant (Roeder *et al.*, 1968). He noticed that a slight hiss or the jangling of keys caused the moths to fly away, only to return and continue feeding a few seconds later. He confirmed these observations on caged moths although the moths did not respond to sound when at rest.

Sphingid moths do not possess tympanic organs on the thorax or on the abdomen, but Roeder, Treat and Vande Berg (1968) found that when a moth was exposed to ultrasonic stimuli, nerve impulses could be recorded from an interneurone in the cervical nerve tracts connecting the brain with the first thoracic ganglion. These nerve impulses originated in the head of the moth and were produced in response to frequencies of

7 kHz to 100 kHz, with the lowest threshold at 20–40 kHz. Roeder and his co-workers found that amputation of the antennae or the proboscis did not affect the acoustic response but amputation of the labial palps reduced the sensitivity roughly in proportion to the amount amputated. But even complete amputation did not abolish the response altogether. Deflection of the palps from their normal position also greatly reduced the response but it returned to normal when the palps were replaced. The receptor mechanism therefore appeared to be in the base of the palp but no acoustic response could be found in the palpal nerve. The hawk

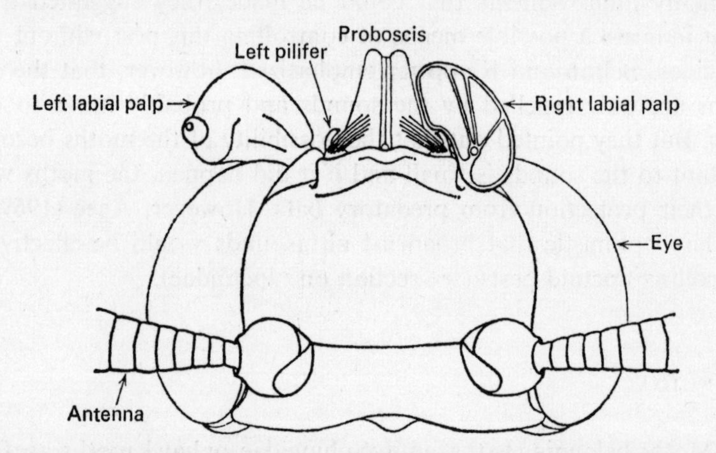

FIGURE 4.3 A diagram of the dorsal view of the denuded head of *Celerio lineata*. The right labial palp is in its normal position and has been sectioned to show its apposition to the pilifer and also the internal airsac. The left labial palp has been deflected laterally to expose the distal lobe of the left pilifer. (After Roeder *et al.*, 1970)

moths in which these behavioural and acoustic responses occurred, belonged to the sub-family Choerocampinae which is distinguished from other sphingids by the bulbous form of the second segment of the labial palp. Both the first and second segments are almost entirely filled by an air sac (Fig. 4.3) but Roeder and his colleagues could find no sensory structure in this palp.

Later, Roeder and Treat (1970) investigated the acoustic response further by recording from an interneurone in the prothoracic ganglion while manipulating the palps in various ways. When they deflected both of the palps laterally by 10°, the sensitivity of the acoustic response fell by about 40 dB, but when they deflected only one palp, the sensitivity of the response was reduced by a few decibels at the most. Any given interneurone in the prothoracic ganglion responded similarly to deflection of either the right or the left palp. This indicated that the interneurone was excited by impulses coming from either side of the head and so it could not relay information about the direction of the sound source. Although no acoustic response was detected in the palpal nerve, such a signal, identified by its short latency, was detected in the suboesophageal region of the brain, close to the root of the palpal nerve. Single-peaked impulses were produced at all sound intensities, suggesting that only a single sensory cell was involved in each acoustic organ.

Roeder, Treat and Vande Berg (1970) at last found the probable site of the acoustic receptor, not in the palp itself but in an adjacent structure on the head, the pilifer. In most butterflies and moths, the pilifers are small stiffened projections which lie on either side of the base of the proboscis. They generally bear long, sensory bristles which probably sense movements of the proboscis during feeding. In the Choerocampinae, however, the pilifers have in addition thin-walled and somewhat elastic distal lobes that flex when pressure is applied towards the midline. The enlarged second segment of the labial palp has a small ridge on its inner wall and normally the pilifer touches this ridge and is completely concealed by the overlying palp (Fig. 4.3). Airborne ultrasounds of suitable frequency apparently set up vibrations in the inner wall of the palp which acts as sounding board and transmits vibrations to the distal lobe of the pilifer. It is thought that impulses produced by a receptor within the pilifer are then transmitted to the brain via the labral nerve.

When contact between the palp and the pilifer lobe was broken, the 40 db decrease in sensitivity could be almost completely restored (Fig. 4.4) by placing a small 'flag' of mylar (plastic film) so that its tip just touched the convex surface of the pilifer lobe. Later Roeder (1972) showed that the distal lobe of the pilifer was sensitive to mechanical vibrations as small as 0·02–0·1 nm at 30–70 kHz. This suggested that the frequency response of the intact organ, which decreases above 30–40 kHz, is limited by the physical characteristics of the labial palp and not by the pilifer sense organ itself.

The neural responses of the whole organ, that were recorded from either side of the thoracic or suboesophageal ganglia, contained little information about the frequency, intensity or the direction of the source of acoustic stimulus. Roeder therefore suggested that it must be regarded as an acoustic alerting system stripped to the bare essentials. Nevertheless this acoustic organ, like that of noctuid moths, probably serves to alert choerocampine hawk moths to the presence of predatory bats. This particular hearing organ, however, is unique in several ways. It is formed by interaction between two different appendages of the head, the labial palps and the pilifers, which are derived from labial and labral structures respectively. The sensitivity of the organ depends only on the contact of these two appendages, which are not fused and so can easily be separated and re-aligned. Also, this 'ear' is restricted to the one sub-family Choerocampinae. Roeder examined many other sphingid species but failed to find any interneurone responses to acoustic stimuli,

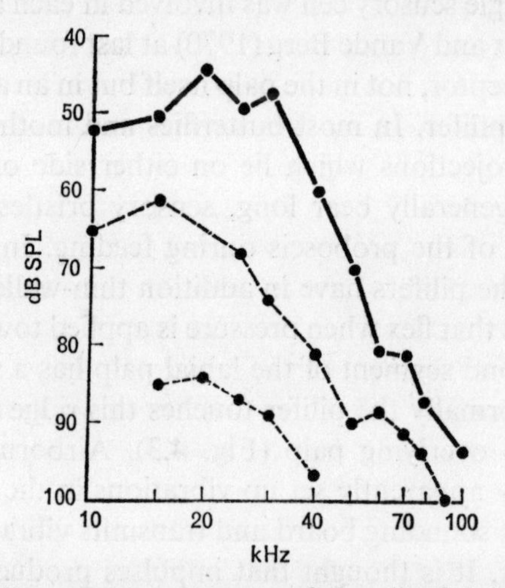

FIGURE 4.4 Auditory response curves of the distal lobe of the pilifer of *Celerio lineata*. (Top curve) The intact organ. (Bottom curve) After the labial palp has been deflected to one side. (Middle curve) After a small mylar flag has been brought into contact with the distal lobe. (After Roeder, 1972)

except in one species which was less sensitive than any of the Choerocampinae. This unusual acoustic mechanism therefore seems to have evolved in only one out of five sub-families of the Sphingidae although many members of the other four sub-families are similar in size, flight pattern and feeding habits. As Roeder points out, it is possible that they have other types of auditory organ not yet discovered.

It was mentioned earlier that one sphingid, not a choerocampine, emits sound. This is *Acherontia atropos*, the death's head hawk moth which produces two different sounds alternately. The first is a broad-band sound that resembles a grinding noise and it is followed immediately by a shrill whistle (Plate VIIa–e). Busnel and Dumortier (1959) studied these sounds using equipment that had an upper frequency limit of 25 kHz. They reported that the broad-band sound had a frequency spectrum of 5–20 kHz and a peak of energy at 6–8 kHz. The whistle began with components at 5 kHz, 10 kHz and 15 kHz which decreased in frequency to about 2, 4 and 6 kHz respectively at the end of the call. The sonagram shown by Busnel and Dumortier had an upper frequency limit of 16 kHz but it showed indications of higher harmonic components. East African specimens, recorded by Pye (unpublished) using ultrasonic equipment, produced broad-band sounds of 8–20 kHz with a peak at 13–16 kHz and these were followed by a whistle with a fundamental component at 12–16 kHz and a second and possibly a fourth harmonic component (Plate VIIa).

Busnel and Dumortier studied the sound-producing mechanism of *Acherontia* and verified an earlier suggestion by Prell that the sound had

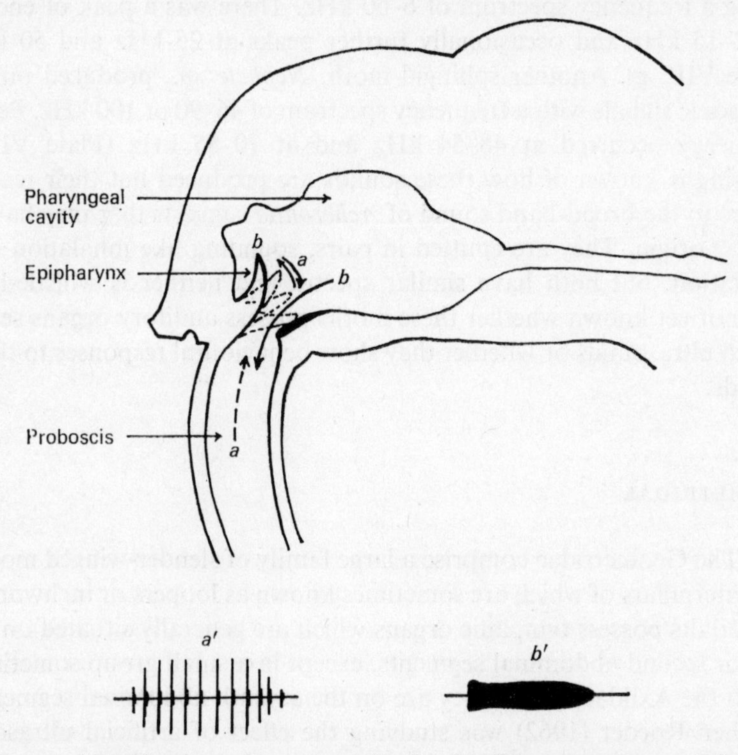

FIGURE 4.5 Mechanism of sound production in *Acherontia atropos.* Diagram of a sagital section through the head showing the action of the epipharynx. *a*, air is drawn into the pharyngeal cavity, the epipharynx vibrates and the broad-band sound *a'* is produced. *b*, the whistle *b'* is produced as air is forced out of the pharyngeal cavity while the epipharynx is raised. (Modified from Busnel and Dumortier, 1959) An analysis of these two sounds is shown in Plate VII.

a dual origin. They found that the broad-band sound was produced when the roof of the pharyngeal cavity was raised by the pharyngeal dilator muscles and air was sucked in to the cavity through the proboscis. A crescent-shaped flap, the epipharynx, which can be lowered to close off this cavity, vibrated as the air was sucked in and this was thought to be responsible for producing the broad-band sounds. When the roof of the cavity relaxed, the epipharynx was raised and air was forced out again. This coincided with the shrill whistle (Fig. 4.5).

Some other sphingid moths also produce sounds that extend into the ultrasonic range. Dr A. Pye noticed that, when handled, the East African *Coelonia mauritii* emitted sounds which were later analysed (Pye, J. D., unpublished) and found to consist of a series of spiky pulses giving a frequency spectrum of 6–60 kHz. There was a peak of energy at 12–15 kHz and occasionally further peaks at 25 kHz and 50 kHz (Plate VIIf, g). Another sphingid moth, *Nephele sp.*, produced purely ultrasonic signals with a frequency spectrum of 46–90 or 100 kHz. Peaks of energy occurred at 48–54 kHz and at 70–85 kHz (Plate VIIh). Nothing is known of how these sounds are produced but their resemblance to the broad-band sound of *Acherontia* suggests that they have a similar origin. They are emitted in pairs, sounding like inhalation and exhalation, but both have similar spectra and neither is whistle-like. It is not yet known whether these moths possess auditory organs sensitive to ultrasounds or whether they show behavioural responses to these sounds.

GEOMETRIDAE

The Geometridae comprise a large family of slender-winged moths, the caterpillars of which are sometimes known as loopers or inchworms. The adults possess tympanic organs which are generally situated on the first or second abdominal segments, except in a small group sometimes called the Axiidae, where they are on the seventh abdominal segment.

When Roeder (1962) was studying the effect of artificial ultrasonic pulses on the behaviour of free-flying moths, he found that most of the moths taking avoiding action belonged to the Noctuidae or Geometridae. Roeder examined the acoustic responses of the noctuid tympanic organ in detail but so far no one appears to have studied that of geometrids. From their behavioural actions it seems likely that these moths too, are able to detect and evade bats. Perhaps by the time this book is published yet another 'bat-detecting' system will have been demonstrated experimentally.

NEUROPTERA

Another group of nocturnal flying insects which apparently can hear the ultrasonic cries of bats, belongs to the order Neuroptera or lacewings. These are not related to the moths and they have biting mouthparts instead of the long coiled proboscis of adult moths. Roeder noticed that when artificial ultrasonic pulses were relayed through a loudspeaker in the open, many green lacewings behaved in a similar manner to some of the moths; they closed their wings and fell passively towards the ground.

Lacewings of the family Chrysopidae have a small swelling near the base of each forewing where two veins fuse (Plate VIIIa). In 1966, Miller and MacLeod showed that in *Chrysopa* these swellings contained sensory cells resembling those of typical insect tympanic organs and that the insects possessing them were sensitive to sound frequencies of up to 100 kHz. Later (1970, 1971) Miller made a more detailed study of both the structure and the physiological responses of this acoustic organ. He reported that the ventral cuticle of each swelling acts as a tympanic membrane and that the sensory units are grouped together to form two chordotonal organs, each of which makes contact with the tympanic membrane. One chordotonal organ contains five to seven sensory units, the other about 20. Their nerve axons combine to form a chordotonal nerve which eventually joins the radial nerve of the wing and runs to the mesothoracic ganglion.

Miller (1970) pointed out that although the tympanic organ of green lacewings resembles that of many other insects, it differs in two main respects. Firstly the cavity of the swelling, in which the sensory cells lie, is part of the body cavity or haemocoele and so it is filled with fluid, unlike the air-filled tracheal sacs of other tympanic organs; the tracheal air sac which does run through the swelling is not expanded. Secondly, the cuticle forming the tympanic membrane is rippled in lacewings and consists of only one sheet of material. In most other insects the tympanic membrane is smooth and composed of two layers, the external cuticle and the wall of the tympanic air sac. Despite these features, Miller concluded that this tympanic organ probably acts as a sound pressure receptor.

Miller (1971) then recorded the acoustic responses of the tympanic organ in *Chrysopa* from both the radial nerve and from one or a few sensory units within the swelling itself. The tympanic organ responded to acoustic stimuli between 13 kHz and 120 kHz, but within this range it appeared to be about 20 dB less sensitive than the noctuid tympanic organ. The acoustic response did not adapt to pulses up to 30 ms long and only showed slight adaptation to pure tones lasting up to one minute. When *Chrysopa* was exposed to artificial ultrasonic pulses, the flexor muscles of the forewing twitched and from the electrical activity of these muscles it appeared that the acoustic response of the tympanic organ provided the input for this reflex contraction. Miller suggested that although lacewings might possibly use their tympanic organs for intraspecific communication, it was more probable that this was yet another group which had adopted a means of detecting and evading bats.

Miller and MacLeod (1966) found that tympanic organs were present in many species of the family Chrysopidae but were absent from five genera which are sometimes placed in a separate sub-family together with the fossil Chrysopidae. It seems, therefore, that within this family, the evolution of the tympanic organs is relatively recent. Miller and MacLeod suggested that the recent success of the Chrysopidae was possibly a direct result of their advantage over their relatives by virtue of their protection from bats.

EVOLUTION

This chapter has examined the variety of acoustic sense organs that have been evolved by many different nocturnal flying insects. In all cases these organs appear to be used to detect and evade bats, yet they have not evolved uniformly. They occur on different parts of the body in different families and even within a family they may not be possessed by all the members. This suggests that these tympanic organs have evolved independently in each of the different nocturnal insect groups possessing them and that they arose in response to a special selection pressure after these groups had diverged during evolution.

The first fossil insects are found in Devonian strata which were probably laid down about 300 million years ago and the first Lepidoptera (butterflies and moths) seem to have arisen in Permian times, about 200 million years ago. Echo-locating bats probably did not appear until the

Eocene, about 150 million years later, that is 'only' 50 million years ago. By this time the various lepidopteran groups were all firmly established and so each apparently responded in its own way to this new predator that constituted such an important nocturnal hazard. This view is supported by Miller and MacLeod's observations on the distribution and probable time of evolution of tympanic organs in the green lacewings. The probable evolutionary story is summarized in the following verses, composed in 1968 after Roeder discovered the ultrasonic sensitivity of choerocampine sphingids by jingling his keys near a jasmine hedge in Texas.

In days of old and insects bold
(Before bats were invented),
No sonar cries disturbed the skies—
Moths flew uninstrumented.

The Eocene brought mammals mean
And bats began to sing;
Their food they found by ultrasound
And chased it on the wing.

Now deafness was unsafe because
The loud high-pitched vibration
Came in advance and gave a chance
To beat echo-location.

Some found a place on wings of lace
To make an ear in haste;
Some thought it best upon the chest
And some below the waist.

Then Roeder's keys upon the breeze
Made sphingids show their paces.
He found the ear by which they hear
In palps upon their faces.

Of all unlikely places!

(Pye, J. D., 1968, *Nature*, **218,** 797)

CHAPTER FIVE

The Songs of Bush Crickets (Tettigoniidae)

Members of the family Tettigoniidae, the katydids or bush crickets, are found in both tropical and temperate regions. They produce sounds by rubbing their forewings together, a method often known as 'stridulation'. This term, however, is also used more generally to describe sound production in insects by any mechanical means. The more specific term 'strigilation' will therefore be used here to describe a particular method of sound production that occurs in tettigoniids and many other insects; the friction of differentiated parts (Dumortier, 1963a).

Many of the insect songs that are produced by strigilation contain ultrasonic components and in the Tettigoniidae most of the sound energy of the song is often within the ultrasonic band. So far ultrasounds have been detected mainly from this family of insects but they have also been detected from the family Gryllidae, the crickets, as well as from the family Acrididae, the locusts and grasshoppers, and they are probably emitted by many other groups of insects which produce sounds by strigilation. The most recent studies of these ultrasonic components of insects songs have made use of the equipment that was developed for the study of bat echo-location cries. Previously, however, both Pielemeier (1946) and Pierce (1948) had detected the ultrasonic components of insect songs in the field using equipment that had been developed for artificial ultrasounds in the laboratory. Pierce in particular made a detailed study of these songs but his work does not appear to have been fully appreciated by many recent workers in this field.

A study of the mechanism of strigilation shows that it is not really surprising that some ultrasounds are produced. What may seem paradoxical is that the songs of many species of tettigoniids are mainly ultrasonic and yet they can be heard by the human ear. This also, can be explained by the mechanism of strigilation and so, before the songs of these insects are described, the mechanism and the physics of strigilation must be discussed briefly.

THE PHYSICS OF STRIGILATION IN GENERAL

The strigilatory apparatus is made up of two parts, the 'pars stridens' or file or strigil and the 'plectrum' or scraper or strigilator. The file is a part of the exoskeleton which has become modified to form a

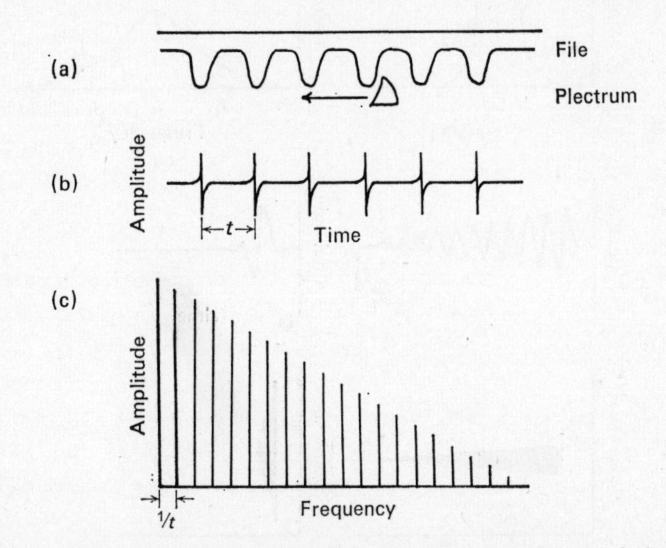

FIGURE 5.1 The physics of strigilation (i). The interaction of the file and plectrum. (a) The plectrum moves along the file, striking each tooth in turn. (b) Each tooth-strike produces a brief click. (c) A series of such clicks has a frequency spectrum with a fundamental at the tooth-strike rate and many harmonics.

series of teeth or ridges. Another part of the exoskeleton, often a protruding peg or hard edge, forms the plectrum. Sound is produced by the plectrum and file moving rapidly across each other. In insects in general, any two parts of the body that can be brought together may be used for strigilation, and different insects use a wide variety of areas. For example strigilation may occur between two mouth parts, two legs or two wings, between one leg and one wing or between one appendage and a region of the thorax or abdomen or between adjacent segments of the body (Dumortier, 1963a).

As the plectrum strikes each tooth of the file it produces a click, a brief sound which consists of a wide range of frequencies (Fig. 5.1). A train of such clicks has a fundamental frequency, equivalent to the rate at which the plectrum strikes the teeth, and a series of harmonics, giving a wide total band of energy. But the actual sound that is emitted is

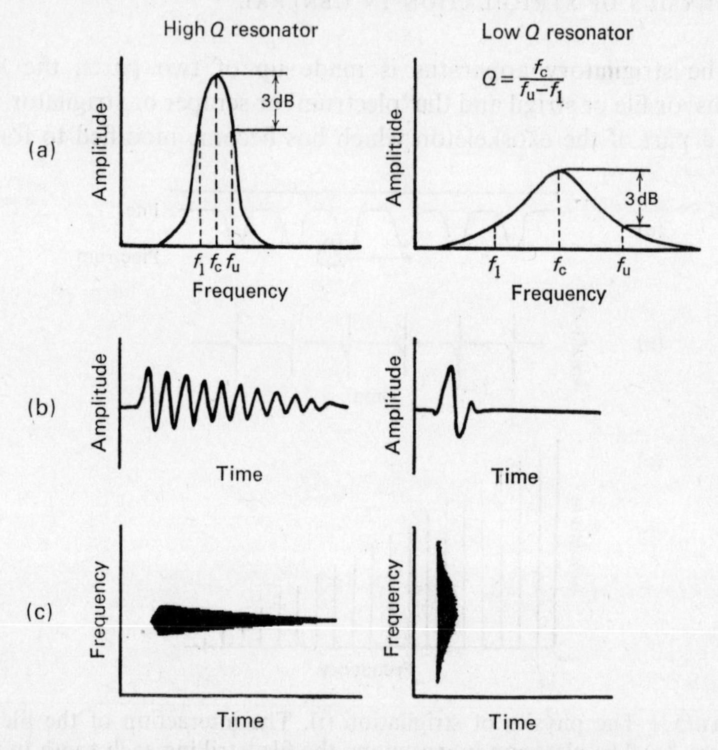

FIGURE 5.2 The physics of strigilation (ii). The responses of two resonators. Left: sharply tuned resonator showing (a) the frequency response curve, (b) the waveform produced when the resonator is excited by a sudden shock of broad-band energy and (c) the sonagram analysis of such a response. Right: the same for a broadly-tuned resonator showing that the response dies away more quickly and covers a wider frequency band.

generally affected by the presence of a resonator coupled to the strigilation mechanism. Familiar resonators include objects such as bells or wine glasses which 'ring' when they are struck. The energy that they receive from the blow is dissipated slowly within a narrow frequency band around a centre frequency (Fig. 5.2 left). A resonator is therefore said to be sharply tuned or to have a high 'quality' or *Q* value. A non-resonating object, such as a wooden table, has a low *Q* value and when it is struck it dissipates the energy rapidly over a wide frequency band (Fig. 5.2 right). The numerical value of Q for a resonator can be calculated from its frequency response curve, that is the relative amplitudes of the different frequencies that the resonator radiates when it is struck. Q is equal to the frequency of maximum amplitude divided by the range of frequencies that are within 3 dB of this amplitude (Fig 5.2a).

Many insects have a resonator that is mechanically coupled to the file or plectrum. This resonator is generally a modified part of the exoskeleton. It receives energy from the clicks produced by the file and plectrum and dissipates it again within a restricted frequency band. This is illustrated in Fig. 5.3. A stylized waveform of a train of clicks was shown in Fig. 5.1b. The frequency spectrum, the relative amplitudes of

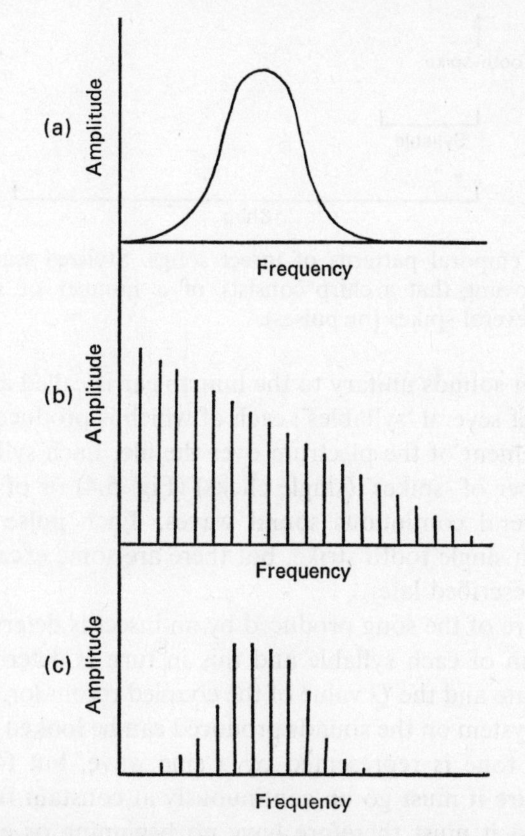

FIGURE 5.3 The physics of strigilation (iii). The production of a formant. (a) The frequency response of a resonator (as in Fig. 5.2a) acting on (b) a harmonic spectrum (as in Fig. 5.1c) results in (c) a new spectrum consisting of a group of harmonics called a formant.

all the frequencies present from Fig. 5.1c, is repeated in Fig. 5.3b. The effects of using these clicks to excite a resonator is shown in Fig. 5.3c. It can be seen that the resonator acts as a filter and the resulting frequency spectrum with a single 'hump' is called a formant (see Scroggie, 1955).

The insect resonator is excited by each tooth-strike in turn and so a train of sounds is produced as the file and plectrum move over each other. Longer bursts of sound are produced by several consecutive movements of the file and plectrum. The different temporal divisions of insect song have been defined by Broughton (1963). A continuous burst

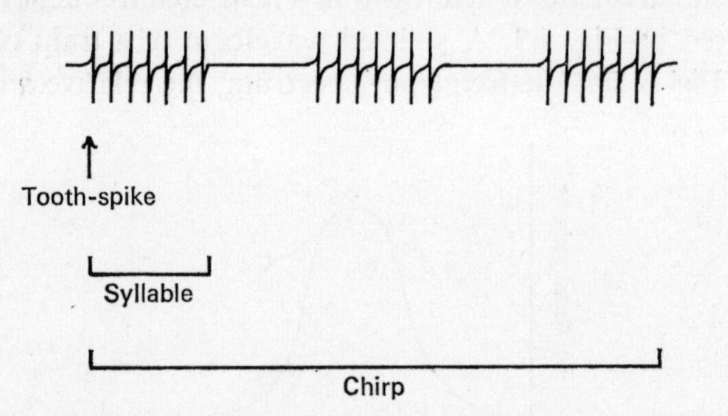

FIGURE 5.4 Temporal patterns of insect songs. Stylized waveform of an insect song showing that a chirp consists of a number of syllables each composed of several spikes (or pulses).

of activity that sounds unitary to the human ear is called a 'chirp'. This is composed of several 'syllables', each of which is produced by a single cycle of movement of the plectrum over the file. Each syllable is made up of a number of 'spikes' (single clicks) (Fig. 5.4) or of 'pulses' consisting of several continuous sound waves. Each pulse is generally produced by a single tooth strike, but there are some exceptional cases that will be described later.

The nature of the song produced by an insect is determined by the final waveform of each syllable and this in turn is determined by the tooth-strike rate and the Q value of the coupled resonator. The effect of this coupled system on the sounds produced can be looked at in another way. A pure tone is represented by a sine wave, but for this to be completely pure it must go on continuously at constant frequency and amplitude and it must therefore have no beginning or ending. If the amplitude or the frequency are changed at all, energy at other frequencies is introduced. These additional frequencies are both above and below the original 'carrier' frequency and are called 'side-bands'.

The more sudden the amplitude change, or modulation, the wider is the range of additional frequencies that are produced. Side-bands that occur only at the beginning or at the end of a sound are called 'transients'.

This has of necessity been a very brief and simplified discussion of the formation of side-bands. Readers wishing for a more detailed discussion are referred to Scroggie (1955) and to Brown and Glazier (1964).

If an insect resonator has a low Q value, as in the tettigoniid *Conocephalus conocephalus*, each tooth-strike results in a broad-band sound which decays rapidly before the next tooth is struck. A typical syllable of this type is illustrated in Fig. 5.5a. Each spike has a rapid onset and decay and so a wide range of side-bands must be inferred. The waveform of such a spike (seen in Fig. 5.5a) shows only the 'carrier' or 'resonant' frequency. As this is amplitude modulated, side-bands must be present but they are not seen as superimposed waves on the oscilloscope trace, although they do appear on the sonagram trace. In the case of a very low Q system an additional low frequency component will be present due to the asymmetry of the syllable envelope. In a high Q system this component will be fainter.

When the insect resonator has a high Q value as in some other tettigoniids, the sound produced by one tooth-strike does not die away completely before the next tooth is struck (Fig. 5.5b). The amplitude of each spike changes less rapidly than in *Conocephalus* and so the side-bands that are produced cover a much narrower frequency band. The sound produced by a syllable of this type would therefore have only a small frequency spread.

The side-bands in these examples are very real for, even though they do not appear as separate waves in the waveform, they can be seen as amplitude changes in the carrier. So even though the carrier frequency may not be detected by the human ear, some of the lower side-bands may be. In the same way a low frequency carrier may have side-bands that extend into the ultrasonic range.

Two further cases must be mentioned briefly. When the resonant frequency is only twice the tooth-strike rate, one tooth strike will produce only two sound waves before the next tooth is struck (Fig. 5.5c). The waveform of the sound produced is equivalent to that of a fundamental tone (at the frequency of the tooth-strike rate) and its second harmonic at twice this frequency, the resonant frequency of the filter. On the other hand, if the tooth-strike rate is such that only one wave of sound is produced by each tooth-strike, each 'syllable' will consist of a continuous and coherent train of waves (Fig. 5.5d). The whole syllable

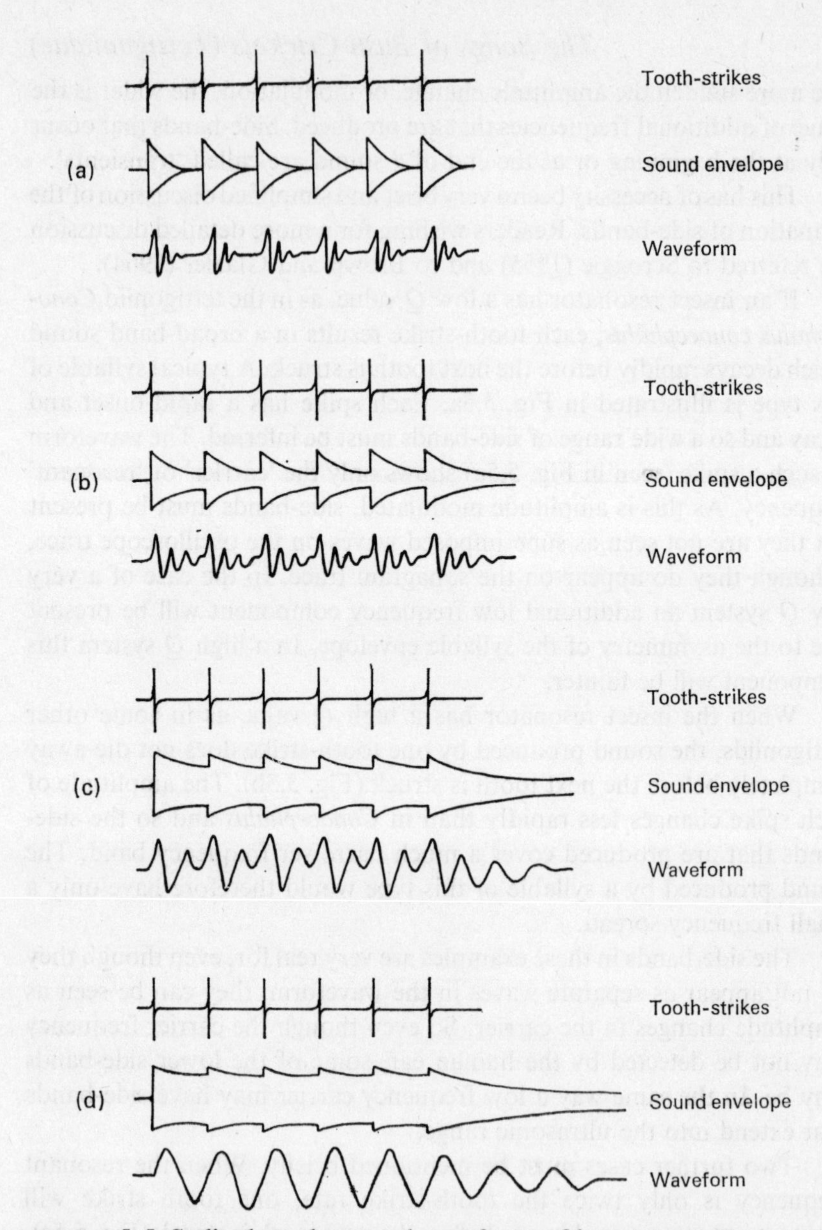

FIGURE 5.5 Stylized envelopes and waveforms of insect songs. Each tooth-strike initiates a pulse with a rate of decay that depends on the associated resonator. (a) A broadly tuned, low Q resonator gives a rapid decay and a spike-like waveform. (b) Medium Q resonator gives a slower decay so that each pulse runs in to the next. (c) A high Q resonator tuned to twice the tooth-strike rate gives a smooth waveform including a strong second harmonic. (d) A high Q resonator tuned to the tooth-strike rate gives a 'pure' sine wave, the fundamental, with a long decay at the end of the syllable.

is then equivalent, acoustically, to a 'pulse' and an almost pure tone is produced with side-bands present only as onset and offset transients, as well as the weak component due to the asymmetry of the syllable envelope. Calls of this type are produced by the tettigoniid *Drepanoxiphus modestus* (Suga, 1966), and during the calling song of crickets (p. 132).

In order for the sound energy to be radiated effectively, the resonator must be coupled to air by a reasonably large surface area acting as a sounding board. This structure is not necessarily the same as that of the resonator and in tettigoniids there is evidence that the two are separate, although of course they are closely coupled together.

It can be seen that several factors are involved in the production of sounds by even a single movement of the plectrum over the file and that insects can produce ultrasounds either as the main carrier frequency, possibly with audible side-bands, or as the upper side-bands of an audible carrier frequency. If such insect songs are recorded with a good quality 'audio' microphone and tape-recorder, they sound exactly right when replayed at normal speed. But examination of the waveform or acoustic analysis will not determine whether the carrier frequency is present in the recording or whether it has been filtered off by the apparatus. Obviously in the latter case several features of the original insect song will not be represented in the remaining waveform. Many insect songs have been recorded and studied using 'audio' equipment. If the frequency of one of these songs is high and if it is faint to the human ear when the animals appear to be calling over long distances, it would be worth re-examining the song with ultrasonic equipment.

MECHANISM OF SOUND PRODUCTION IN TETTIGONIIDS

In tettigoniids the file and plectrum are found near the base of the forewings, which are thickened to form protective covers, called tegmina, for the hindwings. The plectrum is situated on the upper surface of the right tegmen and consists of a sharp edge of thickened cuticle on the medial edge of the tegmen. The file is on the under surface of the left tegmen and is developed from one of the wing-veins as a series of folds in the cuticle (Pierce, 1948; Anstee, 1971). These folds form the teeth of the file which runs approximately at right-angles across the long axis of the tegmen (Plate IX a, f, g).

The mechanism of sound production in these animals has been known in outline for many years. The wing covers are raised so that the plectrum is roughly at right-angles to the file and they are then moved rapidly across each other. Pierce filmed the actions of the wings and measured the time taken for them to open and close. He concluded that in several species of tettigoniids sounds were produced only during the closing movement of the wings, the opening movement being silent. More recently, Suga (1966) has shown that in at least one species of tettigoniid, sounds are also produced during the opening movements of the wings. To start with Suga demonstrated the relationship between the structure of the sounds and the teeth of the file by studying the sounds that were produced both before and after several teeth had been removed from the file. Suga found that in two different species of *Phlugis* and in *Conocephalus saltator* each of the spikes making up a syllable was produced by one tooth strike, for when a few teeth were eliminated from the file, the corresponding number of spikes was absent from the syllable (Plate VIIIb). A fourth species produced syllables in which the tooth-strikes formed a continuous waveform (Fig. 5.5d), resulting in almost pure-tone syllables. When a few teeth were removed from the file of this species, the pulse contained some sound waves which were reduced in amplitude. The number of these smaller waves nearly corresponded to the number of teeth removed, indicating that each tooth-strike probably produced one normal large sound wave. In some pulses the total number of waves was greater than the number of teeth in the file and so two waves may have been produced by each tooth-strike.

Then by removing teeth from one end of the file only, for example the medial end, Suga found that in both species of *Phlugis* and in *Drepanoxiphus* the gap in the syllable always occurred in the same place, in this case the beginning of the syllable (Plate VIIIb). This showed that sounds were produced only during the closing movements of the tegmina. *Conocephalus saltator* produced long and short syllables alternately and when teeth were eliminated towards the medial end of its file, several spikes were missing from near the beginning of the longer syllables, and from near the end of the shorter syllables. These long and short syllables were therefore produced by the closing and opening movements of the wing respectively. In this case the sounds produced on each stroke of the wing movement are often known as 'hemisyllables'.

Amblycorypha uhleri produces complex songs with four different types of sounds that differ in the rate and amplitude of spike production and to some degree in the frequency spectrum (Alexander, 1960). Recently, Walker and Dew (1972), using sound-synchronized cine-photography, found that each type of sound was produced by a different movement of the wings and that sound was produced during both the opening and closing strokes of the wing, either of which could have two phases, one silent and one producing sound.

During strigilation the wings move very rapidly and rates of up to 200 strokes per second have been reported (Josephson and Elder, 1968; Heath and Josephson, 1970). In *Neoconocephalus* the muscles that are responsible for moving the wings during strigilation are also those that are used for flight. Before strigilation begins, the body temperature of the insect is increased by up to 5–15°C above that of the surroundings, by the simultaneous contraction of all the flight muscles (Heath and Josephson, 1970). The ultrastructure of the muscles is well adapted to contract and relax very rapidly although, unlike the rapidly contracting muscles of some other insects, each contraction of these flight muscles is caused by a single muscle action potential (Josephson and Elder, 1968; Elder, 1971; Josephson and Halverson, 1971).

The identification of the resonator in tettigoniids has not been as straightforward as the identification of the file and plectrum. In many species there is an area of thin, transparent cuticle near to the base of the right tegmen which is known as the 'mirror'. Some species of tettigoniids have a similar area on the left wing as well, although this may be smaller and thicker than the right mirror. The mirror on the right tegmen is surrounded by a rigid ring of sclerotized wing veins. The most anterior of these is the vestigial file which corresponds to the true file on the left tegmen. In the *Conocephalus* group of tettigoniids, the frame forms a complete rigid boundary to the mirror whereas in the *Homorocoryphus* group the frame is more 'horseshoe' shaped as there is no rigid connection between the outer edge of the vestigial file and the distal part of the mirror frame (Bailey, 1970). The mirror frame itself is surrounded and separated from the rest of the tegmen by thin, membranous tissue which slopes up towards the frame on all sides except at the end nearest to the base of the tegmen. Here the vestigial file is connected to the plectrum by a fairly rigid and heavily sclerotized region (Plate IX, b, d).

Pierce (1948) reported that the plectrum was set in vibration by the teeth of the file passing over it and that the vibrations were communicated to the nearby mirror region. He believed that all of the wing in this region took part in the vibration and when he removed part of the tegmen of one species, *Conocephalus spartinae*, leaving the mirror region, which he called the disc, intact, he found that the main frequencies produced by this animal rose from 40 kHz and 31 kHz to 56 kHz and 42 kHz respectively. This, he suggested supported the idea that the frequency of the sound was determined by wing resonance. But Pierce also suggested that the mirror caused 'a more pronounced sound radiation of a frequency which is resonant with the disc's natural frequency'. Making certain assumptions, he calculated the resonant frequencies of the mirror membranes of several species of tettigoniids and these calculated frequencies fitted the recorded frequencies of the natural sounds fairly well in some cases.

Pierce's suggestion that the mirror membrane was the principal resonator in tettigoniids remained unchallenged until 1964, when Broughton expressed doubts as to the acoustic function of this region. Broughton pointed out that as the frame was surrounded by a thin membrane, the frame and the mirror could oscillate as whole, relative to the rest of the tegmen. A few years later Morris and Pipher (1967) found that puncturing the mirror membrane of *Conocephalus nigropleurum* did not affect the frequency spectrum of the sounds produced but only decreased the amplitude of these sounds. In the same year Bailey (1967) showed that the frequency spectra of *Conocephalus discolor* and of *Homorocoryphus nitudulus* were affected if the mirror frame as well as the mirror membrane were damped by applying a thin layer of latex. All of these workers therefore concluded that the mirror frame was the important structure in sound production rather than the mirror membrane. Morris and Pipher suggested that in *C. nigropleurum* the rigid structure of the plectrum, the vestigial file and the mirror frame together act as a cantilever with the whole of the vestigial file forming the fulcrum. The mirror frame would then vibrate about this fulcrum when the plectrum is set in motion by the teeth of the real file. These authors suggested that the mirror membrane acts as a sounding board, radiating all frequencies equally.

More recently Bailey (1970) and Bailey and Broughton (1970) have studied this lever mechanism in some detail. They developed a method of activating pairs of isolated tegmina of *Homorocoryphus* to produce sounds that were very close to the natural song of the species. Bailey found that in *Homorocoryphus* the region of maximum vibration of the mirror frame was the part most distal to the vestigial file. This supported Morris and Pipher's suggestion that the whole frame vibrates about a pivot. But Bailey pointed out that the frame of *Homorocoryphus* differed from that of the *Conocephalus* studied by Morris and Pipher. As described earlier, the frame of *Homorocoryphus* does not form a complete rigid ring around the mirror. In this group therefore the frame resembles a horseshoe that is clamped along one arm (the vestigial file) and struck at the mid-point. In both of these groups of tettigoniids the thin membrane around the frame allows this to vibrate freely and so decouples the resonator from the rest of the more rigid tegmen.

Bailey and Broughton (1970) showed that in *Homorocoryphus*, when the plectrum encounters a tooth of the file, it is forced upwards. The lever formed from the plectrum, vestigial file and mirror frame pivots about the fulcrum and so the end of the frame is depressed and the thin membrane around the frame is carried down with it. Once the

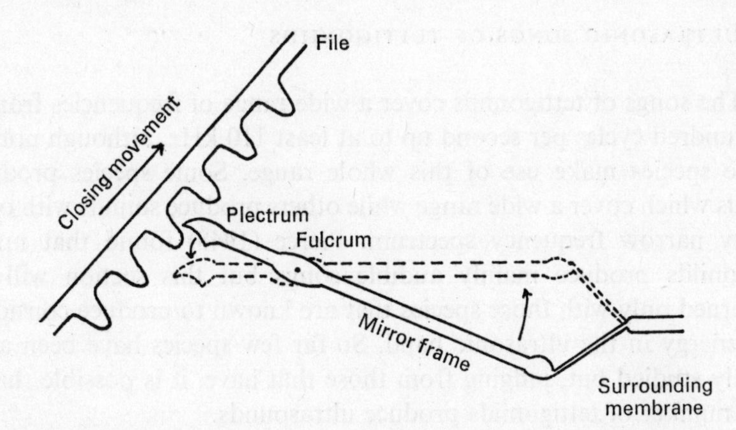

FIGURE 5.6 The cantilever action of a tettigoniid mirror-frame and plectrum, seen in transverse section.

plectrum has left the tooth, the frame springs back to its original position before the next tooth is struck, due to the elasticity of the system (Fig. 5.6). In *Homorocoryphus* the mirror frame is made to vibrate 12,000–15,000 times a second when the tooth-strike rate itself approaches these values. Each syllable is therefore composed of a continuous waveform at 15 kHz and few side-bands are produced. In *Conocephalus* the tooth-strike rate does not approach the resonant frequency of the mirror frame and so each tooth impact merely sets the mirror frame vibrating at its natural frequency and the pulse dies away before the next tooth is struck, resulting in a broad-band sound.

Bailey (1970) found that the dominant frequencies produced by both *Homorocoryphus* and *Conocephalus* fitted well with a theoretical consideration of the lever model (see later) but he only considered the frequency components below 40 kHz. *Conocephalus* emits songs with a second band of energy above 40 kHz although the lower band does contain the major energy component. The theoretical consideration is therefore successful but the source of this second frequency component is unexplained and should be sought. Dumortier (1963b) suggested that the wide-band sounds produced by some species of tettigoniids indicate that several resonators are involved. Possible sources of other resonances are the column of air beneath the tegmina (Bailey and Broughton, 1970), the right mirror membrane or the frame of the left tegmen or even that of the right tegmen vibrating in a different mode.

THE ULTRASONIC SONGS OF TETTIGONIIDS

The songs of tettigoniids cover a wide range of frequencies from a few hundred cycles per second up to at least 110 kHz, although not all of the species make use of this whole range. Some species produce sounds which cover a wide range while others produce sounds with only a very narrow frequency spectrum. Pierce (1948) found that many tettigoniids produce mainly audible songs but this section will be concerned only with those species that are known to produce considerable energy in the ultrasonic band. So far few species have been adequately studied but, judging from those that have, it is possible that a large number of tettigoniids produce ultrasounds.

The species from which ultrasounds have been recorded are listed in Table 5.1 together with two species whose songs are mainly audible but have not been described previously. Different authors give different sorts of information about the sounds. Pielemeier (1946) and Pierce (1948), who used tunable detectors, report the frequency of the major components, including those in the ultrasonic band. Other authors show oscillograms of either whole chirps or of syllables or pulses. These give some indication of the amplitude modulation of the sounds but the waveforms will be distorted if the full bandwidth is not included. Total frequency spectra are given by some authors, for example Morris and Pipher (1967) while others, including Suga (1966), present sonagrams which show the distribution in time of the sound energy at different frequencies. These forms of display are complementary and some are illustrated in Plate X.

It can be seen from Table 5.1 that there does not appear to be any correlation between the taxonomic position of the insects and the frequency patterns of the sounds that they produce. For this reason the ultrasonic songs of tettigoniids will not be described taxonomically but instead the range of frequency patterns involving ultrasound will be reviewed briefly.

One of the widest frequency spectra found in tettigoniids is probably that produced by *Pterophylla camellifolia*, a member of the sub-family Pseudophyllinae. Pierce (1948) described the song of this species as having 'a practically continuous spectrum from 18 kHz to 63 kHz... also a considerable amount of energy in the pitches below 18 kHz which probably extends all the way down to include the harmonics and fundamental of the teeth impact frequency of 254 Hz'. It seems probable that a very low Q system, or possibly more than one resonant frequency, is involved in sound production in this species. Many other tettigoniids produce wide-band calls in which the sound energy is distributed in distinct frequency bands and there are often one or more peaks of energy within these bands. These frequency peaks are generally not related harmonically and any one of them may be more intense than the others. For example in *Conocephalus conocephalus*, a member of the sub-family Conocephalinae, the higher frequency peak, at about 52 kHz, is more intense than the lower peak, at about 30 kHz, but in *C. maculatus* the lower frequency peak, between 27 kHz and 35 kHz is more intense than the upper peak of 43–52 kHz (Plate X a,b). Pierce reported that the sounds produced by five species of *Conocephalus* were modulated at the frequency of the tooth impact rate and more recently oscillograms of the sounds of another five related species within this genus have shown that each tooth-pulse begins sharply and decays rapidly (Pye and Sales, unpublished). It therefore appears that the sounds of this genus also are produced by a low Q system and moreover that in many species of *Conocephalus* at least two carrier frequencies appear to be involved. In some species of tettigoniids such as *Metrioptera brachyptera* (sub-family Dectinae), either or both of the two major frequency peaks may be subdivided to give a total of three or four peaks of energy (Lewis *et al.*, 1971). Such peaks may remain at the same frequency throughout the pulse or the syllable or they may vary in frequency. The lower main peak of energy in the songs of *M. roeselii* sweeps from 25 kHz down to 18 kHz and then up to 30 kHz within each syllable while the upper peak sweeps from 65 kHz to 55 kHz (Plate X d). In *C. saltator* and *M. roeselii* which also produce a hemisyllable on the opening stroke of the wing, this minor hemisyllable has a narrower frequency spread than the opening or major syllable.

Some tettigoniids may produce different frequency spectra at

TABLE 5.1 Emission of ultrasound by tettigoniids

Species and location	Total energy bands kHz	Main energy (and spectral peaks) kHz	Remarks	Authors
Sub-family Tettigoniinae				
Tettigonia viridissima (France)	7–100			Busnel (1953)
Tettigonia viridissima (France)	9–30 45–58	12, 22 47, 55	Rounded tooth-pulses without rapid decay.	Pye (Unpub).
Sub-family Dectinae				
Decticus verrucivorus (France)	5–100	Variable with circumstances		Busnel & Chavasse (1951) Busnel (1953)
D. albifrons (France)	1·6–95	Variable		Busnel & Chavasse (1951)
	6–100	Variable		Busnel (1953)
Platycleis denticulata (U.K.)	15–50 65–110	15–50 (22, 30, 40) 70–100 (73, 83, 93)	Energy widely distributed in each band, no tooth-pulses visible.	Pye (Unpub).
Metrioptera brachyptera (U.K.)	15–85	25 (22, 30) 67 (65, 77)	Rapidly decaying tooth-pulses.	Howse *et al.* (1971) Lewis *et al.* (1971)
M. roeselii (U.K.)	13·5–40	25		Broughton (1965)
		26		Bailey (1970)
	Major hemisyllable 12–40 50–70 80–110	25→18→30 65→55	'U-shaped' energy sweeps, rounded tooth pulses. At end of last hemisyllable.	Pye (Unpub).

	Minor hemisyllable			
	25–40 55	33 55		
M. sphagnorum (Canada)	10–70	33, 15, 63	Two modes of strigilation.	Morris (1970)
Gampsocleis buergeri (Japan)	Up to 40	8–12		Katsuki & Suga (1960)
Sub-family Conocephalinae				
Conocephalus conocephalus	Major and minor hemisyllables			Pye & Sales (Unpub).
(Uganda and Nigeria)	20–70 max. 90	25–45 (30) 43–65 (52) max. 80	Two main peaks sometimes double and of varying frequency, rapidly decaying tooth-pulses.	
C. maculatus	Major and minor hemisyllables			
(Uganda and Nigeria)	20–60 max. 90	27–35 (28) 43–52 (48)	Moderately rapidly decaying tooth-pulses.	Pye & Sales (Unpub).
C. brevipennis (U.S.A.)		47, 34		Pierce (1948)
C. spartinae (U.S.A.)		40, 31		Pierce (1948)
C. discolor (France)	Major hemisyllable			
	15–70	20–35 (32) 50–55 (50)	Lower peak sometimes double.	Pye (Unpub).
	Minor hemisyllable			
	15, 30–45, 55		Very faint.	
C. discolor (U.K.)	Major hemisyllable			
(recorded in lab.)	20–90	30–35 (31) 80–85 (80)	Broad spectra with slight peaks, rapidly decaying tooth-pulses.	Pye (Unpub).
	Minor hemisyllable			
	30–60	40	Short and faint with weak peak.	

TABLE 5.1—*continued*

Species and location	Total energy bands kHz	Main energy (and spectral peaks) kHz	Remarks	Authors
C. dorsalis (U.K.)		28		Bailey (1967), Bailey & Broughton (1970)
C. dorsalis (U.K.)	Major hemisyllable			
	20–90	40, 52, 63	Rapidly decaying tooth-pulses.	Pye (Unpub).
	Minor hemisyllable			
	{30–55 {15–30	{45 {23	Tooth-pulses not distinct.	
C. saltator (Trinidad)	Major hemisyllable			
	32–120	40, 60–65	Rapidly decaying tooth-pulses.	Suga (1966)
	Minor hemisyllable			
	14–76	18, 42		
C. nigropleurum (Canada)	22–55	30, 36, 42		Morris & Pipher (1967)
C. attenuatis (Canada)	26–60	{34–35, 36–38 {48, 58		Morris & Pipher (1967)
C. fasciatus (U.S.A.)		40, 20–25		Pielemeier (1946)
C. fasciatus (U.S.A.)		40, 16		Pierce (1948)
C. gracimilis (U.S.A.)		40, 20–25		Pielemeier (1946)
C. strictus (U.S.A.)		40, 20–25		Pielemeier (1946)
Orchelimum vulgare (U.S.A.)		14·2, 18–20, 7·1		Pielemeier (1946)

Orchelimum vulgare (U.S.A.)		7·6, 16, 27		Pierce (1948)
O. concinnum (U.S.A.)		20		Pierce (1948)
Sub-family Copiphorinae				
Homorocoryphus nitidulus (Uganda)	9–32	15·5	No distinct tooth-pulses, syllable of coherent sound waves.	Bailey (1967, 1970)
Neoconocephalus ensiger (U.S.A.)		13·7		Pierce (1948)
Neoconocephalus ensiger (U.S.A.)	4·5–50			Borror (1954)
Sub-family Ephippigerinae				
Ephippiger ephippiger (France)	6–88	Variable with circumstances		Busnel & Chavasse (1951)
Ephippiger ephippiger (France)	7–75	Variable		Busnel (1953)
E. bitterensis (France)	5–45	Variable		Busnel & Chavasse (1951)
E. bitterensis (France)	6–50	Variable		Busnel (1953)
E. cuneii (France)	5–70	Variable		Busnel & Chavasse (1951)
E. cuneii (France)	6·5–40	Variable		Busnel (1953)
E. provincialis (France)	6·4–42	Variable		Busnel (1953)
Sub-family Pseudophyllinae				
Pterophylla camellifolia (U.S.A.)	0·25–63			Pierce (1948)
Drepanoxiphus modestus (Trinidad)	18–28	22–24	Syllable of coherent waves.	Suga (1966)

TABLE 5.1—*continued*

Species and location	Total energy bands kHz	Main energy (and spectral peaks) kHz	Remarks	Authors
Zabalius ophthalmicus (Uganda)	4–28	8–11, 22, 33	No distinct tooth-pulses ('handling' call).	Sales (Unpub).
Sub-family Listroscellinae				
Phlugis sp. (1)	33–66	44–50		Suga (1966)
Phlugis sp. (2)	40–100	40–70		Suga (1966)
Sub-family Mecopodinae				
Anoedopoda lamellata (Uganda)	1–50	15, 23, 36, 42	('handling' call).	Sales (Unpub).
Mecopoda elongata (Japan)	1–40	10		Katsuki & Suga (1960)
Sub-family Meconematinae				
Anepitacta egestoides (Nigeria)	45–60	50–55 (52)	Syllable of coherent waves.	Pye (Unpub).
Sub-family Phaneropterinae				
Barbitistes fischeri (France)	12–90			Dumortier (1963b)
Unknown species (France) possibly *Leptophyes punctatissima*	30–56	35–48 (42)	Tooth-pulses not distinct.	Pye (Unpub).

different times (Busnel and Chavasse, 1951; Morris, 1970). Morris (1970) reported that the Canadian *Metrioptera sphagnorum* produced prolonged calls that alternated between two intensity levels. The higher intensity sounds were associated with a low tooth-strike rate and contained predominantly ultrasonic frequencies at 10–70 kHz with peaks at about 30 kHz and between 60 kHz and 70 kHz. The lower intensity sounds were associated with a higher tooth-strike rate and had predominantly lower frequencies, 10–65 kHz, with a peak at 15–20 kHz.

Some tettigoniids produce sounds with a single frequency peak. This may be within a fairly wide frequency spread as in *Phlugis sp.* or within a very narrow frequency band as in *Drephanoxiphus modestus* which produces almost pure-tone sounds. The highest pure-tone sound that is known in tettigoniids is produced by *Anepitacta egestoides*. The song of this species has a frequency spread from 45 kHz to 60 kHz with a frequency peak at 50–55 kHz (Plate X f). Even though the frequency is so high, the waveform suggests that there is only one or at the most two waves per tooth-strike. Nevertheless the teeth of the strigil are spaced about 5 μm apart, which is not especially small (Plate IX g).

The song of this small nocturnal species (Plate I) was first detected while listening for bats in Ibadan, Nigeria (Pye, unpublished). The insects were traced to the middle of a thick hedge and were later found to have been previously undescribed (Beier, 1967). *Anepitacta egestoides* is therefore the first new species to be discovered by a bat detector and it is likely that many more unobtrusive insects might be found in this way.

THE ACOUSTIC BEHAVIOUR OF TETTIGONIIDS

The acoustic behaviour of tettigoniids has been studied in detail in only a few species. Unlike the studies of behaviour in moths (Chapter 4), those of tettigoniids have seldom been coupled with physiological investigations into the neural responses of the auditory system. Many of the behavioural observations have been made on species whose songs contain most energy in the audible band, but as it is likely that similar behaviour occurs in at least some of the mainly ultrasonic insects, the studies on the behaviour of both 'audible' and 'ultrasonic' tettigoniids will be reviewed briefly here.

The songs of tettigoniids appear to have two main functions. The 'calling song' and the 'aggressive song' serve to space out rival males, and the calling song also attracts females towards males so that mating can occur. These songs differ in their temporal patterns. A third type of call, a 'disturbance' or 'warning' call has also been reported in a few species.

The calling song is produced by individual male tettigoniids in their territories and it is the characteristics of this song that are given for most of the species in Table 5.1. The calling song consists of multisyllabic or occasionally monosyllabic chirps separated by intervals of silence, and it may be produced in this way for many hours at a time. Many species of tettigoniids are nocturnal singers but some also sing during the day. Internal factors are important in strigilation and in many species males do not sing for 2–3h after they have successfully copulated with a female. The calling song of male tettigoniids stimulates other males to sing (Busnel, 1963b; Jones, 1966a), and two males may enter into 'competition' with each other by producing their songs alternately. This alternation is achieved by each male modifying its own rate of sound emission, but this depends on auditory feedback and alternation only occurs if the tympanic organs of both the strigilating males are intact (Busnel, 1963b).

Several authors have reported that, in the laboratory at least, male tettigoniids can be induced to alternate with the calling songs of other species of tettigoniids, with some of the songs of some acridid grasshoppers and also with artificial signals such as the tapping of a typewriter key, blasts on a Galton high frequency whistle, or pulses produced by a signal generator (Alexander, 1960; Busnel, 1963b; Jones, 1964; 1966b). Pierce (1948) found that a male *Pterophylla camellifolia* responded to loud raucous shouts and that the insect replied to two, three, four or five shouts in a row by producing a similar number of bursts of sound, even though this insect normally produced only two chirps at a time. Busnel (1963b) induced male *Ephippiger ephippiger* to alternate with a Galton whistle, and these males never sang during the artificial signal however prolonged this was. Such signals therefore appear to have a lasting inhibitory effect on strigilation and Jones (1964, 1966a) has suggested that in *Pholidoptera griseoaptera* the alternation of singing in males is due to the song of each male inhibiting strigilation by the other in turn. After studying song alternation in *Pterophylla camellifolia*, Shaw (1968) concluded that this acoustic behaviour could be explained in terms of inhibition of a spontaneously firing neurone (an acoustic pacemaker), together with increased activity after such inhibition (post-inhibitory rebound). The alternation of singing in males could give each male the opportunity of assessing the loudness, and so the proximity of, the other singer (Jones, 1966a).

In some species of tettigoniids, populations of territorial males produce the calling song together in synchrony. *Orchelimum vulgare*, a North American species, produces a calling song consisting of a series of syllables that are emitted singly at intervals and sound like 'ticks'. These are followed by a rapid series of syllables which form a 'buzz'. On a warm sunny day all the males in a population first 'tick' together and then 'buzz' together (Alexander, 1960). Alexander has suggested that such chorusing behaviour enhances certain rhythms of the specific calling song which would be less apparent in individual songs. In this way they possibly could attract more females of that species to the area. Alternatively the behaviour may form an epideictic display, giving an indication of population density that in turn controls the rate of reproduction (see Wynne Edwards, 1962).

Sexually receptive female tettigoniids are attracted by singing males and move towards the source of sound, a response known as positive phonotaxis (Busnel, 1963b). The intensity of the song is important in this response and a female will move towards the louder of two signals, even if this is an artificial signal and even if she had previously been moving towards the source of the weaker signal. As the female approaches the sound source, she moves more rapidly and Busnel has shown that there is a linear relationship between the intensity of the sound and the speed of the female. Only an acoustic stimulus is necessary for the female to find the male and blinded females are able to move towards a singing male as rapidly as sighted females. Just as males will respond to the calling songs of other species and other groups by singing, so females will move towards singers of other species. But if they are given the choice, they move towards a conspecific male rather than to a male of a different species. There appears to be no special courtship song in tettigoniids, as there is for example in crickets, and females seem to recognize non-specific calling songs only when they are close to the singer. Some attempts at interspecific mating have been observed in the laboratory (Busnel, 1963b) but hybrid forms are rare in the wild.

Within a genus, species recognition appears to depend on the chirp, syllable, or even the spike repetition rates. Bailey and Robinson (1971) reported that four different species of *Homorocoryphus* emit songs in which the main energy is at the same frequency, 15–16 kHz. They found that the activity of females of two of these species, *H. vicinus* and *H. flavovirens*, was increased in response to artificial signals of 15 kHz but only if these signals were delivered at the syllable repetition rate characteristic of the species. The activity of *H. flavovirens* was inhibited to some extent by the song of *H. paraplesius*, whose syllable repetition rate is nearest to that of its own.

Generally it is the male tettigoniids that produce the calling song but sound production is not uncommon among females. The females of several species within the sub-family Phaneropterinae can strigilate but they use a rather different strigilatory mechanism. A region of the inner margin of the left tegmen is bent downwards to form a plectrum and this moves over an adjacent patch of minute, stout spines projecting from veins on the upper surface of the left tegmen (Fulton, 1933). These spines are particularly well-developed in female *Microcentrum rhombifolium* which, when sexually receptive, are known to sing in response to the male's calling song. The male in this case then moves towards the female (Fulton, 1933; Alexander, 1960). Dumortier (1963c) reports that female *Ephippiger ephippiger* of the sub-family Ephippigerinae, also sing when highly motivated sexually, and this advertises their willingness to mate.

Aggressive songs are less common in tettigoniids than they are in other insects such as crickets, but they occur in some species when two males of the same species are close together. These songs are sometimes called disturbance calls, a rather misleading term which has also been applied to the response to disturbance by animals of other species. In both *Pterophylla camellifolia* and *Pholidoptera griseoaptera*, the aggressive songs contain a greater number of syllables per chirp than the calling song and the chirps are consequently much longer (Jones, 1966a; Shaw, 1968). Jones (1966a) heard the aggressive song from a pair of caged male *Pholidoptera griseoaptera* while they were facing each other and lashing out with their antennae, and he suggested that it might be an 'intimidation' call. The emission of this song by one animal stimulates another either to produce this song alternately or to leave the vicinity (Jones, 1966a; Shaw, 1968). The function of the aggressive song therefore may be to space out individual males and so increase the probability of all the available females mating successfully (Jones, 1966a).

A disturbance in the environment causes some tettigoniids, both male and female, to emit a 'disturbance' or 'protest' call. This call is also produced by some species when they are captured or handled (Busnel, 1953; Busnel and Chavasse, 1951). It is again less common than the calling song. Alexander (1960) found that only four out of the 90 species of North American tettigoniids that he studied produced this song. Three of these species were large and flightless. In *Pterophylla camellifolia* the disturbance call consists of monosyllabic chirps that are produced irregularly (Shaw, 1968). Alexander (1960) reported that the production of a characteristic disturbance call often causes other males to stop calling. A similar cessation often occurs if just one male stops calling. While tape-recording the echo-location cries of bats in Nigeria, Pye noticed that strigilating *Conocephalus conocephalus* and *C. maculatus* stopped calling whenever an echo-locating bat flew overhead, an observation that was later confirmed by Sales for the same species in Uganda.

One further acoustic behaviour pattern is known in tettigoniids and that is an 'aggregation' call. Dumortier (1963c) reported that groups of male *Pholidoptera griseoaptera* were attracted to the cage of a single singing male. This particular type of song apparently differs from the calling song in that it does not attract females. It may be important in the formation of groups of males and so enhance the chorusing behaviour mentioned earlier.

THE ABILITY OF TETTIGONIIDS TO HEAR ULTRASONIC SOUNDS

There is good physiological evidence that tettigoniids can hear the ultrasounds that they produce. These insects have complex tympanic organs which are situated on the fourth segments, the tibiae, of the first pair of legs. The tibiae are dilated to accommodate these organs (Fig. 5.7a) which were first described in detail by Schwabe in 1906. Two enlarged branches of the tracheal air system run down the centre of each tibia and the two air sacs are closely apposed in the midline so that a thin membrane is formed between them (Autrum, 1963). The anterior and posterior faces of these air sacs are in connection with the soft tissues of the leg, but their lateral surfaces bear the twin tympanic membranes. These lie within pockets of cuticle (Fig. 5.7b) and are exposed to the exterior by narrow, forwardly-facing longitudinal slits in the cuticle or by wide oval apertures. There are between 100 and 300 sensory cells in the tympanic organ of tettigoniids but these are situated along the anterior edge of one of the tracheal air sacs and are apparently not directly associated with either of the tympanic membranes. The cells are arranged in a row down the surface of the trachea (Fig. 5.7c), the size of the cells decreasing from above downwards. The reason for the multiplicity of these cells is not known.

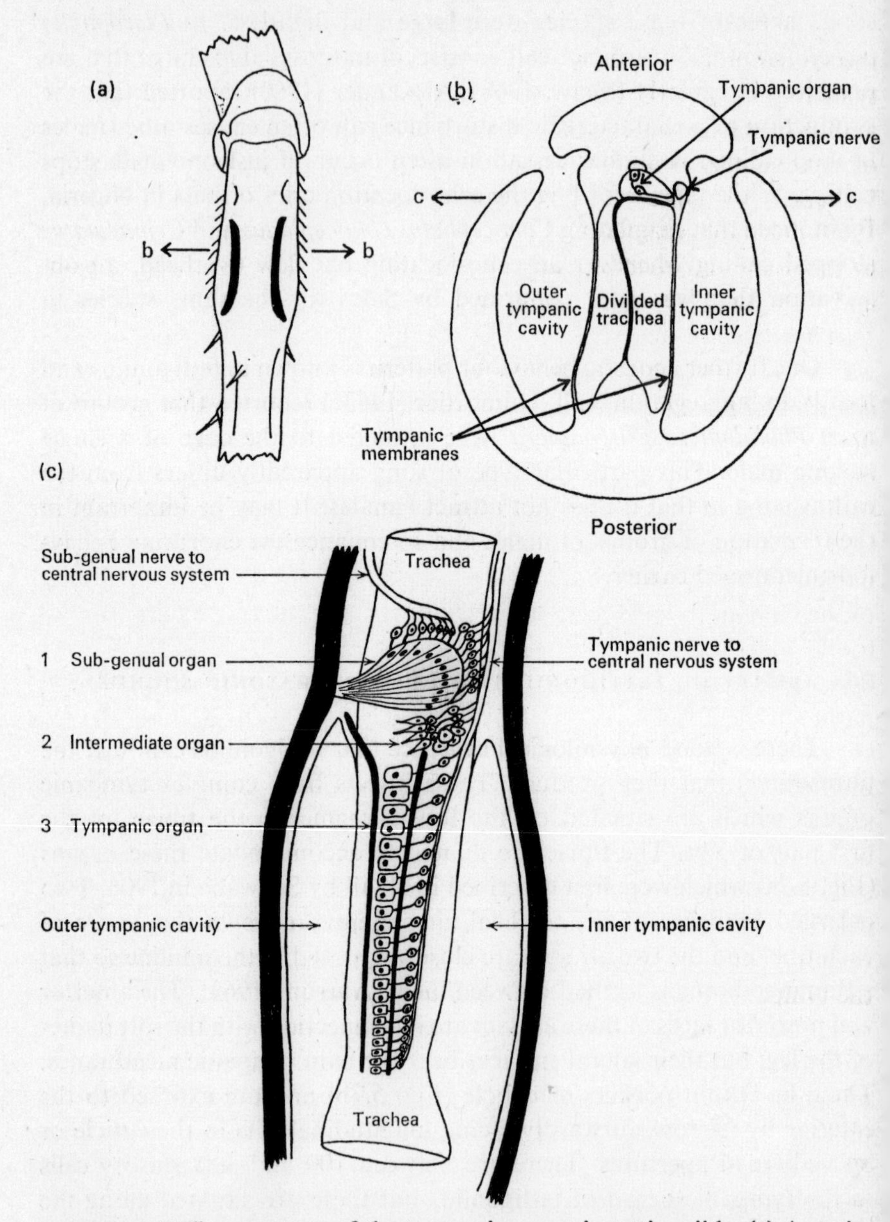

FIGURE 5.7 The structure of the tympanic organ in tettigoniids. (a) Anterior view of the upper part of the tibia of the foreleg of *Tettigonia viridissima* showing the tympanic slits. (b) Transverse section through the tympanic organ of *Decticus verrucivorus* (i.e. the plane b–b in (a)). (c) Anterior view of the exposed tympanic organ of *Decticus verrucivorus* (the plane c–c in b). (b and c redrawn after Schwabe, 1906)

The nerve fibres, one from each of the sensory cells, together form the tympanic nerve which runs to the prothoracic ganglion. Here the tympanic nerve from each fore-leg is thought to synapse with several auditory interneurones, some of which then run to the brain. The largest of these interneurones are called *T* neurones as they are believed to be T-shaped. There is as yet little morphological evidence for this (McKay, 1969) but these interneurones do conduct nerve impulses both rostrally to the brain and caudally to the meso- and metathoracic ganglia (Suga and Katsuki, 1961a; Rheinlaender *et al.*, 1972).

Probably the first attempt to study the hearing ability of tettigoniids was made in 1926 by Regen. He studied the behaviour of *Thamnotrizon apterus* (= *Pholidoptera aptera*) and noted that when two males were in the same vicinity they tended to sing alternately. Regen was able to induce young males to alternate with artificial signals and they appeared to be sensitive to signals of up to about 28 kHz. Later Wever and Bray (1933) inserted electrodes into the third segment of the leg of an unnamed species, possibly *Pterophylla camellifolia*, and studied the response of the tympanic nerve to tones produced by an oscillator. These authors found that impulses were produced in response to stimulus signals of 0·8–45 kHz. As in moths (Chapter 4) within this frequency range the nature of the response was not affected by the frequency of the stimulus.

In 1940 Pumphrey reviewed the evidence then available for hearing in insects. He concluded that although insects with tympanic organs were sensitive within certain frequency ranges to the amplitude modulation of signals (by which he apparently meant the pattern of the chirp and syllable envelopes and their repetition rates), they were not able to discriminate between signals of different carrier frequencies within these ranges. This view was also adopted by Autrum (1960) who found that the tympanic organ of the large green tettigoniid, *Tettigonia viridissima* responded to all sounds up to 100 kHz, independent of their frequency. Autrum (1940, 1941, 1963) also reported that two other species of tettigoniids, *T. cantans* and *Decticus verrucivorus* could detect signals up to 90 kHz but they were most sensitive to frequencies of 10–60 kHz. Responses to signals up to 80–120 kHz were recorded from the tympanic nerve of *Conocephalus strictus* by Wever and Vernon (1959). This species showed maximum sensitivity at 10–40 kHz but another tettigoniid, *Neoconocephalus retusus* had a narrower range of sensitivity with a peak at 15 kHz and an upper limit of 60 kHz.

Katsuki and Suga (1960) also concluded that the tympanic organ of tettigoniids was insensitive to frequency although in the tympanic nerve of *Gampsocleis buergeri* they found two types of auditory neurones which had different frequency responses. These authors recorded the acoustic responses of the whole tympanic nerve using fine silver wire electrodes and they sampled single auditory neurones with capillary micro-electrodes. Impulses were recorded from the whole nerve in response to sounds between 0·6 kHz and 75 kHz. Some individual neurones, however, responded only to signals between 0·6 kHz and 30 kHz and were most sensitive to frequencies of 6–7 kHz, while others responded to signals from 3 kHz to 60 kHz and were most sensitive to signals of 10 kHz. These 'higher frequency' neurones were more sensitive than the 'lower frequency' ones.

Suga and Katsuki (1961a) then studied acoustic responses in the ventral nerve cord of *Gampsocleis* using a fine silver wire electrode. Here they recorded impulses of large amplitude that were attributed to a single large *T* fibre. This *T* fibre responded to a narrower range of frequencies than did the tympanic nerve. Its total range was from 2 kHz to 70 kHz but its threshold was lowest to signals of 10–20 kHz. Up to a certain limit, a great number of impulses was produced in response to signals of higher intensity. Suga and Katsuki found that in both *Gampsocleis* and *Homorocoryphus lineosus* the tympanic nerve nearer to the source of sound had an excitatory effect on the ipsilateral *T* fibre but it also excited interneurones that had inhibitory effects on the activity of the contralateral *T* fibre (Suga and Katsuki, 1961a, b; Suga, 1963). The *T* fibres on each side of the ventral nerve cord therefore responded very differently to a stimulus signal that was presented from one side of a preparation; the ipsilateral fibre produced a large number of impulses and the contralateral one produced very few impulses. This difference, however, decreased as the source of sound was moved through an angle of 90° to a position directly ahead of the animal. The differential response of the two *T* fibres could therefore provide the insect with information about the direction of the source of sound. In *Gampsocleis* the difference in the responses of the two *T* fibres was greatest when the stimulus was 17 kHz, the most dominant frequency in the species song. This probably means that this insect can localize the species song very sharply.

It appears, however, that not all species of tettigoniids have this ability to localize sound sharply and not all species have *T* fibres that are most sensitive to the dominant frequency of the species song. Two other species that Suga and Katsuki studied, *Mecapoda elongata* and *Phaneroptera falcata*, showed no inhibitory interaction between the activity of the tympanic nerves and that of the contralateral *T* fibres. As might be expected, there was very little difference in impulse production by the two *T* fibres of *M. elongata* when a preparation of this insect was presented with sound stimuli from different directions. Suga (1966) also found that the auditory systems of two species of *Phlugis* and of *Conocephalus saltator*, as measured by their *T* fibre responses, were most sensitive to frequencies of 20–25 kHz (Fig. 5.8) whereas the dominant frequency of the song of *Phlugis* is 44–56 kHz and that of *C. saltator* is 40–66 kHz.

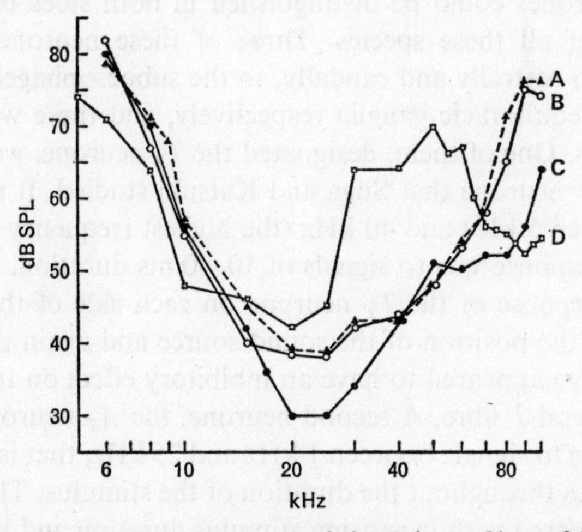

FIGURE 5.8 Auditory response curves of the *T* large fibre of four species of tettigoniids. A; *Phlugis* sp. 2. B; *Phlugis* sp. 1. C; *Conocephalus saltator*. D; *Drepanoxiphus modestus*. (After Suga, 1966)

An indication that the auditory system of tettigoniids is capable of some degree of frequency discrimination comes from the work of McKay (1969, 1970) who studied the auditory responses of the tympanic nerve, of the auditory *T* fibre and of three other auditory interneurones in three species of *Homorocoryphus*. The dominant frequency of the song of these species is 15–16 kHz and McKay found that the threshold of the tympanic nerve was lowest to signals of 10–15 kHz and was also low to signals above 30 kHz. The behaviour of the *T* fibre, however, was very different. It responded preferentially to signals of 30 kHz but its threshold to signals of 15 kHz was high and its activity appeared to be inhibited by the species song. McKay therefore suggested that this *T* fibre was not important in species song recognition but that it might have a warning function, possibly against bats which are known predators of

Homorocoryphus or against other predatory small mammals which would create ultrasounds by moving through the grass. Unfortunately there is no evidence as yet to suggest that *Homorocoryphus* reacts behaviourally to such ultrasonic signals. One of the smaller interneurones that McKay studied did respond to the species song but this neurone was difficult to find repeatedly and so was not studied in detail.

Recently Rheinlaender and his colleagues (1972) studied four species of tettigoniids, *Tettigonia viridissima*, *Decticus albifrons*, *Eupholidoptera chabrieri* and *Ephippiger sp.* and found that four different auditory neurones could be distinguished in both sides of the ventral nerve cord of all these species. Three of these neurones conducted impulses both rostrally and caudally, to the suboesophageal and to the meso- and metathoracic ganglia respectively, and these were classified as T neurones. One of them, designated the T_T neurone, was equivalent to the large T neurone that Suga and Katsuki studied. It responded to signals between 5 kHz and 40 kHz (the highest frequency studied) and its greatest response was to signals of 10–40 ms duration. As in *Gampsocleis* the response of the T_T neurone on each side of the nerve cord depended on the position of the sound source and again the ipsilateral tympanic nerve appeared to have an inhibitory effect on the activity of the contralateral T fibre. A second neurone, the A_T neurone, showed a tonic response to signals between 1 kHz and 25 kHz, that is, it produced nerve impulses throughout the duration of the stimulus. The number of impulses increased with increasing stimulus duration and intensity. The D_T neurone produced only one impulse per stimulus regardless of the intensity, duration or the frequency of the stimulus. This neurone therefore appears to be comparable to the pulse marker neurone in the ventral nerve cord of noctuid moths (Chapter 4). The fourth type of neurone was found only in the suboesophageal ganglion and it showed a tonic reaction to stimuli of 1–35 kHz. It responded most readily to signals of 60–80 dB but it only responded after a fairly long latency of 20–30 ms.

It has recently been shown that in at least one species of tettigoniid, *Metrioptera brachyptera*, the ultrasonic components of the species song are necessary to provide an adequate stimulus for the tympanic organ (Howse *et al.*, 1971; Lewis *et al.*, 1971). This species produces syllables consisting of spikes which decay very rapidly and the main energy of the song lies between 15 kHz and 85 kHz. When the live song of a caged animal or a high frequency tape-recording of the song was used as a stimulus for a tympanic nerve preparation of this species, the number of impulses recorded from the tympanic nerve showed a high correlation with the number of intra-syllabic spikes. A similar correlation was obtained when the live song or a high frequency tape-recording of the acridid grasshopper, *Chorthippus parallelus* was used as the stimulus signal. *Chorthippus* also produces syllables with rapidly decaying spikes and the main energy of the song lies between 7 kHz and 80 kHz. However, when the natural song of *Metrioptera* was recorded on an 'audio' tape-recorder with an upper frequency limit of approximately 15 kHz, the song was distorted and the tympanic nerve response showed a very poor correlation with the intra-syllabic spikes, although there was still a close correspondence with the duration of the stimulus signal. When a high frequency recording of the species song was replayed through a narrow band-pass filter, the song was again (differently) distorted but fairly good correlations were obtained at all filter settings above 20 kHz. From these results the authors concluded that the important feature of the song of *M. brachyptera* was the intra-syllabic spike structure, particularly its ultrasonic components. But they pointed out that the tympanic organ of this species also responded well to the song of *Chorthippus* which had a different overall frequency spectrum. The tympanic organ of *M. brachyptera* therefore is not 'tuned' to respond only to the song of the species.

It is clear that the studies of hearing in tettigoniids that have so far been carried out raise many problems on both the method of functioning of the tympanic organ and on the adaptiveness of the auditory system to respond to specific signals. At the present time these questions are unresolved and they offer a wide field for further study.

SOME FURTHER CONSIDERATIONS

Compared with the story of the importance of ultrasound in the lives of nocturnal moths, the story of ultrasonic communication in tettigoniids is far from complete. There are many questions, particularly about the nature of sound production and hearing, that remain unanswered. Some of the acoustic analyses given in Table 5.1 are of interest in connection with the nature of the resonator and therefore the mechanism of sound production. As described earlier, Pierce (1948) attempted to predict the resonant frequency of the mirror membrane from the formula of Rayleigh (1894) for vibrations of a circular disc clamped at its periphery. Several assumptions had to be made for the physical properties of insect cuticle, and the dimensions (area and thickness) were measured for the wings of several species. In some cases a fairly close agreement was found between these calculated frequencies and the frequencies recorded from the insects, but in other cases they were much too high or too low.

Broughton's suggestion that the mirror frame determines the resonant frequency led Morris and Pipher (1967) to consider the frame as a simple cantilever, using another formula given by Rayleigh. Morris and Pipher assumed that the physical properties of cuticle were similar for different insects and Rayleigh's formula then reduced to:

$$f_{res} = \frac{K}{l^2}$$

where f_{res} is the resonant frequency of the cantilever, K is a constant and l is the distance from the vestigial file, acting as the fulcrum, to the furthest side of the mirror frame where the maximum movement occurs. Morris and Pipher then plotted $1/l^2$ against the lowest resonant frequency of *Conocephalus nigropleurum* and *Orchelimum gladiator* and the points fell on a straight line passing through zero. Bailey (1970) replotted this graph and added points for two more species, *Homorocoryphus nitidulus* and *Conocephalus discolor*. These points also fell on the line.

Further points can now be added using the data of Table 5.1 and measurements of l obtained in two ways. Seven measurements have been made from the scale drawings of mirror frames given by Pierce (1948) and these have been related to the lowest resonant frequencies that he recorded for these species. Eight more species have actually been measured and the length of the mirror frame has been compared with the lowest spectral peaks obtained from ultrasonic tape-recordings of the songs. The total of 19 numbered points for 18 species is shown in Fig. 5.9. Most of the points fall very close to the line plotted by Morris and Pipher although some exceptions are obvious. In these cases the second resonance has also been considered. The interesting species *Anepitacta egestoides*, with a single resonance at 52 kHz, cannot be included in this analysis because it has no rigid mirror frame and l is therefore unmeasureable.

The resulting graph is interesting but not highly significant. It does suggest that for a number of species the resonant frequency is closely related to the length of the mirror frame. Morris' assumption that the properties of cuticle are fairly constant is therefore likely to be true. It does not, however, prove the correctness of the cantilever theory; Pierce's disc formula also contained a term similar to $1/l^2$ (although he

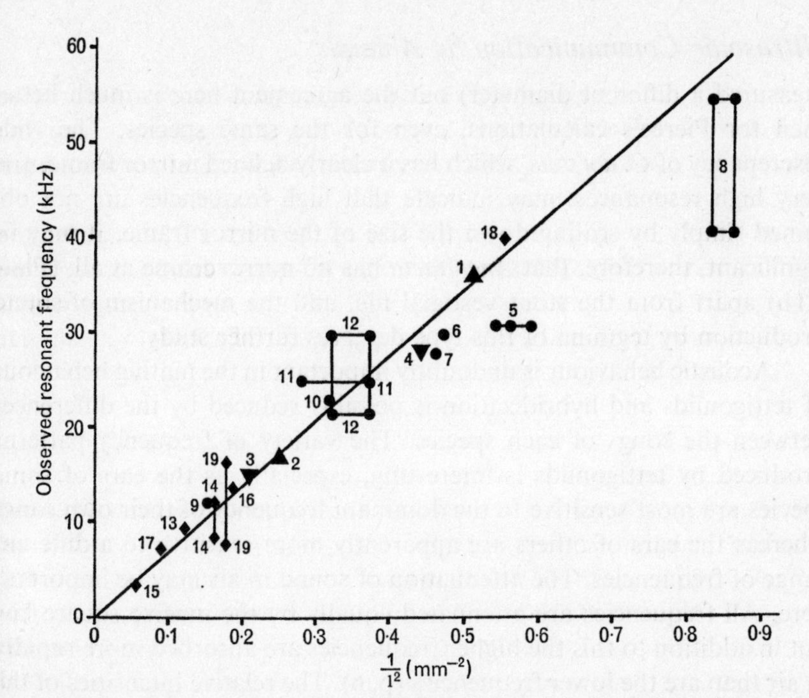

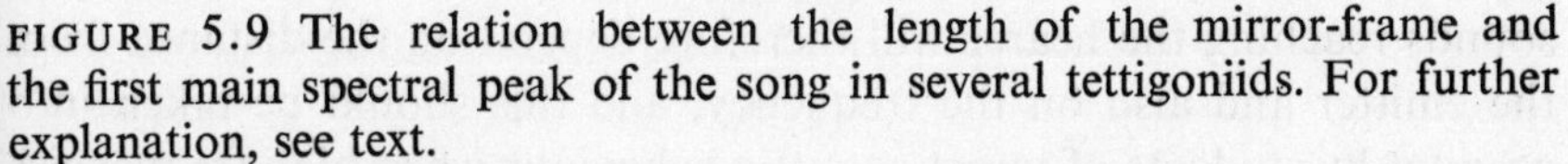

FIGURE 5.9 The relation between the length of the mirror-frame and the first main spectral peak of the song in several tettigoniids. For further explanation, see text.

Key to Fig. 5.9.

1. *Conocephalus nigropleurum*
2. *Orchelimum gladiator*

} Morris and Pipher, 1967. ▲

3. *Homorocoryphus nitidulus*
4. *Conocephalus discolor*

} Bailey, 1970. ▼

5. *Conocephalus discolor**
6. *Conocephalus conocephalus*
7. *Conocephalus maculatus*
8. *Conocephalus dorsalis**
9. *Tettigonia viridissima*
10. *Metrioptera roeselii*
11. *Metrioptera brachyptera**
12. *Platycleis denticulata**

} Pye and Sales, unpublished. ●

13. *Amblycorypha oblongifolia*
14. *Scudderia curvicauda*†
15. *Pterophylla camellifolia*
16. *Neoconocephalus ensiger*
17. *Neoconocephalus robustus*
18. *Conocephalus fasciatus*
19. *Orchelimum vulgare*

} Plotted from measurements and scale drawings given by Pierce, 1948. ◆

* Two or more different sized animals considered.

† Daytime (lower) and night-time (upper) peaks considered.

measured a different diameter) but the agreement here is much better then for Pierce's calculations, even for the same species. The wide discrepancy of *C. dorsalis*, which has a clearly defined mirror frame, and very high resonances, may indicate that high frequencies are not obtained simply by scaling down the size of the mirror frame. It may be significant, therefore, that *Anepitacta* has no mirror frame at all, (Plate XI h) apart from the stout vestigial file, and the mechanism of sound production by tegmina of this type deserves further study.

Acoustic behaviour is undoubtly important in the mating behaviour of tettigoniids and hybridization is possibly reduced by the differences between the songs of each species. The variety of frequency patterns produced by tettigoniids is interesting, especially as the ears of some species are most sensitive to the dominant frequency of their own songs whereas the ears of others are apparently more sensitive to a different range of frequencies. The attenuation of sound in air may be important here. All frequencies are attenuated equally by the inverse square law but in addition to this the higher frequencies are absorbed more rapidly in air than are the lower frequencies (p. 6). The relative intensities of the sounds reaching the hearer will therefore depend on the distance from the emitter and also on the frequency, and this should be taken into account by students of insect acoustic behaviour when high frequencies are involved.

The activity of the auditory *T* fibres also presents a problem. As the single fibres that have been studied in some species do not respond to the species song, they may act as 'escape' fibres to inform the metathoracic ganglion (the 'jumping centre') and the brain of danger. It is possible that the *T* fibres have different functions in different species.

Obviously the study of ultrasonic communication in strigilating insects is still in its early stages and much more needs to be known of the physiology of the auditory system, of the ecology and of the social behaviour of these animals. Further examples of acoustic behaviour in other groups of insects will be described briefly in Chapter 6.

CHAPTER SIX

Other Insects

Insects produce sounds in a variety of ways. The most common method is by friction of one part of the hard exoskeleton against another part, but other methods are also used. Dumortier (1963a) summarized the methods of sound production in insects as (1) friction of differentiated parts as in tettigoniids, crickets and grasshoppers, (2) vibration of membranes, for example the buckling of tymbal organs in cicadas and some moths, (3) the expulsion of a fluid (gas or liquid) as in the death's head hawk moth, (4) shocks to the substrate, well known in the death watch beetle and (5) vibration of appendages such as the wings which produces the 'droning' sounds of bees and mosquitoes when in flight. So far ultrasounds are known to be produced by at least the first three of these methods.

In the previous chapter it was pointed out that the production of ultrasound by insects is not really surprising and it was suggested that many insects, other than tettigoniids, probably also emit ultrasound. Saby and Thorpe (1946) found that the ambient noise in the jungles of Panama contained a high proportion of ultrasound, all of which could be attributed to insects such as cicadas, whose songs were otherwise audible. Several groups of insects are now known to produce songs that are clearly audible but that also contain energy at ultrasonic frequencies. This chapter will discuss some of these.

The study of ultrasound production and reception in many of these insects is even less complete than in tettigoniids and much of the available information comes from isolated reports like that of Saby and Thorpe above. Ultrasound is probably of varying importance to these groups but the picture is still unclear. This chapter will attempt to review the available literature and will also describe a few new examples. Earlier work on this subject was briefly reviewed by Schaller (1952).

GRYLLIDAE, CRICKETS

Gryllids or crickets are found throughout the world. In Britain they can often be heard singing in the summer and early autumn and they sometimes enter houses for the winter. Like tettigoniids, they produce sounds by rubbing their tegmina or wing covers together (Chapter 5) but, although closely related, they differ from tettigoniids in several ways. Male crickets possess a perfect strigilatory apparatus on each tegmen but the right tegmen almost always overlaps the left, so that the plectrum of the left tegmen moves over the file of the right one (Huber, 1963). Each tegmen bears two areas of thin cuticle, the 'harp' which is near to the base of the tegmen and the more distal 'mirror' (Plate IX c). In crickets these sound-producing structures are confined to the males and are absent in females (Dumortier, 1963a).

Male crickets produce songs with different temporal patterns in different situations; a male alone in its territory produces the 'calling song', in the presence of a rival male it gives the 'aggressive song' and a sexually excited male produces the 'courtship song' when close to a female. The calling song generally consists of regularly produced chirps, each containing three to five syllables. As in tettigoniids, the calling song attracts females, even when it is relayed over a telephone (Regen, 1913) and it also advertises the male's presence to rivals with which the calling male may sing alternately. The calling song of crickets is often a rather 'pure' note, each tooth-strike producing one sound wave (Fig. 5.5d). The dominant frequency is about 3–5 kHz but Nocke (1972) has described a secondary peak of energy at about 14 kHz in *Gryllus campestris*, a field cricket, and Lottermoser (1952) found components up to 100 kHz in the calling song of this species. Pielemeier (1946) reported that the calling song of *Nemobius fasciatus*, the little brown cricket, contained frequency peaks at 8·3 kHz, 17·4 kHz and 28 kHz (Table 6.1). Pierce (1948) found that during the calling song of both *Gryllus assimilis*, another field cricket, and *Gryllus* (now called *Acheta*) *domesticus*, the house cricket, sound is produced only during the closing stroke of wing movement.

In tettigoniids, the mirror frame was found to be important in determining the frequency spectrum of the song but Nocke (1971) has shown that in some gryllids there is a close correlation between the resonant frequency of the harp membrane and the main, low frequency peak (4 kHz) of the calling song. The harp frame has no resonator function and the mirror region does not appear to be involved in the production of the calling song.

If a male cricket is approached by a rival male, the calling song is replaced by the aggressive song. This is similar to the calling song but it consists of a greater number of syllables per chirp and it has a higher intensity. It often results in the withdrawal of the intruder (Dumortier, 1963c). Recently Phillips and Konishi (1973) have suggested that the acoustic signals of dominant male crickets inhibit the aggressive tendency of subordinates. They found that subordinate males became much more aggressive after they had been deafened, possibly because they were freed from this inhibition.

Tactile stimuli from a female cricket may stimulate the male to produce the courtship song. This varies between different forms; in some species the courtship song consists of monosyllabic chirps, but in others the chirps are very long and may end in a single rapid 'tick' (Alexander, 1957). The courtship songs of crickets have often been described as more subdued or quieter than the calling songs (Alexander, 1961; Huber, 1963; Ragge, 1965). The change in the male's song in the presence of a female was perhaps first described in 1836 by Goureau: 'Lorsqu' une femelle se presente . . . il . . . modifie ses accents, son chant devient beaucoup plus doux et plus tendre'!

The courtship songs are characterized by a broadening of the frequency spectrum which may result from the changed position of the wings: during the calling song the wings are held at an angle of about 45° to the body but during the courtship song they are lowered so that they are only just raised above the body. Several groups of muscles are involved in closing the tegmina and Bentley and Kutsch (1966) found that, in three species of gryllids, all of these 'closing' muscles contract simultaneously when the tick is produced. Recently Nocke (in preparation) has shown that during the tick the inward movement of the wings, and therefore the tooth-strike rate, is about three times faster than during the calling song. This could well account for the differences in the predominant frequencies of these songs (see below).

It seems that in most gryllids 'courtship' syllables are at first emitted between normal or modified calling syllables as the pre-courtship song and that the full courtship song is produced only gradually. Pierce (1948) reported that the courtship song of *Gryllus assimilis*

TABLE 6.1 Emission of ultrasound by 'other insects'

Species and location	Total energy bands kHz	Main energy (and spectral peaks) kHz	Remarks	Authors
Order Orthoptera				
Sub-order Ensifera				
Family Gryllidae				
Nemobius fasciatus (U.S.A.)		8·3, 28, 17·4		Pielemeier (1946)
Nemobius fasciatus (U.S.A.)		7·5		Pierce (1948)
Gryllus campestris (Germany)	3–100 14–100	5 17	Calling song. Courtship song.	Lottermoser (1952)
Gryllus campestris (Germany)	4–40 4–50	4 8–17 (14)	Calling song. Courtship song.	Nocke (1972)
G. assimilis (U.S.A.)		17	Courtship song.	Pierce (1948)
Acheta domesticus (U.S.A.)		9, 24	Courtship song.	Pierce (1948)
Acheta domesticus (U.K.)	13–26 13–23		Pre-courtship song. True courtship tick.	Lenahan (Pers. comm.)
Sub-order Coelifera				
Family Acrididae				
Chorthippus biguttulus (Germany)	4–50	5–8, 18–40		Dumortier (1963b)
C. brunneus (Germany)	4–30	7–13, 14–25		Dumortier (1963b)
C. parallelus (U.K.)	7–80	10–15, 20–25 37–40, 60–70		Lewis *et al.* (1971)
Arcyptera fusca (France)	3–45	12, 16, 21		Dumortier (1963b)

Arcyptera fusca (France)	20–42	40, 32, 24	Moderately decaying tooth-pulses.	Pye (Unpub.)
Locusta migratoria (France)	3–18			Busnel & Chavasse (1951)
Encoptolophus sordidus (U.S.A.)		19		Pierce (1948)
Order Dictyoptera				
Sub-order Blattaria				
Family Blattidae				
Giant cockroach, unknown form (Uganda)	{8–32 {54–64	{14–20 (14, 17) {26–31		Pye (Unpub.)
Order Hemiptera				
Sub-order Homoptera				
Family Cicadidae				
Unknown form (Panama)	Up to 30			Saby & Thorpe (1946)
Unknown form (Trinidad)	12–100	20–25, 40–45	Tooth-pulses show sudden onset, moderate decay.	Pye (Unpub.)
Order Coleoptera				
Sub-order Polyphaga				
Family Silphidae				
Necrophorus sp.	Up to 28			Autrum (1936)
N. vespilloides	3–30			Dumortier (1963b)
Family Geotrupidae				
Geotrupes sp.	2–40			Autrum (1936)
Family Cerambicidae				
Cerambix cerdo	4·2–>20			Dumortier (1963b)
Longicorn beetle, unknown form probably *Petrognatha gigas* (Uganda)	7–30	{10–20, 20–30 {11–16		Pye (Unpub.)

consists of an irregular series of pulses at a frequency of about 17 kHz. Lottermoser (1952) found a similar peak of energy at 17 kHz in the courtship song of *G. campestris* in which he also detected weaker components up to 100 kHz. Huber (1963, 1970) and Nocke (1972) have now shown that the courtship song of *G. campestris* consists of two different sound elements. The most prominent sound is the loud tick in which the main energy is at 8–17 kHz with a peak at 14 kHz and weaker components extending up to 100 kHz. These ticks are separated by between six and 26 less intense 'intertick' sounds which have a main peak of energy at 4–5 kHz and a smaller peak at 14 kHz. The mechanism of intertick production is at present being investigated.

Loud audible tick sounds also occur in the courtship songs of several species of *Acheta* (Pierce, 1948; Alexander, 1961; Ragge, 1965). They may occur regularly between courtship chirps or more or less irregularly. Pierce (1948) reported that the courtship song of *Gryllus* (= *Acheta*) *domesticus* consisted of a continuous trill at about 9 kHz with an occasional chirp at 24 kHz. Alexander (1956, 1961), however, using equipment with an upper frequency limit of 15 kHz, found that the song consisted of a more or less continuous series of syllables at 3–5 kHz which ended in a tick, apparently at about 8 kHz. Recently J. Lenahan (personal communication) has re-examined the courtship song of this species using equipment with an upper frequency limit of at least 100 kHz. He found that in the pre-courtship song some of the syllables contained frequencies up to 26 kHz. The true courtship song consisted of a more or less continuous train of such syllables but the tick, which appeared intermittently, consisted of a rapid frequency sweep from 23 kHz to 13 kHz and often showed a second harmonic component (Fig. 6.1). Although the intertick syllables, including their ultrasonic components were of fairly low intensity, the frequency sweeps were at a high intensity, comparable with that of the calling song.

The courtship song of crickets is believed by some authors to be the necessary stimulus for the female to take up the copulation position on top of the male (Alexander, 1960). As the tick sounds are often much more intense than the rest of the courtship song, it is possible that they form the actual stimulus for the female to mount the male. Lenahan certainly gained the impression that female *Acheta domesticus* generally made no attempt to mount until this sound had been produced, or if they did they were rebuffed.

At present there is little evidence that crickets can hear ultrasonic frequencies. The tympanic organ of crickets, like that of tettigoniids, is on the fourth segment, the tibia, of the prothoracic legs. It may have one or two tympanic membranes which vary in size and shape (Autrum, 1963). The tympanic organ contains many sensory cells and recently Friedman (1972a, b) has shown that in *Gryllus assimilis* it does not contain scolopale organs as in many other insects such as moths and lacewings (Chapter 4). It consists of a highly modified epithelium and is structurally associated with the smaller, anterior tympanic membrane but not to the larger, posterior membrane.

The acoustic sensitivity of the tympanic organ in crickets has been studied by several groups of workers who used equipment with upper frequency limits of 50–100 kHz (e.g. Wever and Bray, 1933; Wever and

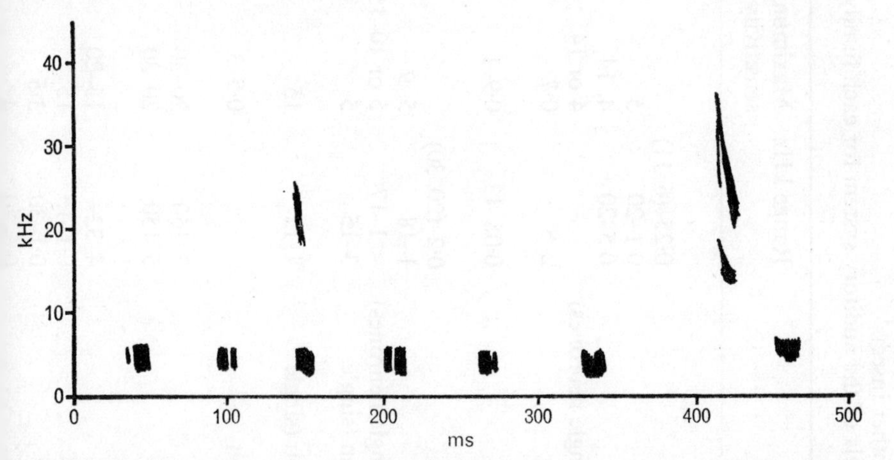

FIGURE 6.1 Sonagram of part of the courtship song of *Acheta domesticus* showing a rapid train of low frequency syllables (one with a marked ultrasonic component) and a single frequency sweep forming the more intense 'tick'.

Vernon, 1959; Katsuki and Suga, 1960; Lenahan, personal communication). These workers obtained acoustic responses to frequencies up to between 6 kHz and 20 kHz and some reported a single peak of sensitivity at about 1 or 5 kHz (Table 6.2). Horridge (1960), however, mentioned that in his experimental preparations the threshold of the tympanic nerve of *Acheta domesticus* did not begin to rise until 20–30 kHz although the prothoracic ganglion appeared to respond preferentially to lower frequencies of 0·5–3 kHz. This indicates that the tympanic organ may possess at least two different types of receptors, one sensitive to low frequencies and the other to higher frequencies up to 30 kHz. Two different receptors were later found in *Acheta* by Popov (1971) who

TABLE 6.2 Auditory sensitivity of some 'other insects' (as measured at ascending levels of the auditory system for each family)

Species	Recording site	Range kHz	Maximum sensitivity kHz	Authors
Family Gryllidae				
Gryllus assimilis	Tympanic nerve	0·25–(6–11)		Wever & Bray (1933)
G. abbreviatus	Tympanic nerve	0·1–20	5	Wever & Vernon (1959)
G. campestris	Tympanic nerve	0·5–20	4, 14	Nocke (1972)
	Tympanic nerve (single neurones)		4 or 14	
Homeogryllus japonicus	Tympanic nerve	2–8	0·7	Katsuki & Suga (1958, 1960)
Xenogryllus marmoratus	Tympanic nerve	0·08–13	0·9–1	Katsuki & Suga (1958, 1960)
Acheta domesticus	Tympanic nerve	0·2–(20–30)		Horridge (1960)
Acheta domesticus	Tympanic nerve	1–18	5, 9	Lenahan (pers. comm.)
Acheta domesticus	Tympanic nerve (single neurones)	<1–17	5 or 10–15	Popov (1971)
	Prothoracic ganglion (single neurones)	3–15	5	
	Prothoracic ganglion (single neurones)	4–17	15	
Acheta domesticus	Prothoracic ganglion		0·5–3	Horridge (1960)
Family Gryllotalpidae				
Gryllotalpa hexadactyla	Tympanic nerve	5–150	20–30	Suga (1968)
Scapteriscus didactylus	Tympanic nerve	5–150	20–30	Suga (1968)
Family Acrididae				
Acridium aegyptium	Behaviour	4–32	15–20	Auger & Fessard (1920)
Paroxya atlantica	Tympanic nerve	0·1–15	15	Wever & Vernon (1957)
P. a. paroxydes	Tympanic nerve	0·1–30	3·5	Wever & Vernon (1959)
Oxya japonica	Tympanic nerve	0·6–30	4	Katsuki & Suga (1958)

Oxya japonica	Tympanic nerve	0·6–30	4–10	Katsuki & Suga (1960)
Locusta migratoria danica	Tympanic nerve	0·6–40	3–5	Katsuki & Suga (1958)
Locusta migratoria danica	Tympanic nerve	0·6–45	4–9	Katsuki & Suga (1960)
Locusta migratoria danica	Tympanic nerve	0·6–45	4–9	Suga (1960)
	Tympanic nerve (single neurones)	Variable	4–9	
L. migratoria	Tympanic nerve	0·2–50	10	Horridge (1960, 1961)
L. migratoria	Tympanic nerve	0·5–20	3–5	Adam & Schwartzkopff (1967)
L. m. manilensis	Tympanic nerve	0·5–40	3–4	Yanagisawa *et al.* (1967)
Schistocerca gregaria	Tympanic cells			
	(a)	2–9	3·7, 8	
	(b)	1–5	3·5, 5	Michelsen (1966, 1968, 1971a)
	(c)	1–14	1·5, 2–3, 8	
	(d)	6–40	12, 19	
Arphia sulphurea	Metathoracic ganglion	0·3–21	10 or over	Wever (1935)
Sphyngonotus sp.	Mesothoracic ganglion	Up to 35		Benedetti (1950a, b)
Locusta sp.	Mesothoracic ganglion	Up to 35		Benedetti (1950a, b)
Locusta migratoria	Mesothoracic ganglion (single neurones)	4–40	6–7	Horridge (1960, 1961)
L. m. manilensis	Prothoracic-mesothoracic connectives (single neurones)	0·5–40	4	Yanagisawa *et al.* (1967)
L. migratoria	Metathoracic-suboesophageal connectives (single neurones)	3–(30–40)		Kalmring *et al.* (1972)
Gastrigmarus africanus	Prothoracic-mesothoracic connectives (α neurones)	5–40	40	Rowell & McKay (1969a, b)
L. migratoria	Suboesophageal ganglion (single neurones)	1–20		Kalmring (1971)
L. migratoria	Protocerebrum (gross response)	Up to 50	6–7, 20	Adam & Schwartzkopff (1967)
	Protocerebrum (single neurones)	4–30, 4–8, or 12–30		

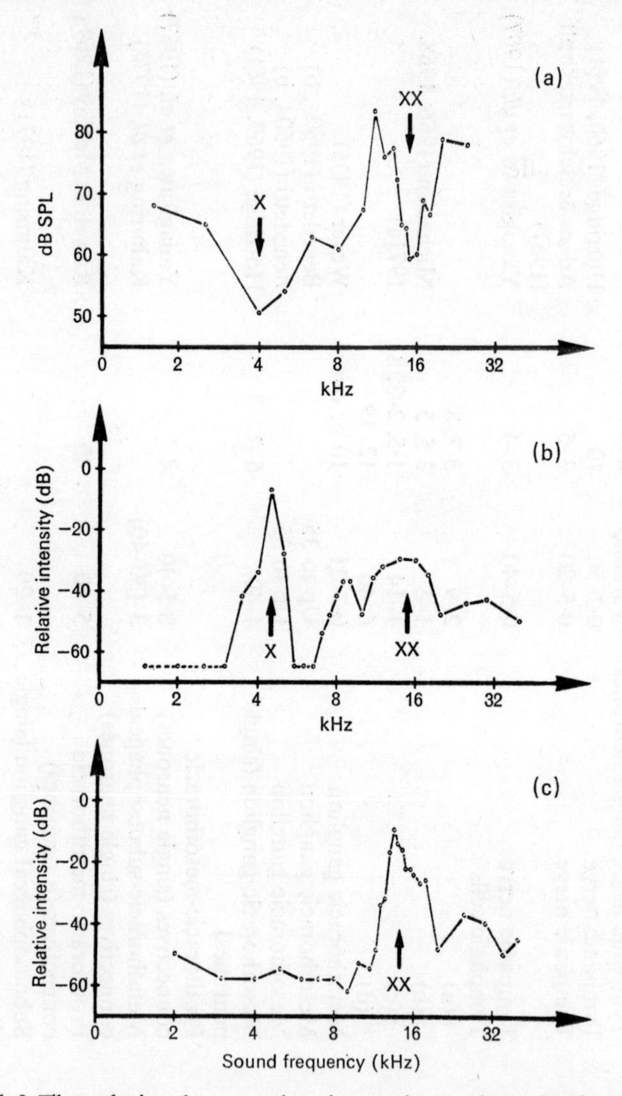

FIGURE 6.2 The relation between hearing and sound production in *Gryllus campestris*. (a) Auditory response curve of the tympanic nerve. (b) The spectrum of the calling song. (c) The spectrum of the courtship song. The maximum auditory sensitivities correspond closely with the spectral peaks as shown by the arrows X and XX. (After Nocke, 1972)

reported that the low frequency receptors were most sensitive to frequencies of 4–5 kHz, the dominant frequency of the calling song, whereas the high-frequency receptors were most sensitive to frequencies of 10–15 kHz. Unfortunately Popov does not appear to have investigated frequencies above 17 kHz. Each kind of tympanic receptor apparently connects with different interneurones in the thoracic ganglion. Popov found that the low frequency interneurones had a peak of sensitivity at 4–5 kHz and were spontaneously active but they were inhibited during the activity of the high frequency interneurones which had a peak of sensitivity at about 15 kHz.

Nocke (1972) has shown that the threshold curve of the whole tympanic nerve of *Gryllus campestris* matches the spectrum of the males' songs. He found a peak of sensitivity at 4 kHz, which corresponds to the main component of the calling and aggressive songs, and a second peak at 14 kHz, corresponding to the main frequency of the courtship song (Fig. 6.2). Many single neurones within the nerve responded to signals of 4 kHz, but only one, or at the most three, neurones responded to 14 kHz signals and these few neurones were connected functionally to the smaller of the two tympanic membranes. Nocke concluded that *G. campestris* could distinguish between sounds near 4 kHz and those near 14 kHz and he suggested that different sound frequencies may affect the two tympana in different ways.

The lower peak of sensitivity in crickets may result directly from the resonant properties of the tympanic membranes. The actual movements of the tympana in response to sounds up to 20 kHz have been studied using the Mössbauer effect in *Teleogryllus commodus* and *T. oceanicus* (Johnstone *et al.*, 1970). Both membranes showed greatest movements to sounds of 5 kHz although the smaller membrane was less sensitive than the larger. But other workers (Loftus-Hills *et al.*, 1971) found that the auditory sensitivity of the tympanic nerve had a sharp peak at 4 kHz in *T. commodus* and a broader peak at 4–5 kHz in *T. oceanicus* and they questioned whether these could result from mechanical tuning of the tympanic membrane at 5 kHz.

GRYLLOTALPIDAE, MOLE CRICKETS

Mole crickets are related to true crickets and they also produce sounds by rubbing the two tegmina together. Their name refers to their habit of burrowing through the ground using powerful, spade-like front legs which also bear the tympanic organs. The sounds emitted by these insects are low in frequency, 1–4 kHz (Busnel and Chavasse, 1951; Alexander, 1960; Bennet-Clark, 1970) and no ultrasounds have been reported. But these animals are interesting for their unique method of radiating their low frequency sounds. They construct burrows in the shape of two exponential horns which join to form a bulb at the base (Bennet-Clark, 1970). An animal sits in the bulb with the body occluding the throat of the horn. The wings are directed backwards, towards the horns which radiate the sound. The two sound sources interact above the

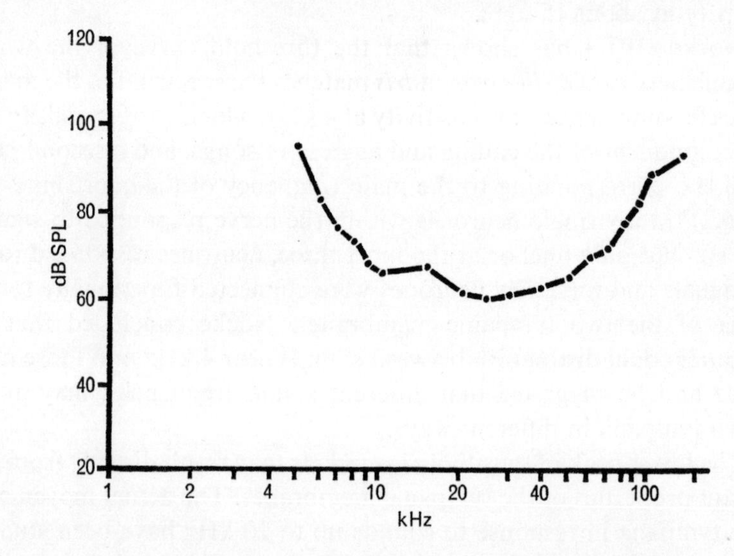

FIGURE 6.3 Averaged auditory response curve of central auditory neurones of some Brazilian mole crickets. (Redrawn after Suga, 1968b)

ground to give a very intense and directed sound. The higher frequencies produced, for example, by tettigoniids can be radiated effectively by the mirror region of the wing, because although this is small, the wavelengths of the sounds are very short.

Suga (1968b) studied the auditory responses of the tympanic organs of two species, *Gryllotalpa hexadactyla* and *Scapteriscus didactylus*, and found that they were sensitive to sounds between 5 kHz and 150 kHz and had a peak of sensitivity at 20–30 kHz (Fig. 6.3). Suga suggested that mole crickets, which are nocturnal, may listen not only to the species song, but also to the ultrasound emitted by some predators. This may therefore be another example of the detection of bats by nocturnal insects. At the moment, however, there is no evidence for this and the significance of the auditory sensitivity of these animals at high frequencies has yet to be explained.

ACRIDIDAE, GRASSHOPPERS AND LOCUSTS

The acridids form a very large group of insects which includes the familiar grasshoppers as well as the notorious locusts. Grasshoppers and locusts strigilate standing on their front two pairs of legs. The back legs are then raised and rapidly moved up and down across the outer edges of the tegmina. In some species the file is on a vein of the tegmen and the plectrum is on the femur of the hind leg, in other species the positions of the file and plectrum are reversed. The mechanism of sound radiation in acridids is not fully understood but it appears that the whole wing may be involved (Pierce, 1948). Acridids produce sounds with a wide frequency spectrum and Dumortier (1963b) has suggested that this could be due to the displacement of the point of contact between the file and strigil as the femur moves over the tegmen. In this way different parts of the tegmen would be brought into resonance. Different sounds are certainly produced by different movements of the femora over the tegmina (Brown, 1955). Sound production is generally confined to the males but in some acridids the females also strigilate (Dumortier, 1963b).

The frequency spectrum of acridid songs often extends into the ultrasonic range (Dumortier, 1963b) and songs with frequency peaks above 20 kHz have now been recorded from several species (Table 6.1). Pierce (1948) found high frequency sounds in only one species, *Encoptolophus sordidus*, the clouded locust, in which the main energy was at 19 kHz. More recently a 'purely ultrasonic' song has been recorded from *Arcyptera fusca*, from the Pyrenees (Pye, unpublished). This had a frequency range of 20–42 kHz and frequency peaks at 40 kHz, 32 kHz and 24 kHz (Plate IX c).

The acoustic behaviour of acridids has been studied in some detail (Busnel, 1955; Busnel and Loher, 1955; Dumortier, 1963c; Alexander, 1960, 1967 for reviews). Acridids show acoustic responses similar to those of tettigoniids and crickets: calling males sing in alternation with other males or with artificial signals up to at least 25 kHz (Busnel, 1956) and they also produce aggressive calls when close to rival males. Again as in tettigoniids, females move towards singing males and in some species they reply to the male's calling song by producing an 'agreement' song which indicates their willingness to mate. The males of some, but not all, species produce a courtship song when the female is near and some males may also give another song just before mounting the female (Dumortier, 1963c). The recorded frequency spectra of the calling song and the courtship song of three species of grasshoppers, *Stenobothrus lineatus*, *Omocestus viridulus* and *Chorthippus brunneus*, are reported to differ very slightly in the audible range (Haskell, 1957) but there do not appear to be any reports of changes in the ultrasonic content of the different acridid songs.

The tympanic organs of acridids are situated one on each side of the first abdominal segment. Each organ consists of 60–80 sensory units, or scolopales, which are directly attached to the tympanic membrane (Gray and Pumphrey, 1958; Gray, 1960). In *Locusta migratoria* and other locusts there are four groups of sensory cells and three of them, groups *a*, *c* and *d*, are orientated in mutually perpendicular planes. The fourth group, *b*, is in the same plane as group *a*.

The ability of acridids to respond to ultrasonic signals has been known for many years. In 1920 Auger and Fessard reported that *Acridium aegyptium* gave a 'reflex-like movement' in response to sounds of up to 20 kHz. Tones of 15–20 kHz were most effective in eliciting the response, which was abolished when the tympanic organs were covered with vaseline or destroyed. Since then there have been many reports of acoustic responses to ultrasound in acridids at all levels of the nervous system from the tympanic organ to the brain. These are summarized in Table 6.2.

The tympanic organs of many species appear to be sensitive to signals up to at least 20 kHz but reports of the upper limit of responsiveness vary between 20 kHz and 50 kHz. The greatest sensitivity is generally reported to be between 3 kHz and 10 kHz, although Wever and Vernon (1957) found that in *Paroxya atlantica*, the American Atlantic grasshopper, the sensitivity increased up to 15 kHz, the upper limit of their apparatus.

Until recently it was thought that insects in general were unable to distinguish between signals of different carrier frequency (Pumphrey, 1940), but this idea is now being re-examined. The first evidence for frequency discrimination in acridids came in 1960–61 when Horridge found that although the whole tympanic nerve of *Locusta* responded to signals of 0·2–50 kHz, an interneurone in the metathoracic ganglion responded preferentially to high frequency signals of 2–30 kHz and rejected those of lower frequency (Horridge, 1960, 1961). He also found that the form of the responses recorded from the tympanic nerve varied with the frequency of the stimulus.

Later Michelsen (1966, 1968, 1971a, b) showed that in locusts this frequency discrimination occurred in the four groups of sensory cells in the tympanic organ, each of which was sensitive to a different range of frequencies. The cells of only one group, *d*, were sensitive to ultrasonic frequencies and they responded to signals of 6–40 kHz. Their maximum sensitivity was at about 12 kHz, but most cells had a second peak of sensitivity at 19 kHz. Michelsen (1971b) showed that this frequency discrimination may have a purely physical explanation. Using

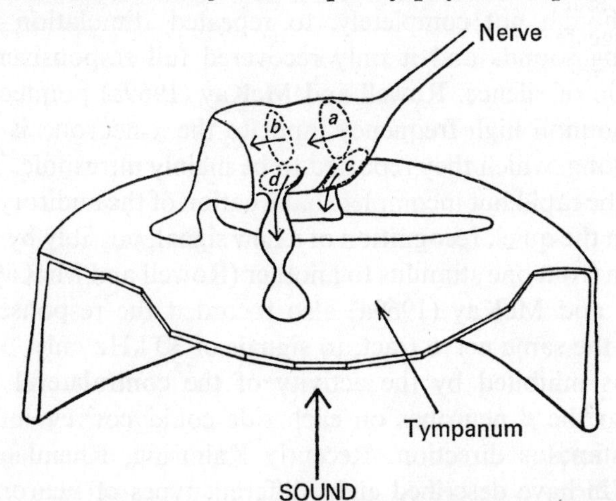

FIGURE 6.4 A diagram of the left tympanic organ of a female locust *Schistocerca gregaria* in the gregarious phase, showing various supporting structures and the four groups of sensory cells *a–d* attached to different parts of the tympanic membrane (tympanum). (After Michelsen, 1971a modified from Gray, 1960)

the technique of laser holography, he demonstrated that the tympanic membrane of locusts can vibrate in at least two different ways. Since the four groups of sensory cells are attached at different points on the membrane (Fig. 6.4), this could account for their different frequency sensitivities.

In some acridids, acoustic responses to signals up to 40 kHz have been recorded from the meso- and metathoracic ganglia and from the ventral nerve cord running both between these ganglia as the thoracic connectives, and also to the brain. However, reports of the frequencies of maximum sensitivity vary between 4 kHz and 40 kHz (Table 6.2). One interneurone that is apparently very sensitive to ultrasound is the 'α' neurone studied by Rowell and McKay (1969a, b) in *Gastrigmarus africanus*, an East African grasshopper. This neurone runs from the mesothoracic ganglion to the brain and Rowell and McKay found that it received impulses from both tympanic nerves. But unlike the auditory *T* fibre of tettigoniids, the activity of the α neurone was not inhibited by that of the contralateral tympanic nerve (p. 124) and so it could not carry information about the direction of the stimulus. The neurone was apparently most sensitive to 40 kHz, the highest frequency tested, and the authors suggested that it would probably respond to even higher frequencies. It adapted slowly to a prolonged tone and habituated fairly rapidly, although not completely, to repeated stimulation by either short or long sounds and it only recovered full responsiveness after about 20 min of silence. Rowell and McKay (1969a) pointed out that the most common high frequency input to the α neurone is probably the species song, which they reported to be mainly ultrasonic. They suggested that the rapid but incomplete habituation of the auditory response could help in the quick recognition of a new signal, possibly by switching the attention from one stimulus to another (Rowell and McKay, 1969b).

Rowell and McKay (1969a) also recorded the response of a 'β' neurone, in the same nerve tract, to signals of 30 kHz only. This interneurone was inhibited by the activity of the contralateral tympanic nerve and so the β neurones on each side could convey information about the stimulus direction. Recently Kalmring, Rheinlaender and Romer (1972) have described eight different types of neurones in the ventral nerve cord of *Locusta* that responded to frequencies up to 30–40 kHz. Three of these neurones responded to movement of the sound source.

There appear to have been comparatively few studies on the acoustic responses of the brain in acridids (Table 6.2). Kalmring (1971) showed that the suboesophageal ganglion of *Locusta* contained at least four different types of acoustic neurones, all of which responded to signals of up to 20 kHz, the highest frequency tested. Adam and Schwartzkopff (1967), however, found that the gross response of the protocerebrum of *Locusta* extended up to 50 kHz and that there were at least three different types of acoustic neurones. Some were sensitive to a wide range of frequencies, 4–30 kHz, others were sensitive to one of two narrower bands, 4–8 kHz or 12–30 kHz. It therefore appears that at least some information on the frequency, including ultrasonic frequencies, reaches the brain in acridids but the importance of the information is not yet known.

INSECTS OF OTHER GROUPS

Many other insects produce sounds in a variety of ways, but so far ultrasound has been reported in only a few species (Table 6.1) and its importance, if any, is not known.

Cicadas, which are plant-sucking insects of the order Hemiptera, possess tymbal organs similar to those of arctiid moths (Chapter 4) and situated on each side of the first abdominal segment. These tymbals consist of a simple membrane (Pringle, 1954) or one with several grooves or striae (Reid, 1971; Young, 1972) (Plate XI a) and the sounds produced during buckling and relaxation are generally reported to be at frequencies between 3 kHz and 10 kHz (Pringle, 1954; Alexander, 1957, 1960; Simmons *et al.*, 1971; Reid, 1971). However, Saby and Thorpe (1946) found that some unnamed Panamanian cicadas produced frequencies up to 30 kHz and the song of a single unidentified species recorded in Trinidad by Pye (unpublished) was mainly ultrasonic and contained frequencies up to 100 kHz (Plate XI b) although it was also very loud to the unaided ear. The songs of cicadas appear to be aggregation calls. Males move towards other singing males and groups of males sing in chorus (Alexander, 1960). Unmated females are attracted to these groups and it has been suggested that bird predators may be repelled by the intense noise that they produce (Simmons *et al.*, 1971).

Cicadas have tympanic organs on the ventrolateral walls of the second abdominal segment (Leston and Pringle, 1963). Each organ contains about 1500 sensory cells and Pringle (1954) found that in *Platypleura capitata*, a cicada from Ceylon, the tympanic organs responded to the species song as well as to 'high pitched sounds' for which, unfortunately, no value was given. The tympanic nerve responses of four other species of cicadas were studied by Katsuki and Suga (1960) who found the highest frequency responses in *Tanna japonensis* which was sensitive to signals up to 20 kHz and *Graptopsaltria nigrofuscata* which responded to signals 'beyond' 20 kHz. However, the maximum sensitivities of all four species lay in the audible range, 1·3–5 kHz.

Ultrasounds have been reported from four families within the order Coleoptera, or beetles (Table 6.1). Frequencies up to 28 kHz have been detected from *Necrophorus*, the burying beetle (Autrum, 1936), which produces sound by rubbing the apical edge of the thick elytra, or wing covers, over two striated ridges on the dorsal side of the fifth abdominal segment (Dumortier, 1963a, c). *Geotrupes*, the dor beetle, has a strigilatory organ on the coxa of the hind leg and frequencies up to 40 kHz have been detected from an unnamed member of this genus (Autrum, 1936). The Cerambicidae, or longicorn beetles, strigilate by rubbing together two plates on the back of their thorax; the hind edge of the pronotum moving over striations on the anterior edge of the mesonotum (Imms, 1957; Dumortier, 1963a). Dumortier (1963b) has recorded sounds extending beyond 20 kHz from *Cerambix cerdo* and frequencies up to 30 kHz were recorded by Pye (unpublished) from an unknown species of longicorn beetle (probably *Petrognatha gigas*) from Uganda (Plate XI d). There appear to be no reports of acoustic responses to ultrasound in these groups.

Acoustic responses to ultrasonic signals have been obtained from one insect in which the strigilation sounds do not appear to have been analysed in detail. This is *Corixa striata*, a water boatman. Male corixids produce sounds by rubbing a spiny area on the femur of the fore-legs over the clypeus, part of the front of the head (Imms, 1957). Schaller (1951) found that male *Corixa striata* responded to tones of 20 kHz, 30 kHz and 40 kHz by making a sharp strigilation sound. Both male and female corixids possess tympanic organs on the mesothorax, near to the base of the wings (Leston and Pringle, 1963). Each contains two sensory cells but the acoustic sensitivity of these organs does not appear to have been studied. In two other water bugs, *Notonecta obliqua*, and *Nepa cinerea*, the tympanic organs are reported to respond to airborne sounds of up to 10 kHz and 7 kHz respectively (Arntz, 1972).

Finally, yet another insect from which ultrasounds have been detected is a giant cockroach from Uganda. The mechanism of stridulation in this animal was not known but sounds that were mainly ultrasonic and contained frequencies up to 64 kHz were emitted when the insect was picked up (Pye, unpublished) (Plate XIe).

Plates IX–XIV

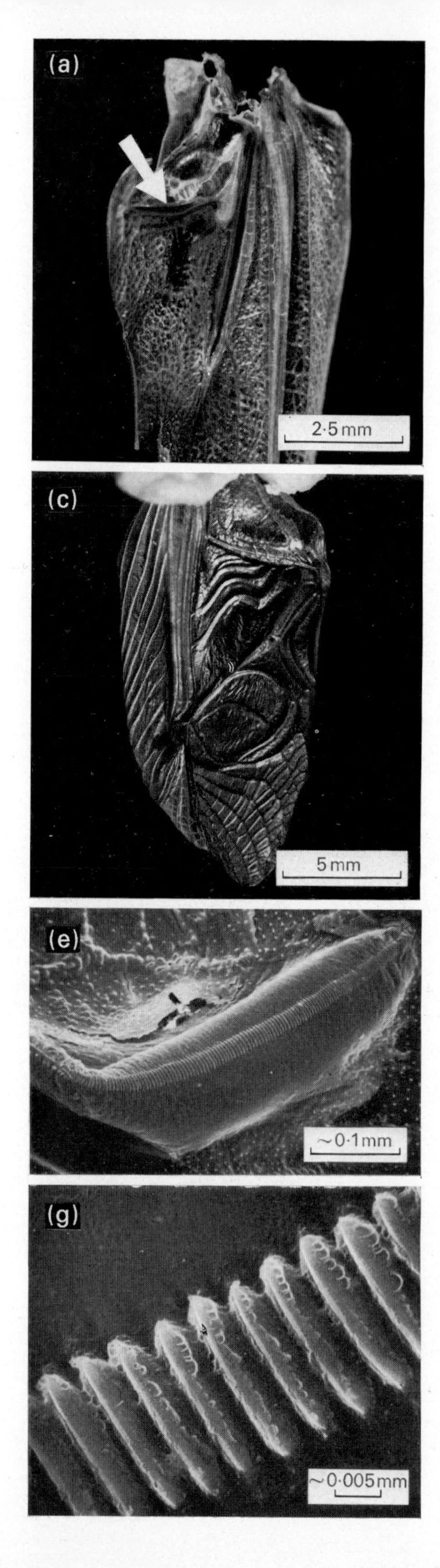
(a)
2·5 mm
(c)
5 mm
(e)
~0·1 mm
(g)
~0·005 mm

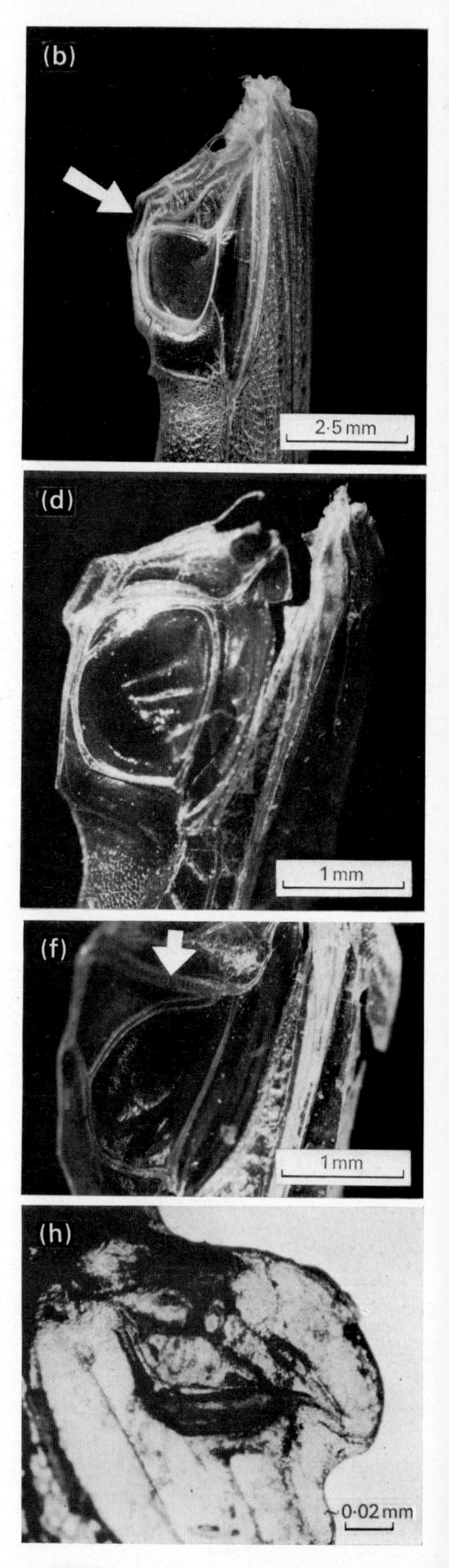
(b)
2·5 mm
(d)
1 mm
(f)
1 mm
(h)
~0·02 mm

PLATE IX Strigilatory structures of three bush crickets and a cricket.

(a and b) *Homorocoryphus nitidulus*, Tettigoniidae.

(a) Underside of the left tegmen showing the true file as a dark line (arrowed) running across the wing and the absence of a mirror. (Bar = 2·5 mm.)
(b) Upper side of the right tegmen showing the mirror frame, thickened on three sides to form a 'horseshoe', and the plectrum (arrowed). (Bar = 2·5mm.)

(c) Upper side of the left tegmen of *Gryllus bimaculatus*, Gryllidae, showing the triangular 'harp' and the more distal mirror. (Bar = 5 mm.)

(d and f) *Conocephalus conocephalus*, Tettigoniidae.
(d) Upper side of the right tegmen showing the plectrum, and the mirror frame forming a complete rigid ring. (Bar = 1 mm.)
(f) Underside of the left wing showing the true file (arrowed) and a region of transparent cuticle corresponding to, but smaller than, the mirror on the right wing. (Bar = 1 mm.)

(e, g and h) *Anepitacta egestoides*, Tettigoniidae.
(e) Underside of the right tegmen showing the L-shaped vestigial file as seen with a scanning electron microscope. (Bar = approx. 0·1 mm.)

(g) The teeth of the true file spaced about 1/200 mm apart as seen with a scanning electron microscope. (Bar = approx. 0·005 mm).
(h) The right tegmen seen from below, showing the L-shaped vestigial file and the absence of a mirror frame. (Bar = approx. 0·02 mm.)

(e and g electron micrographs by John Huxley. Other photographs taken by A. Howard and R. Reed.)

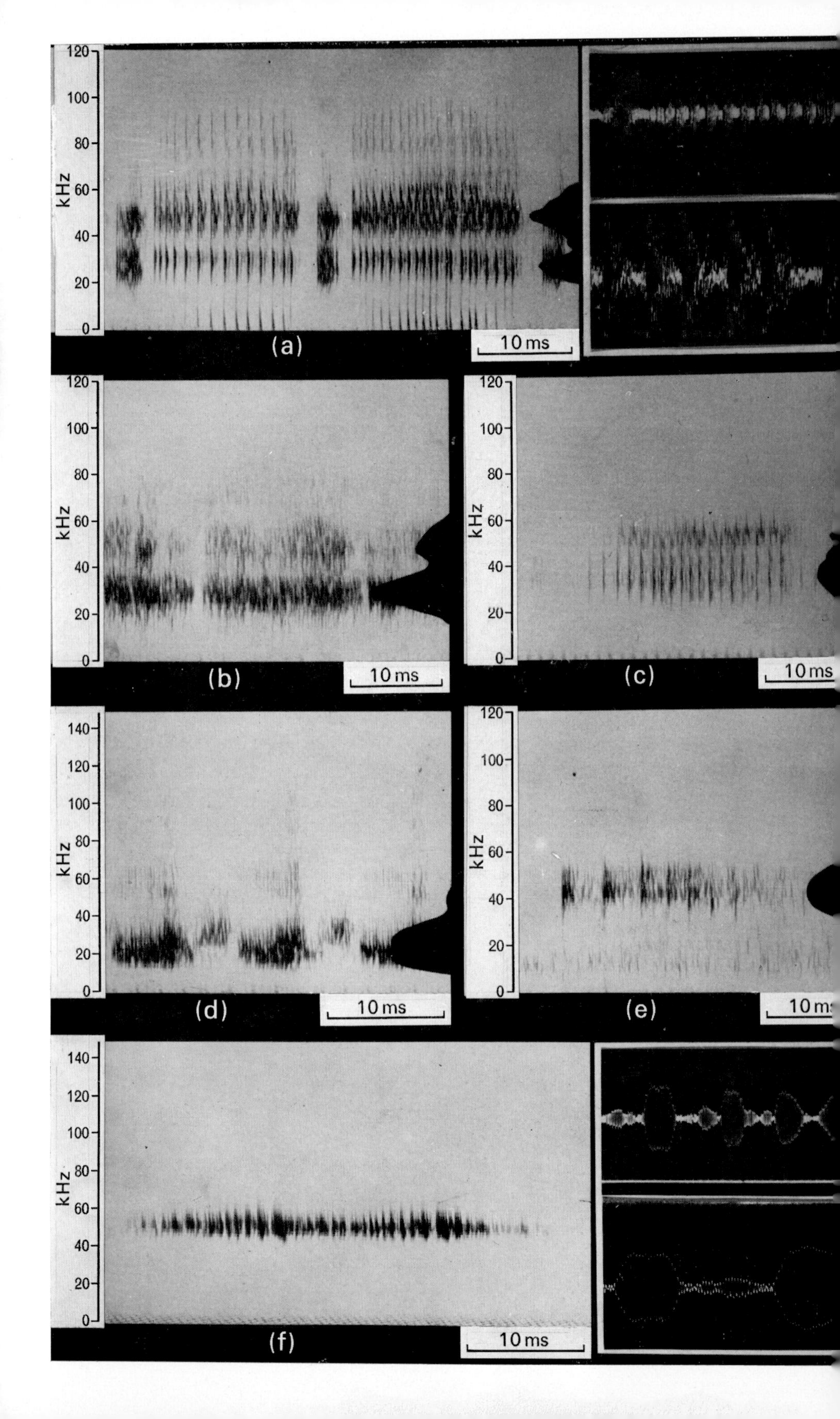

kHz
0
20
40
60
80
100
120
140
(a)
(b)
(c)
(d)
(e)
(f)
10 ms

PLATE X The songs of some bush crickets (Tettiogniidae).

(a) *Conocephalus conocephalus*. A sonagram showing four hemisyllables with two main components at ultrasonic frequencies and an integrated spectrum of the total energy distribution (in black). The waveforms show (above) two hemisyllables and (below) pulses from the longer hemisyllable on an expanded time scale.

(b) *Conocephalus maculatus*. A sonagram again showing two main ultrasonic components but with the lower one more intense [compare with *C. conocephalus* in (a)].

(c) *Conocephalus discolor*. A sonagram showing two main ultrasonic components less clearly separated.

(d) *Metrioptera roeselii*. A sonagram showing frequency changes within the major hemisyllables.

(e) A sonagram of an 'inaudible' song from an unidentified insect, probably a tettigoniid,* recorded from tall trees at Rennes, France.

(f) *Anepitacta egestoides*. A sonagram showing the narrow-band of energy at high frequency. The waveforms at two different time scales show alternating hemisyllables, the rounded syllable envelopes and the sinusoidal wave form.

*Recent recordings in the laboratory strongly suggest that this was *Leptophyes punctatissima*.

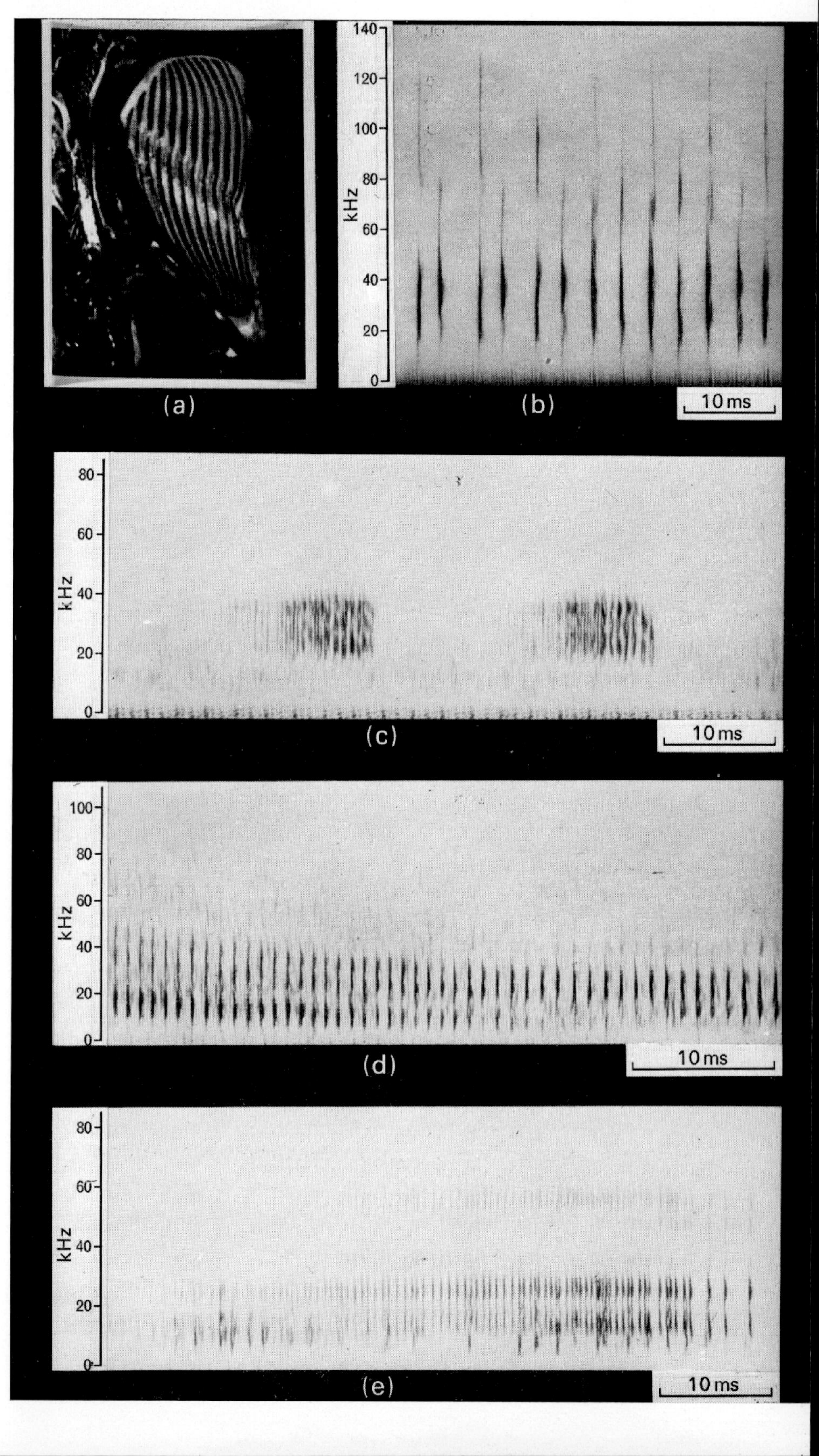

140
120
100
80
60
40
20
0
kHz
(a)
(b)
10 ms
80
60
40
20
0
kHz
(c)
10 ms
100
80
60
40
20
0
kHz
(d)
10 ms
80
60
40
20
0
kHz
(e)
10 ms

PLATE XI Sound production in a variety of insects.

(a) The tymbal organ of the cicada, *Magicicada cassini*, showing vertical grooves which buckle successively from the back forwards to produce one series of clicks and a further series during relaxation, in a similar manner to the arctiid moths described in Chapter 4. (From Reid, 1971.)

(b) A sonagram of the sound produced by an unidentified cicada in Trinidad. Pairs of pulses are produced by alternate buckling and relaxation of the ungrooved tymbal organ as a whole.

(c) A sonagram of two syllables produced by the grasshopper, *Arcyptera fusca*, recorded out of doors in the Pyrenees.

(d) A sonagram of part of a train of clicks produced by a longicorn beetle, probably *Petrognatha gigas*, in Uganda.

(e) A sonagram of a train of clicks produced by a large cockroach in Uganda when it was captured in the open.

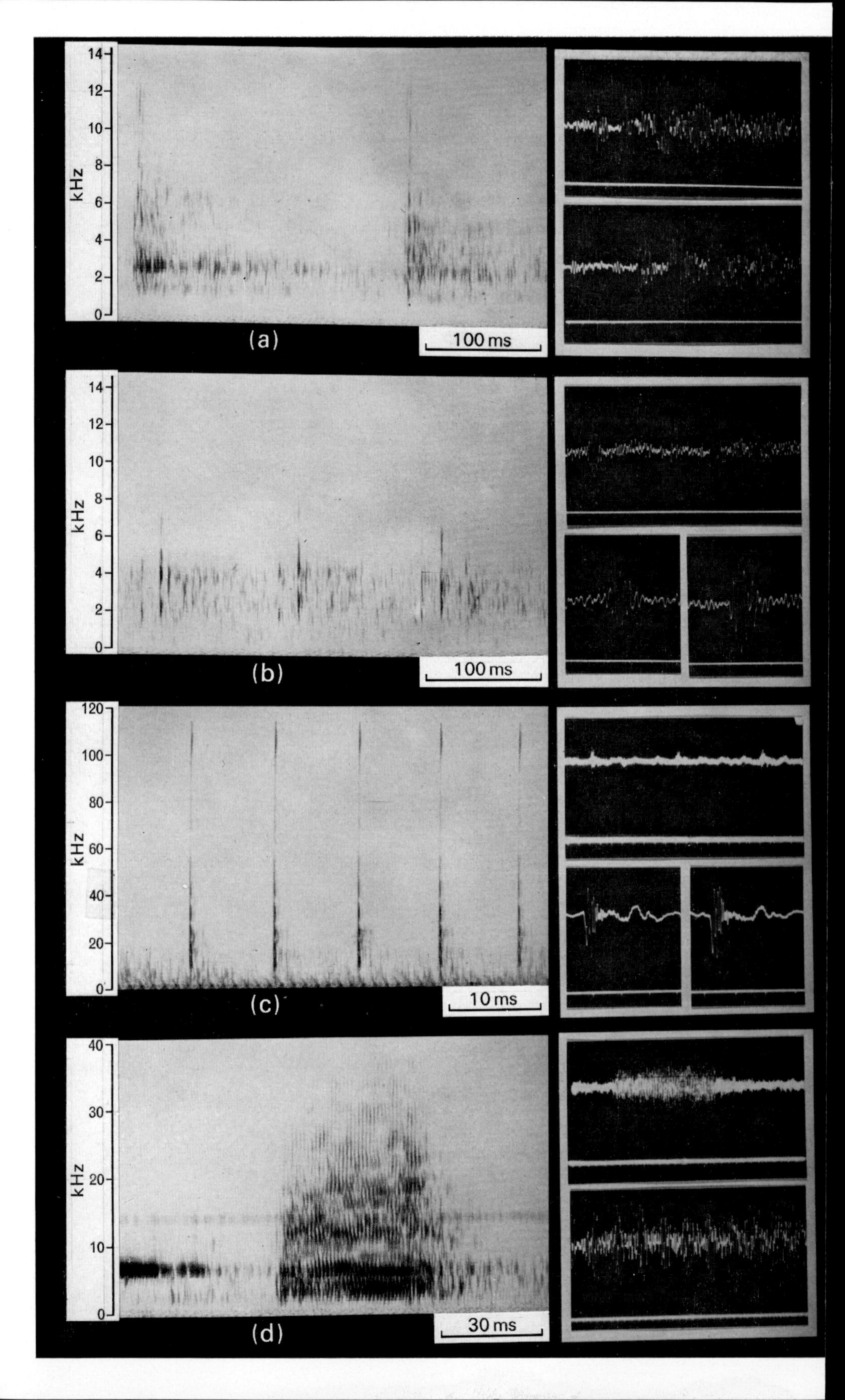

kHz
(a)
100 ms
(b)
100 ms
(c)
10 ms
(d)
30 ms

PLATE XII Clicks of birds and cetacea.

(a) The oil bird, *Steatornis caripensis*. A sonagram of two bursts of clicks recorded from a single bird in a cave in Trinidad. Each waveform (duration 20 ms) shows a burst composed of several rough clicks.

(b) A cave swiftlet, *Collocalia fuciphaga*. A sonagram of clicks recorded in Kuala Lumpur by Lord Medway. The waveforms show (above) several such clicks and (below) the two clearest clicks on an expanded time scale (sweep time 5 ms). (c and d) The bottlenose dolphin, *Tursiops truncatus*, recorded with a hydrophone at Whipsnade Zoo.

(c) A sonagram of five echo-location clicks. The waveforms show (above) three clicks (sweep time 28 ms) and (below) two single clicks (sweep time 0·5 ms).

(d) A sonagram of a low-frequency ‘creak’. The end of a whistle signal is seen just before this. The waveforms show (above) a single low-frequency ‘creak’ (sweep time 115 ms) and (below) part of the creak (sweep time 13 ms).

(a)
(b)
(c)
(d)

(a) Four-day-old spiny mouse, *Acomys cahirinus*. This species is born in a relatively advanced stage; the eyes open and the young can run soon after birth. The pups produce ultrasound and, if they stray, they are retrieved by their mother.

(b) A bushbaby, *Galago crassicaudatus*. Several primitive primates have been shown to be sensitive to sound frequencies as high as 30–40 kHz.

(c) An adult female C_3H laboratory mouse. She is retrieving a four-day-old albino rat pup and her own litter can be seen in the nest. The baby rat is the same age as the spiny mouse in (a) but is much less developed, with the external ears just free and the eyes still closed.

(d) A bottlenose dolphin, *Tursiops truncatus*, demonstrating its echo-location ability. Although its eyes are temporarily covered by rubber suction cups, it is swimming accurately between two iron poles. (From Norris *et al.*, 1961.)

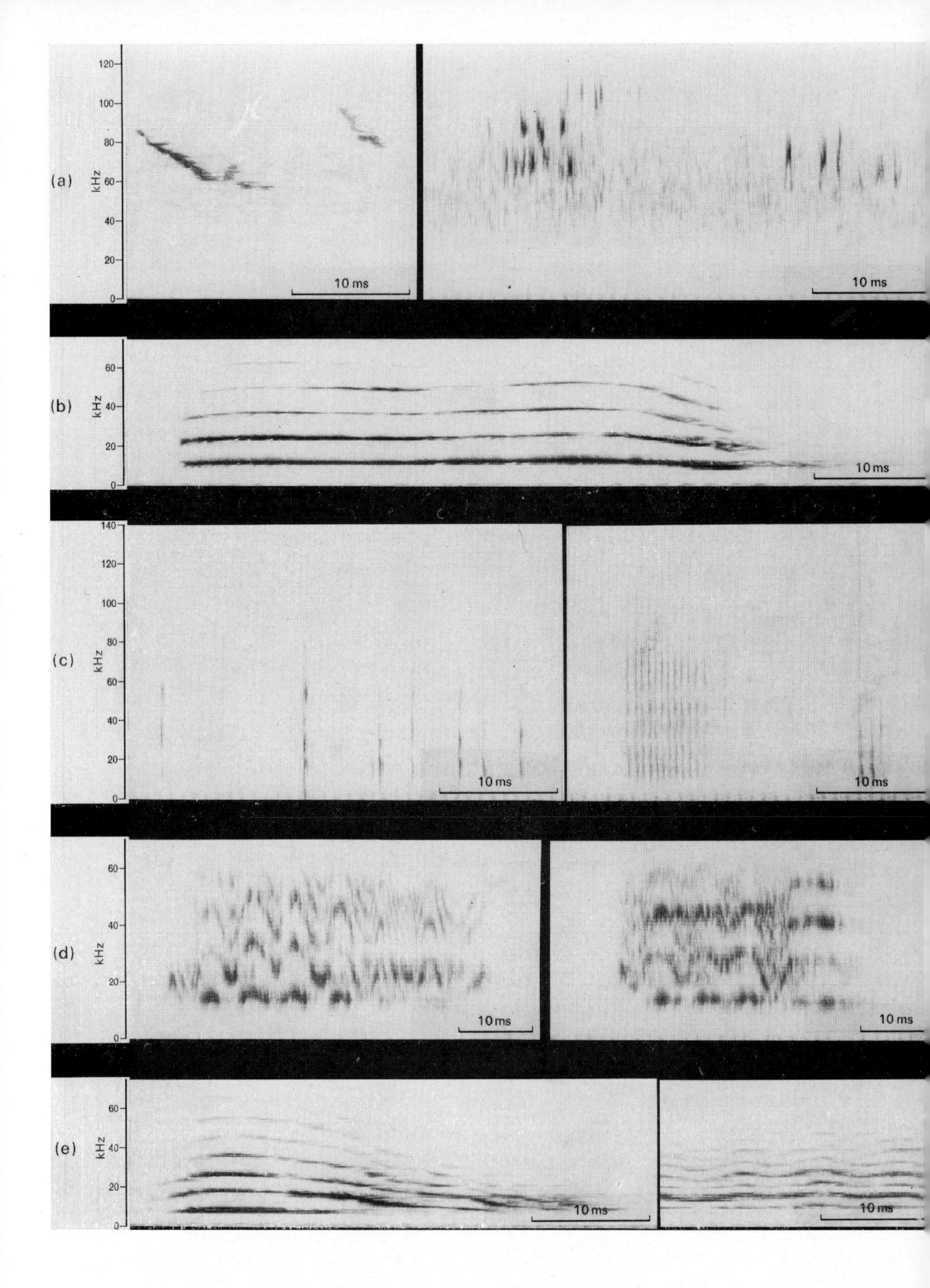

PLATE XIV Some sounds recorded from shrews. (a–b) *Neomys fodiens* (water shrew): short ultrasounds and an audible squeak. (c) *Crocidura russula judaica* (Israeli shrew): clicks produced while examining the microphone. (d) *Crocidura suaveolens* (Scilly Isles shrew): two audible squeaks. (e) *Suncus murinus* (musk shrew): two audible squeaks.

CHAPTER SEVEN

Ultrasound in Rodents

The rodents form a very large and varied group of mammals and are found all over the world wherever food is available. The order Rodentia contains three main sub-orders: the Hystricomorpha with porcupines and guinea pigs; the Scuriomorpha with squirrels and beavers; and the Myomorpha with mice, rats, hamsters and their allies. Pure ultrasounds have so far been found in only one of these sub-orders, the Myomorpha, and in only two families within this group. These are the family Muridae, which includes rats and mice and the family Cricetidae which includes hamsters, voles, gerbils and lemmings.

In 1948 Schleidt reported that *Clethrionomys glareolus*, the bank vole, produced high pitched cries that were barely audible to humans and which, he suggested, were used to contact others of the same species. Later (1951) Schleidt described calls of 15·5 kHz produced by this species and he believed that these inhibited the escape reaction of mates. Sounds extending up to 30 kHz, but with clearly audible components, were detected from *Mesocricetus*, golden hamsters, and *Glis*, dormice, of the order Rodentia as well as from *Crocidura,* shrews, of the order Insectivora by Kahmann and Ostermann (1951) who suspected that these sounds might be used for echo-location.

The first report of purely ultrasonic calls in rodents appears to have been made in 1954 when Anderson found that adult *Rattus norvegicus*, laboratory rats, alone in their cages often produced calls at frequencies of 23–28 kHz which lasted for 1–2 s. He suggested that these signals might be used for communication between individuals or for echo-location. Anderson did not follow up his suggestion but a little later echo-location in rats was studied by Rosenzweig, Riley and Krech (1955) and by Riley and Rosenzweig (1957). These authors found evidence that rats can use echo-location to detect objects but they concluded that the echoes of incidental sounds such as toe scratching and footsteps were used rather than the few ultrasonic calls that they detected.

In 1956 Zippelius and Schleidt published their important discovery that the young of three species of myomorph rodents, *Apodemus flavicollis*, the yellow-necked mouse, *Mus musculus domesticus*, the house mouse, and *Microtus arvalis*, the field vole, emitted ultrasounds when they were removed from the nest and so were probably cold and hungry. The calls, which were about 100 ms long and at frequencies of 70–80 kHz, were produced by pups from the day of birth until their eyes opened, on about the thirteenth day of age. Zippelius and Schleidt suggested that these 'distresss' calls elicited the retrieving response of the mother and guided her to stray young which were then restored to the nest and so to their source of warmth and food.

Later, Noirot found that adult albino mice displayed maternal behaviour readily when presented with newborn young but that their responsiveness decreased progressively towards older pups (Beniest–Noirot, 1957, 1958; Noirot, 1964a, b, c, 1965). Retrieving and nest building in particular were seldom exhibited with pups of 13 days or older and Noirot therefore suggested that the performance of these behaviour patterns might be related to the ultrasonic cries detected by Zippelius and Schleidt. Subsequently Noirot studied the ultrasonic cries of baby mice and proposed that, as well as eliciting the retrieving response, these cries also inhibited the aggressiveness of adult rodents towards stray young which might otherwise be treated as strange animals outside the nest (Noirot, 1965, 1966a). Since then rodent ultrasound has formed an ever-expanding field of research and ultrasounds have been detected not only from infants but also from adults in a variety of different social situations (Sewell, 1967, 1969; Sales *née* Sewell, 1972a, b).

THE ULTRASONIC CALLS OF INFANT RODENTS

It is now known that the young of a wide variety of myomorph rodents emit ultrasonic calls when they are removed from the nest. So far these calls have been detected from all of the species that have been studied from the family Muridae and the family Cricetidae.

The physical characteristics of the calls

The ultrasounds produced by infant *Mus musculus*, laboratory mice, were first studied in detail by Noirot (1966a, b) and by Noirot and Pye (1969). These authors removed young albino mouse pups from the nest and isolated them in a dish or on the bench top for 5 min daily from birth onwards. The calls produced by the pups in this 'isolation' condition were mainly 10–140 ms in duration and at frequencies between 45 kHz and 88 kHz. The frequency generally changed slowly over the length of the pulse but sometimes small instantaneous jumps in frequency occurred (Fig. 7.1). The rate of calling increased markedly on the fourth day after birth and then decreased to zero on the day on which the eyes of the young opened, usually the thirteenth day. More ultrasonic calls were detected when the young were being retrieved by their mother than when they were isolated. Noirot and Pye found that the length of the calls, and the sound pressure level decreased as the pups grew older and they suggested that this might decrease their effectiveness in eliciting maternal behaviour.

Sewell (1968, 1969) also studied the distress calls of the strain of albino mice used by Noirot and Pye (E.N. mice), as well as the albino T.O. Swiss strain and the grey C_3H strain. She found no great differences in the frequency patterns of the pulses produced by these three strains although Nitschke, Bell and Zachman (1972; Bell *et al.*, 1972) reported significant differences between the peak frequencies and between the durations of the distress calls of three other standard strains of laboratory mice. Sewell found that ultrasound emission did not necessarily cease when the eyes of the young opened but that both the intensity of the calls and the time of their cessation depended on how the young were treated. After about the thirteenth day the babies no longer emitted calls when they were isolated, but if they were handled, rolled on their backs, or gently pinched by the scruff of the neck, they produced ultrasonic calls until the age of 16 to 20 days in E.N. mice or until at least 39 days in C_3H mice. On any one day the 'handling' calls produced by a pup were always louder than the 'isolation' calls. Sewell suggested that these two different levels of intensity may reflect the two different functions of the calls previously put forward by Noirot. The less intense 'isolation' calls may be motivated by cold and hunger and initiate the retrieving response of the mother, whereas the louder calls produced on handling may inhibit the mother's

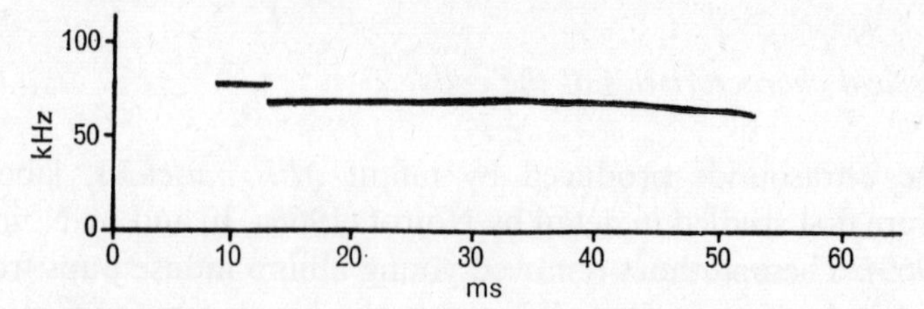

FIGURE 7.1 A call of a four day old albino mouse pup showing a single component with drifts in frequency and also a step-like frequency change.

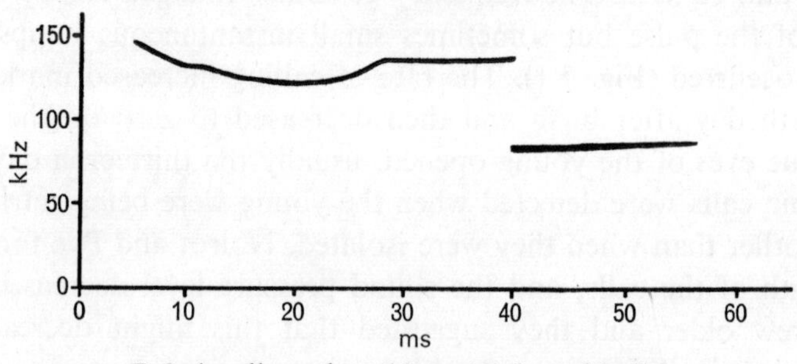

FIGURE 7.2 A call produced by a 17-day-old albino mouse pup showing a large frequency step downwards.

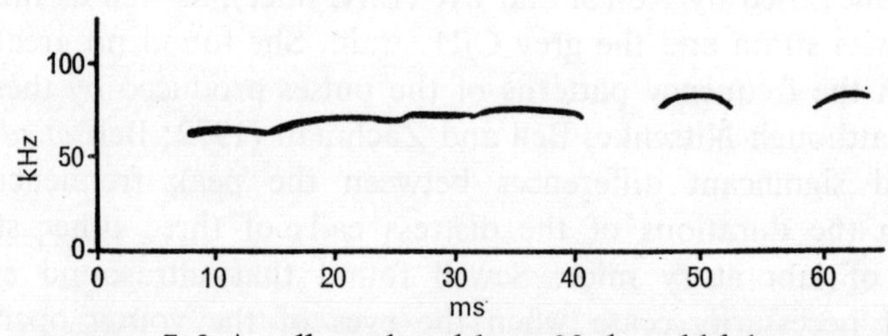

FIGURE 7.3 A call produced by a 20-day-old laboratory rat pup showing rapid frequency changes.

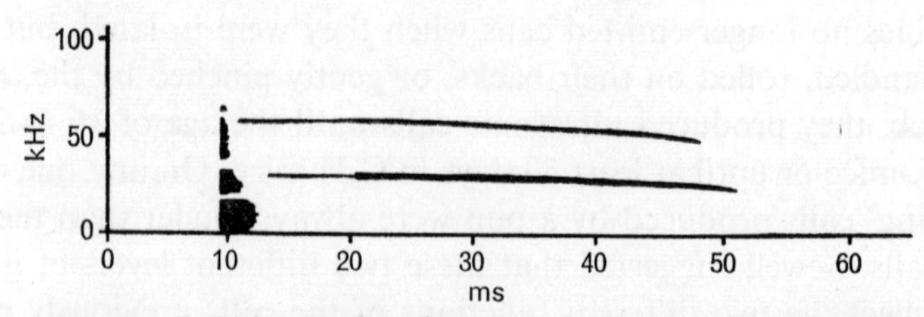

FIGURE 7.4 A purely ultrasonic call of a six-day-old *Clethrionomys glareolus* showing a fundamental and a second harmonic component and also a broadband click at the onset of the call.

aggression and so persist to a greater age, until the young can fend for themselves.

The calls produced before the thirteenth day were similar to those described by Noirot and Pye but the 'handling' calls produced after this time had a wider frequency range of 45 kHz to 148 kHz and contained more complex patterns such as frequency fluctuations and large instantaneous step-like frequency changes of up to 75 kHz (Fig. 7.2). Most of these large frequency steps were in a downwards direction but occasionally an upward step occurred near the end of a pulse. The frequencies on either side of the step were often almost in a simple ratio of 2:1 or 3:2. This may provide a clue to the way in which the sounds are produced and its significance will be discussed in the section on sound production.

The distress calls of two sub-species of *Peromyscus maniculatus*, the North American deer mouse, have been studied by Hart and King (1966), who used equipment with an upper frequency limit of 30 kHz. *Peromyscus* is a member of the family Cricetidae although it is similar in both form and habit to *Apodemus sylvaticus*, the European wood mouse, a member of the family Muridae.

Hart and King subjected infant deer mice *Peromyscus maniculatus bairdi* and *Peromyscus maniculatus gracilis*, to cold stress by placing them in an aluminium dish over iced water for 1 min. The calls elicited during this period were tape-recorded daily from the third day after birth until the young no longer produced distress calls. The calls were longer than those of albino mouse pups, 100–200 ms in duration, and they consisted of two or more harmonic components, the most intense of which was rarely the fundamental. The dominant component ranged from 4·9 kHz to 23·7 kHz for *P. m. bairdii* and from 3·6 kHz to 26·5 kHz for *P. m. gracilis.* These calls were therefore audible with some ultrasonic components and the sonagram traces that Hart and King published showed indications of the presence of even higher harmonics.

The maximum frequency of the pulses produced by *P. m. bairdii* decreased with age from birth onwards but that of *P. m. gracilis* showed a marked increase until 6 days of age and then decreased. *P. m. bairdii* ceased to produce calls between 10 and 12 days of age and *P. m. gracilis* between 14 and 16 days. Hart and King suggested that since this is probably about the time when the young begin to regulate their own temperature effectively, the calls induced by cold stress may guide the mother to her young, and so restore the pups to their primary source of heat. Once the pups are able to thermoregulate, cold stress apparently no longer elicits distress calls.

Bell and his colleagues (1971) studied the calls emitted by infant *P. m. bairdii* immediately after they had been 'handled' by being removed from the nest for 3 min and then replaced. These 'handled' litters produced longer calls with higher peak frequencies (22·2 kHz) than 'non-handled' controls (11·9 kHz). But the authors reported that pups of both groups only produced calls when they were being groomed or retrieved by their mother, a situation comparable to the 'handling' condition mentioned above for laboratory mice.

Recently Smith (1972) has recorded the distress calls of infant *P. maniculatus* (sub-species uncertain) using equipment with an upper frequency limit of at least 100 kHz. She studied the calls emitted by pups both in the isolation condition and when 'handled' at various ages from birth to 21 days. Some pups were removed individually from the nest and placed in a dish at room temperature, the isolation condition, others were 'handled' between finger and thumb for two periods of 30 s within 5 min. Two different types of pulse were emitted by these pups. One was an 'audible' call, 15–200 ms in duration, with a fundamental frequency of 8–29 kHz, and several harmonic components, and showed drifts in frequency. This was equivalent to the calls studied by Hart and King (1966) and by Bell and his colleagues (1971). The second type of pulse was short, 3–45 ms in duration and was purely ultrasonic, with a single component at 60–140 kHz, that consisted of a rapid frequency sweep. Smith found that up to the fourth day after birth pups in the isolation condition produced mainly 'audible' calls. The rate of calling increased to a maximum at day 7 and decreased markedly after day 12. Isolated pups produced few ultrasonic calls before day 6 but these calls were commonly produced by older pups. On handling, pups of all ages showed an increase in the rate of ultrasound production and a decrease in the production of 'audible' calls, but when handling ceased, the rate of 'audible' calling increased immediately although the high rate of ultrasonic calling decreased more slowly. These results indicate that the 'audible' calls emitted by infant *Peromyscus* are elicited by cold stress, but that these calls appear to be suppressed by handling which elicits ultrasonic calls instead.

The ultrasonic calls of other infant rodents have not been studied in such detail as those of *Mus* and *Peromyscus*. Sewell (1969, 1970a) studied the ultrasonic calls of 14 other species, six from the family Muridae and eight from the family Cricetidae. The physical characteristics of these calls are summarized in Table 7.1 together with those of the species studied by other authors.

Sewell (1969) found that in most of the species she studied, the calls elicited by handling were louder and were produced at later ages than isolation calls, but their frequency patterns did not appear to differ significantly. During the first few days after birth, isolated pups generally produced only a few fairly low intensity calls although more vigorous calling could be induced by handling. Acoustic responses to both isolation and handling usually increased rapidly after the second to fourth day. The 'isolation' calls then decreased in intensity over subsequent days and ceased at, or soon after the age when the eyes of the young opened. 'Handling' calls remained at higher intensities until just before the young began to leave the nest at the age of 6 to 16 days depending on the species. In some, such as laboratory rats, low intensity calls could often be elicited by handling until testing ceased, which in rats was after one year. In other species, for example *Acomys cahirinus*, the spiny mouse, low intensity ultrasounds were elicited until the animals could no longer be restrained when handled.

Changes with age have been reported for the rate of ultrasound production in several species of infant rodents when isolated. Noirot (1968) found that the rate of calling in isolated rat pups increased markedly on the fourth day after birth and decreased after the eyes opened, on about day 16, to zero on about day 21. Similar results were obtained by Brooks and Banks (1973) for *Dicrostonyx groenlandicus*, the collared lemming, and by Okon (1971a, b, 1972) for five other species of rodents. Okon studied the changes with age of both 'isolation' and 'handling' calls as part of an investigation into the motivation of ultrasound emission in infant rodents which will be described in the next section.

Sewell (1969) found that the ultrasonic calls produced by different species of murid rodents differed somewhat in their length and frequency patterns although there was much overlap; the same was also true for cricetid rodents. For example the calls of new-born *Apodemus sylvaticus* were commonly at frequencies of 40–75 kHz and lasted for 50–135 ms whereas those of another murid species, *Acomys cahirinus*, were between 30 kHz and 50 kHz in frequency and were often 100–200 ms in duration.

A much greater difference existed between the frequency patterns of the calls of infant murid rodents and those of cricetid rodents (Table 7.1). Murid rodents generally produced purely ultrasonic calls with a single component. The calls of new-born pups were commonly longer

TABLE 7.1 Ultrasonic calls emitted by rodent pups when newborn and when beginning to leave nest

Species	Day eyes open	Day isolation calls cease at 20°C	Duration (ms)		Frequency (kHz)		Frequency pattern	Authors
			Newborn	Older pups	Newborn	Older pups		
Family Muridae								
Mus musculus (E.N.)	13	13	10–140		45–88		Single component	Noirot & Pye (1969)
Mus musculus (E.N.)			8–120	30–105	40–120	45–148	Single component	Sewell (1969)
Mus musculus (C_3H)			3–100	10–90	50–85	50–120	Single component	Sewell (1969)
*Mus minutoides**					70–90			Sewell (1969)
Rattus norvegicus	15–16	16	4–65 (2–205)	5–65 (1–150)	40–75 (35–112)	40–90 (30–100)	Single component	Sewell (1969)
Apodemus sylvaticus	16–18	19	50–135 (10–140)	2–30	40–75	65–85	Single component	Sewell (1969)
Acomys cahirinus	1	6	100–200 (30–200)	2–70 (2–200)	30–50	50–95	Fundamental, sometimes + 1 harmonic component	Sewell (1969)
Praomys natalensis†					60–90			Sewell (1969)
Thamnomys sp.‡			60–150	5–30	35–60	30–80	Single component	Sewell (1969)

Family Cricetidae								
Mesocricetus auratus	16–17	18 (ultrasonic)§	2–180	2–50	28–55	50–80	Single component	Sewell (1969)
			60–200	34–200	20–55	30–60	Several components	
		9§	30–200	5–200	10–60	Up to 80	Many components audible and ultrasonic	
Meriones shawi	18	18–20	5–150	3–30 (3–150)	45–80	50–130	Single component	Sewell (1969)
				80–200		5–148	Fundamental +1 or more harmonic components	
Meriones unguiculatus‖			20–200		50–80		Single component	Sewell (1969)
			80–200		6–70		Several Harmonic components	
	19–21		60–100	30–150	37–58	37–75	Single component	Smith (pers. comm.)
			80–200		6–70		Fundamental and several harmonic components	
Gerbillus sp.	24–25		Few detected	5–200	Few detected	50–70	Single component	Sewell (1969)
				5–200		48–85	Several components	
				25–200		Up to 60	Many components audible and ultrasonic	
Peromyscus maniculatus	12–13		100–200		3–26·5		Several harmonic components	Hart & King (1966)
			Few ultrasonic	3–45	Few ultrasonic	60–140	Single component	Smith (1972)
		12–13 ('audible')	100–200 (15–200)	Few 'audible'	8–70	Few 'audible'	Several harmonic components	

TABLE 7.1—*continued*

Species	Day eyes open	Day isolation calls cease at 20°C	Duration (ms)		Frequency (kHz)		Frequency pattern	Authors
			Newborn	Older pups	Newborn	Older pups		
Clethrionomys glareolus	13–14	10–12	30–70 (3–100)	3–30	20–55	60–110	Fundamental, sometimes +1 harmonic component	Sewell (1969)
Microtus agrestis	11		40–120 (5–120)	3–30 (3–140)	25–60 (20–125)	30–60 (20–110)	Fundamental, sometimes +1 harmonic component	Sewell (1969)
Calomys callosus	8		Few ultrasonic	3–30	Few ultrasonic	60–145	Single component	Smith (1972)
		12 (audible)	80–200	Few 'audible'	15–70	Few 'audible'	Several harmonic components	
Lagurus lagurus			15–70		40–65		Single component	Sales (Unpub).
Dicrostonyx groenlandicus		11–16	100–800		17–44			Brooks & Banks (1973)
Sigmodon hispidus					25–60			Brown (1971b)

Calls of 'newborn' include those produced from day of birth until isolation calls cease.
Calls of 'older pups' indicates handling calls produced after isolation calls ceased.
* One litter less than one week old studied with bat-detector only.
† One litter, day 5, studied with bat-detector only.
‡ One litter studied on day 3 and day 17 only.
§ Okon (1971).
‖ One litter studied up to day 5 only.

and at lower frequencies than those of older pups. As the pups became older the calls became shorter and higher in frequency, and were often between 50 kHz and 110 kHz, some even extending to the limit of the recording apparatus at 150 kHz. These calls also showed more complex frequency patterns, such as rapid frequency changes, frequency fluctuations and instantaneous frequency steps (Fig. 7.3). These changes in frequency and duration were seen in most of the pulses produced by young murid rodents, but in all species some very young pups produced a few short, high frequency pulses and some older pups, particularly rats, produced long, low frequency pulses.

Cricetid rodents produced calls with a variety of frequency patterns and which often contained both audible and ultrasonic components. Many new-born cricetid rodents produced calls with two or more harmonically related frequency components. These were sometimes purely ultrasonic as in the voles *Clethrionomys glareolus* and *Microtus agrestis*, which emitted calls with a fundamental frequency of 20–55 kHz and a second harmonic component (Fig. 7.4), or they had an audible fundamental with a series of harmonic components extending into the ultrasonic range. This latter type of frequency pattern has already been described for *Peromyscus* and it also occurred in some of the distress calls of the gerbils *Meriones unguiculatus* and *Meriones shawi.* These gerbils also emitted pulses in which a further series of harmonics appeared intermittently along the pulse and at half the frequency spacing of the first series (Fig. 7.5). Wide-band structures consisting of many apparently harmonically unrelated components were also found for short periods within some of these pulses (Fig. 7.5). Most of the calls produced by baby *Mesocricetus auratus*, golden hamsters, consisted entirely of such wide-band structures. They generally contained both audible and ultrasonic components and often one ultrasonic component was more intense than the others and was continued as a single ultrasonic component for a few milliseconds after all the other components has ceased (Fig. 7.6). Similar calls were also produced by young *Gerbillus* (species uncertain). Some infant hamsters emitted calls that were purely ultrasonic and consisted of several frequency components that were not harmonically related (Fig. 7.7). A few species of cricetids, particularly *Clethrionomys* and *Microtus*, produced broad-band clicks either at the beginning of an ultrasonic pulse (Fig. 7.4) or during the pulse when step-like frequency changes occurred. These clicks extended from the audible to the ultrasonic range, often up to 70 kHz.

All of the cricetid rodents that were studied also produced some

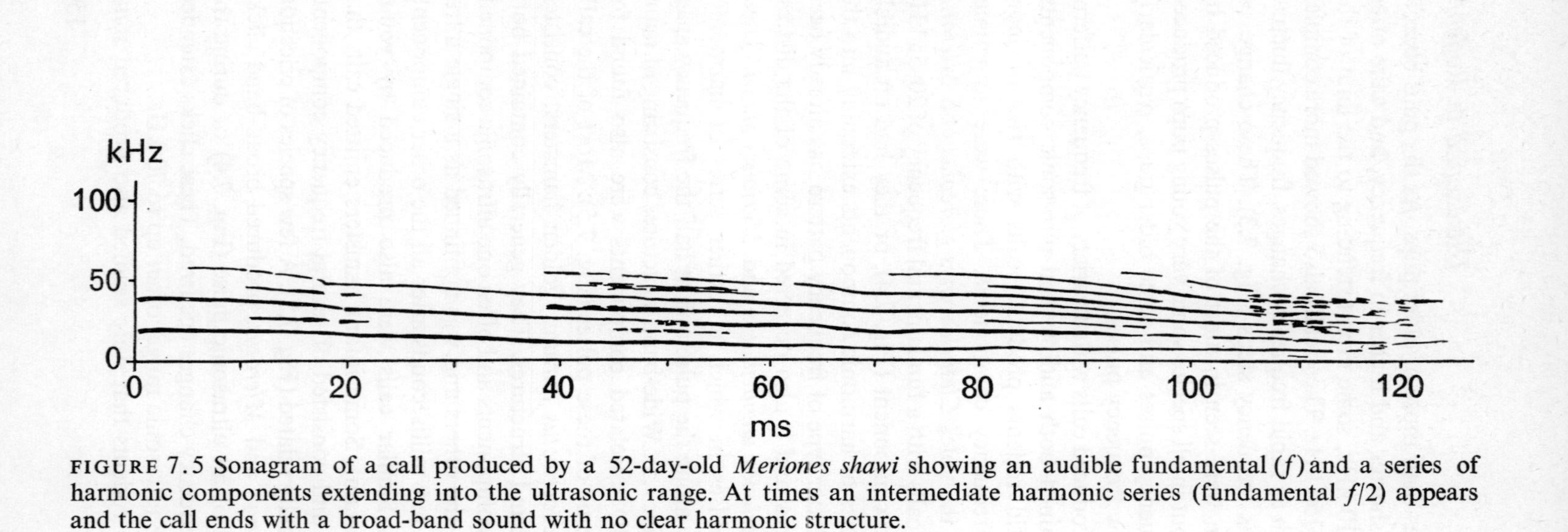

FIGURE 7.5 Sonagram of a call produced by a 52-day-old *Meriones shawi* showing an audible fundamental (f) and a series of harmonic components extending into the ultrasonic range. At times an intermediate harmonic series (fundamental $f/2$) appears and the call ends with a broad-band sound with no clear harmonic structure.

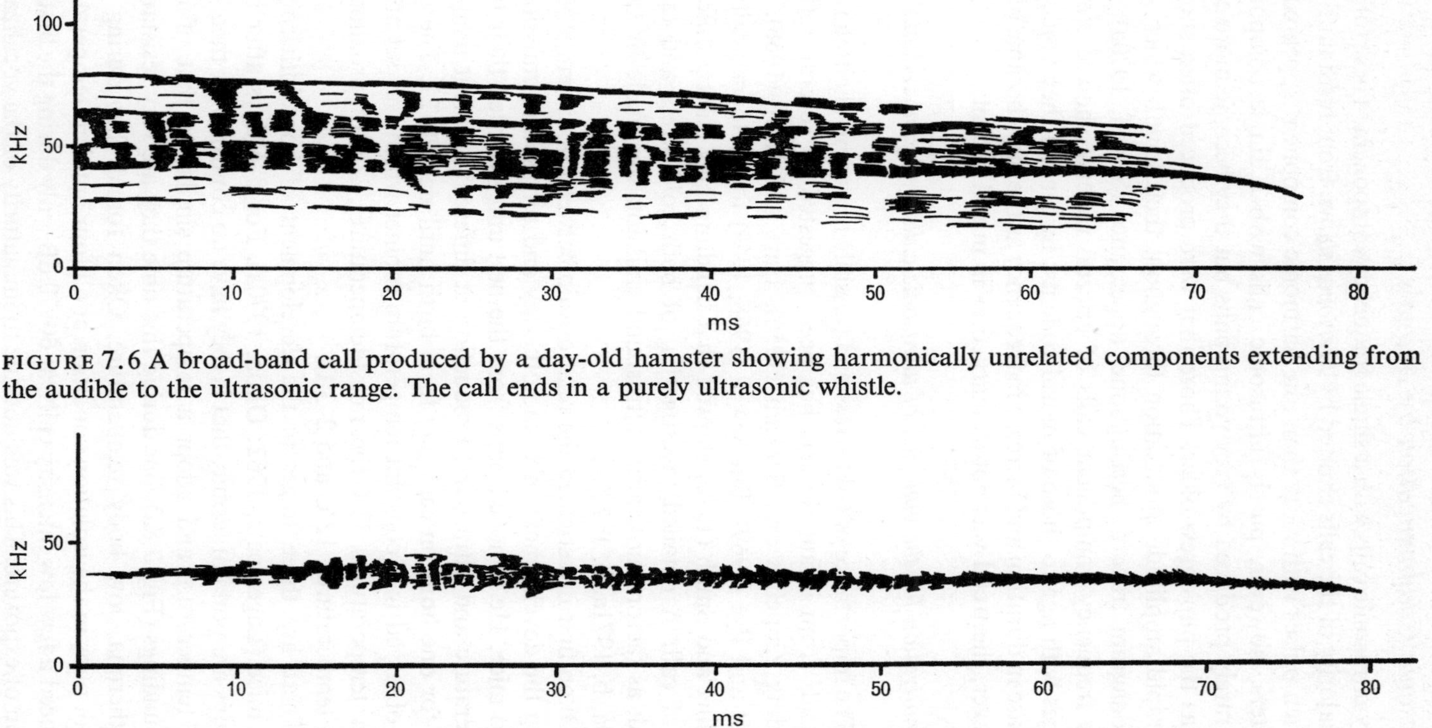

FIGURE 7.6 A broad-band call produced by a day-old hamster showing harmonically unrelated components extending from the audible to the ultrasonic range. The call ends in a purely ultrasonic whistle.

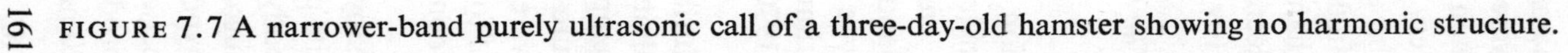

FIGURE 7.7 A narrower-band purely ultrasonic call of a three-day-old hamster showing no harmonic structure.

purely ultrasonic calls with a single frequency component. These formed the majority of the calls emitted by *Meriones shawi*, from birth until day 18 when pulses with more than one harmonic component appeared. In hamsters, however, purely ultrasonic calls with a single component were rarely produced by very young pups but they became more common as the pups grew older. Those that were produced often showed very rapid amplitude modulation throughout their length which gave the sonagram trace a 'herringbone' appearance (Sewell, 1970a). The single frequency component calls of cricetid rodents showed similar changes with age to those of murid rodents: in general they tended to decrease in duration and increase in frequency as the pups became older. However, the trend was not as marked as in murid rodents.

The motivation for the emission of ultrasonic calls by infant rodents

The importance of cold stress and possibly hunger in eliciting isolation calls from infant rodents has been suggested by several authors including Zippelius and Schleidt (1956), Hart and King (1966), and Noirot and Pye (1969). But Sewell (1968, 1969) found louder calls on handling and Smith (1972) showed that handling may elicit a different type of call. A detailed investigation of both cold stress and tactile stimuli as factors motivating ultrasound emission was made by Okon (1970a, b, 1971a, b, 1972).

Okon first investigated the relationship between body temperature during the development of homoiothermy and ultrasound emission in albino mice. He removed pups from the nest and recorded their body temperature and ultrasound production at different ambient temperatures for one hour on each day from birth until day 20–21. The pups were observed in a constant temperature cabinet which was set at one of four temperatures, 33°C (near nest temperature), 22°C (approximately room temperature), 12°C and 2–3°C.

There are three stages in the development of homoiothermy in baby mice (Lagerspetz, 1962; Okon, 1970a). For 5–6 days after birth the pups are poikilothermic, that is they have no control over their own body temperature and adopt a temperature similar to that of their surroundings (Fig. 7.8a), but during this time they are very resistant to hypothermia, low body temperature. Okon found that during this period, pups in the isolation condition at all four ambient temperatures produced a few low intensity calls, at 60–70 dB, only during the first 10–20 min of exposure. This was not due to immaturity of the vocalization

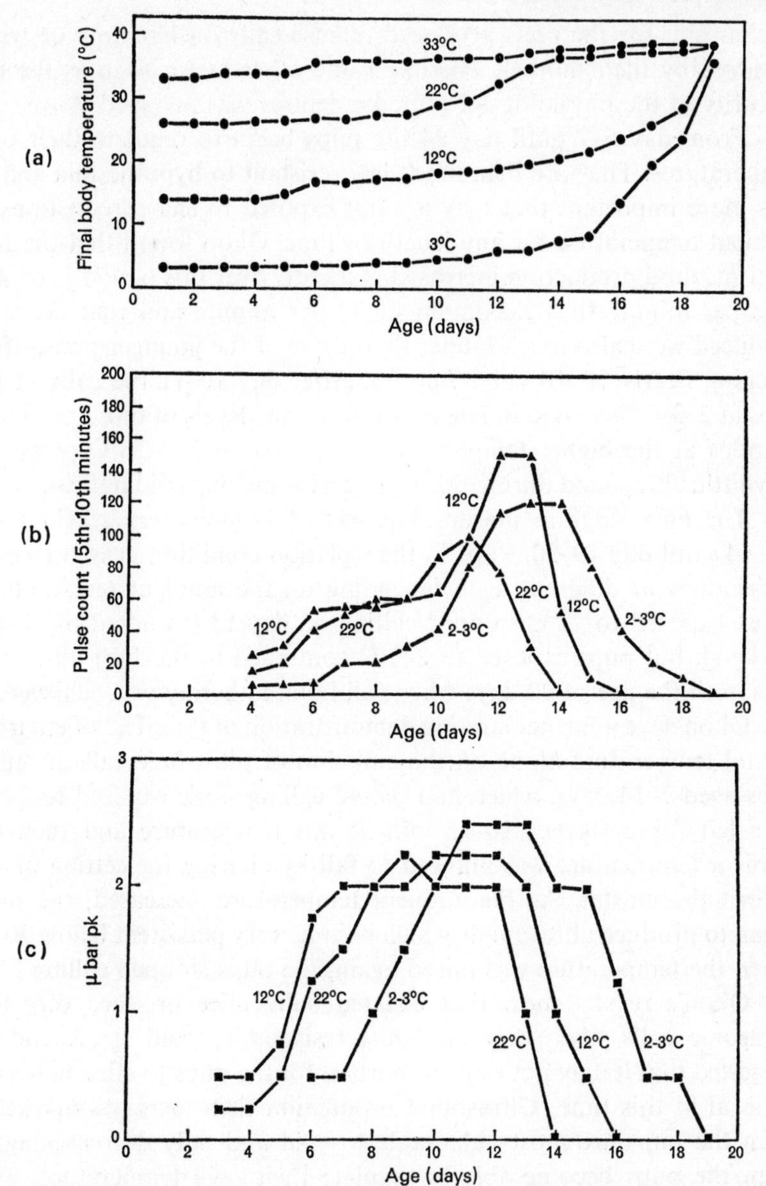

FIGURE 7.8 The relation between body temperature and ultrasound production in infant albino mice of different ages exposed to different ambient temperatures. (a) Final body temperature after 1h exposure at four different temperatures. (b) Rate of calling between the 5th and 10th min of exposure. (c) Peak amplitudes of calls after 10 min of exposure. Few calls were produced at 33°C. All points are averaged from observations on three animals. (From Okon, 1970a)

mechanism, for the pups produced intense calls on handling or when retrieved by their mother. Possibly some other factor such as the immaturity of the physiological response centres was involved.

From day 6–7 until day 14 the pups begin to regulate their own temperatures. They are then much less resistant to hypothermia and so it is more important that they are not exposed to sharp reductions in ambient temperature for any length of time. Okon found that the rate of ultrasound production increased markedly over this period from 4–5 calls per minute to a maximum of 32 per minute and that the calls produced were also much louder than those of the younger pups, often reaching nearly 78 dB (Fig. 7.8b, c). After day 10–11 the calls of the pups at 2–3°C increased in intensity beyond the levels of those produced by pups at the higher temperatures. Pups exposed at 33°C produced very little ultrasound during this stage and some pups did not call at all.

The final stage in the development of homoiothermy lasts from day 14 until day 19–20. Pups in the isolation condition ceased to emit ultrasounds at different ages depending on the ambient temperature. Those exposed to 22°C stopped calling on day 13 (as noted by earlier workers), but pups exposed to 2–3°C continued to produce ultrasonic calls until the age of 20 days when full homoiothermy was achieved.

Okon gave a further striking demonstration of the effect of environmental temperature alone on the emission of ultrasonic calls in mice. Pups aged 7–14 days, which had ceased calling while exposed to 33°C, were left for a further 10–15 min at this temperature and then the ambient temperature was allowed to fall by altering the setting of the cabinet thermostat. As the ambient temperature decreased, the pups began to produce ultrasounds which became very persistent below 30°C. When the temperature was raised again, the pups stopped calling.

Okon's results show that isolated baby mice produce very few ultrasonic calls when they are most resistant to cold stress and he suggested that it may not be so important for the pups to elicit maternal retrieval at this time. Ultrasound production then increases markedly when the pups are most vulnerable to cold and only decreases again when the pups become able to regulate their own temperature. This strong evidence supports the suggestions of earlier workers (e.g. Hart and King, 1966; Noirot and Pye, 1969) that the production of ultrasound by isolated pups is very closely related to the development of homoiothermy.

Okon later (1971a, b) found a similar pattern of association between the stages in the development of homoiothermy and changes in ultrasound emission in rat and hamster pups. Both species showed very little ultrasonic response to cold stress for the first few days after birth. Rats, like mice, showed an increase in ultrasound production from days 6–8, but this did not occur in hamsters until day 9–12. It may be significant that Hissa and Lagerspetz (1964) and Okon (1971a) showed that homoiothermy in hamsters does not begin to develop until this age. Rat pups were able to regulate their own body temperatures by day 20 and ceased to emit ultrasonic calls even at 2–3°C at this age, but hamsters did not achieve full homoiothermy until day 23. This relative delay in the development of temperature regulation in hamsters may be associated with their short gestation period of only 16 days, which means that they are born at an earlier stage of development than rats and mice which have a gestation period of 20–21 days. Hamsters further differed from mice and rats in producing loud, wide-band, audible calls in response to cold stress. These audible calls showed a steady decline from birth onwards. They were strongest soon after birth when ultrasonic calls were weakest and they disappeared during the peak period of ultrasonic calling on days 9–13. Okon suggested that until the ultrasonic responses are fully developed, hamster pups make use of their ability to produce audible calls in order to elicit the retrieving response of the mother and so to be restored to the nest.

Allin and Banks (1971) also studied the effect of temperature on ultrasound emission by albino rat pups. Like Okon, they found that low ambient temperatures, 2°C and 20°C, elicited high rates of ultrasonic calling but these rates decreased 6–8 days after birth. At temperatures near to or above that of the nest, 35°C and 40°C respectively, few ultrasounds were emitted. These authors similarly concluded that cold stress elicits ultrasound production by rat pups and they too suggested that the ultrasonic calls induce the mother to search for the young and to retrieve them to the nest.

Okon obtained less detailed information on ultrasound emission and the development of homoiothermy in *Clethrionomys glareolus* and *Apodemus sylvaticus*. He found that *Clethrionomys* pups show practically no ability to regulate their own temperature before days 10–11. Homoiothermy then develops rapidly and in most pups it is achieved by day 16 although a few pups were not completely homoiothermic when exposed to temperatures of 2–3°C until days 18–19. Very few ultrasonic pulses were produced in the first four days after birth, when presumably the pups were resistant to cold. After this age low intensity calls were emitted by all pups when they were exposed to the ambient temperatures

22°C, 12°C and 2–3°C. These isolation calls generally ceased between days 10 and 16 depending on the ambient temperature. But the few pups that were not able to regulate their own temperature until days 18–19 emitted ultrasounds until this age.

In *Apodemus* the picture is very different. Infant *Apodemus* are very vocal and Okon found that when exposed to low temperatures the pups produced ultrasounds very vigorously from the day of birth onwards. They generally produced much louder calls than the other species studied, up to 96 dB SPL. They also emitted ultrasounds for a longer time on each day and to later ages at the three ambient temperatures used, 22°C, 12°C and 2–3°C. Pups exposed to these temperatures ceased calling on days 23, 24 and 25 respectively. Okon found that the development of homoiothermy in *Apodemus* is apparently much slower than in the other rodents he studied and is probably not complete until after day 24. He suggested that during the first few days after birth this species is less resistant to cold than the other species and so the acoustic responses are developed earlier.

Okon (1970b, 1971b) also studied the ultrasonic responses of baby rodents to different kinds of tactile stimuli: holding and lifting them loosely between two fingers, tapping or pressing the tail lightly against the surface of the dish containing the pups and lightly pinching the scruff of the neck. The pups were also rolled on their backs by tipping the dish in which they were placed, and in addition they were studied while being retrieved by their mother.

Okon confirmed the earlier observations of Sewell (1968, 1969) that the ultrasonic 'handling' calls are louder than the 'isolation' calls of young pups and also that they are produced for some time after the eyes open. The ultrasounds produced by baby albino mice under all the handling conditions were very loud during the first four days after birth and calling ceased on different days, according to the tactile stimulus used. The cries produced by pups falling over onto their backs ceased on day 8, those produced during retrieval ended on day 12, while all other 'handling' calls were no longer produced after days 15–17. The changes with age in the ultrasounds elicited from mice by tactile stimuli thus showed a very different pattern to those elicited by cold stress and so tactile stimuli alone are obviously an effective stimulus for ultrasound production.

Infant rats produced very few calls in response to tactile stimuli during the first few days after birth. Most of these calls were wide-band audible signals but a few were ultrasonic. From day 4 to days 14–16 all methods of handling elicited ultrasonic calls but after this age the response declined rapidly and very few calls were elicited after day 20. Hamsters also produced mainly audible sounds in response to handling during the first few days after birth. As with their acoustic response to temperature changes, the audible calls decreased and the ultrasonic calls increased as the pups grew older. Normal handling failed to elicit any response after days 6–7 but pinching and tapping (of the feet rather than the extremely short tail) elicited calls until day 20 or later. Acoustic responses of hamster pups to maternal retrieving decreased from birth onwards and ceased by day 8. Tactile stimuli elicited very few calls from *Clethrionomys* pups; audible calls were produced in response to tapping and pinching but retrieving and normal handling elicited no response at all.

Okon's work has shown that at least two separate motivating factors are important in eliciting ultrasonic calls from infant rodents: the effect of cold on the physiological mechanisms of temperature regulation and tactile stimuli. Recently (as mentioned earlier), Smith (1972) has shown for the first time that in *Peromyscus* these different motivating conditions elicit very different types of calls, which, as discussed in the following section, may have different functions.

The role of infant distress calls in adult–young relationships

The ultrasonic distress calls of infant rodents appear to play an important role in the survival of the young, particularly through mother–young relationships. Adult rodents are certainly able to hear these sounds (see section on hearing) and so the signals have at least a potential communication value. The earliest workers in this field assumed that the function of the isolation calls was to elicit the retrieving response of the mother and to guide her to the young. In support of this, Zippelius and Schleidt (1956) reported that lactating female mice would retrieve live pups from outside the nest and that one female *Apodemus* retrieved 148 live babies that were scattered successively throughout her cage. Dead or narcotized young which could not emit ultrasonic calls were not retrieved.

Later Noirot (1966a) proposed that the ultrasonic calls also had a negative or inhibitory effect on aggressive behaviour which in female mice increases during the gestation period. In mice this negative effect may be exerted by the 'handling calls' while the positive effect on maternal behaviour may be elicited by the 'isolation calls' (Sewell,

1969). Smith (1972) has suggested that the two completely different calls that she recorded from *Peromyscus* may also reflect these different functions. The more intense 'audible' calls, characteristic of young pups subjected to cold stress, may elicit maternal retrieving (as suggested for the less intense cries of mice), whereas the less intense, and therefore probably shorter-range, ultrasonic calls produced mainly by older pups on handling may serve to inhibit maternal aggression (as suggested for the more intense cries of mice).

Evidence for a positive effect of isolation calls on maternal response comes from a series of studies in which Noirot investigated the cues affecting maternal behaviour in mice. She found that maternal responses such as retrieving, nest building, licking and covering the young in a suckling position were not mainly dependent on the sex or hormonal state of the adults but rather on cues coming from the young themselves (Beniest–Noirot, 1957, 1958; Noirot, 1964a, b). Later (1964c) Noirot presented naïve female and male mice for 5 min with a 'strong' stimulus for eliciting maternal behaviour, a live, one-day-old baby. Then, either immediately or 2–8 days later, these animals were presented with a 'weak' stimulus, a drowned, one-day-old pup. The experimental animals showed an increase in all maternal responses towards the weak stimulus when compared with controls that were presented with the weak stimulus only.

Young pups provide visual, olfactory, tactile and auditory stimuli and Noirot (1969a) showed that this increase in maternal behaviour could be elicited by exposure to only two of these, olfactory and auditory stimuli. It did not appear to depend on previous performance of the maternal responses. Evidence for this was given by an experiment in which mice were exposed for 5 min to a 1–2-day-old pup hidden in a perforated metal box. These animals displayed more retrieving, licking and nest building responses towards a drowned pup than did controls which had not been exposed to such 'priming'. Noirot (1972) next attempted to separate the effects of auditory and olfactory stimuli. Some animals were presented with only olfactory cues, a box containing the nest material used by a litter, and when their maternal behaviour was assessed later they showed more intense licking towards a live one-day-old pup than did controls which had previously been exposed to an empty box. Other mice were exposed to a pup in a non-perforated box which, it was hoped, would allow the maximum auditory cues with the minimum of olfactory cues. These animals performed more intense nest building than did controls. Unfortunately these results did not show whether either olfactory or auditory stimuli preferentially affect retrieving, as the test pup was itself an 'optimum' stimulus for eliciting this response.

A very clear demonstration that isolation calls do affect the searching behaviour and probably also initiate the retrieving response of lactating females was given by Sewell (1970b). She studied the reaction of lactating female *Apodemus sylvaticus* to a purely acoustic 'stimulus' signal, the tape-recorded isolation calls of a 5-day-old *Apodemus* litter. The sound of the tape-recorder motor alone, tape-recorded background noise and artificial 45 kHz pulses were used as control signals. The signals were relayed by an ultrasonic loudspeaker inserted through the side wall of the cage into one of two compartments formed by a 'T' partition and with separate entrances, in the end of the cage away from the nest. The stimulus and control signals were presented in random order and randomly to either the left or the right compartment. Sometimes the stimulus and a control signal were relayed simultaneously into different compartments.

Five lactating female *Apodemus* were observed on a total of 12 different occasions. Out of 54 presentations of different control signals, only three responses occurred, that is the females emerged from the nest. The stimulus signal was presented 56 times. On 48 occasions the females rapidly emerged from the nest and on 38 of these they went immediately to the compartment containing the loudspeaker. The females generally reached the loudspeaker within 5–30 s of the onset of the stimulus signal. On several occasions the females emerged from the nest with the young still attached to the nipples. Some then forcibly removed the pups and returned them to the nest before entering a test compartment, others dragged the pups with them. One female even chewed the foil diaphragm off the loudspeaker!

Allin and Banks (1972) carried out a similar experiment using lactating female, virgin female and male albino rats. A 90 s tape-recording of the ultrasonic cries of a 7-day-old rat pup was used as the stimulus signal and tape-recorded background noise was used as the control. The test area was hexagonal in shape with a nest box in the centre and the signals were relayed randomly from any one side. Allin and Banks reported that 20 out of 40 lactating females left the nest in response to the stimulus signal and most of these orientated correctly towards the sound source. Males and virgin females failed to leave the nest when the pup's calls were relayed but some of the males turned their heads in the direction of the loudspeaker.

There is no direct experimental evidence that the calls produced by pups on handling have a negative or inhibitory effect on maternal behaviour although there are several indications of such an effect. Noirot (1966a) found that naïve, virgin female mice handled young pups rather roughly and many ultrasonic calls were produced by the pups. The adults usually responded to these calls by immediately stopping the activity in which they were engaged and switching to another. For example if calls were produced during retrieving, the pup was often dropped and the adult ran to the nest and engaged in nest building activities before returning to the pup. If calls were produced during licking, this again was often replaced by nest building. Later (1969b) Noirot reported that primiparous females were highly territorial at the time of parturition and often attacked and killed pups over 13 days old whereas younger pups were rarely killed. She suggested that the younger pups were able to exert strong inhibitory effects on the adults' aggression, probably by emitting ultrasonic distress calls. But older pups which no longer produced these calls were morely likely to be attacked. The strong acoustic responses that Okon (1970b) obtained from young pups when handled or retrieved as well as the continuation of 'handling' pulses in many species after the young are capable of leaving the nest and returning to it of their own accord, also indicate the probable importance of these calls for the survival of the young. But ultrasounds are unlikely to be the only cues involved; Gandelman and his associates (1971a, b) have suggested that olfactory cues are essential for eliciting maternal behaviour in mice during the early stage of the pre-weaning period and that olfactory cues from the young may normally serve to inhibit cannibalism at this time.

It is now well known that baby rodents that have been manipulated in some way during infancy, even by being removed from the nest and immediately replaced, tend to develop more rapidly and to display less emotional behaviour than unhandled animals when they become adult (e.g. Levine, 1962; Russell, 1971). This could be due to changes in the mother's behaviour towards the handled pups which in turn affects the pups' later behaviour (Russell, 1971). Bell and his colleagues (1971) have suggested that the differences between the ultrasonic calls produced by young *Peromyscus* that were regularly removed from the nest, and by those that were not, could mediate these effects of handling.

The ultrasonic calls of infant rodents therefore appear to play an important role in determining certain aspects of maternal behaviour and in turn of the pups' own subsequent development. It is probable that these calls are effective interspecifically as well as intraspecifically. Maternal responses are certainly not species specific and there have been many reports of females of one species adopting the young of another (Plate XIIIc). For example Kahmann and Frisch (1952) and Lauterbach (1956) found that it was possible to cross-foster a variety of species of myomorph rodents and Lauterbach reported that in rats the biting response of adults towards a strange object appeared to be inhibited by the young of several species. Ultrasonic calls are probably an important part of a complex pattern of stimuli presented by the young, the whole of which is necessary to elicit the complete pattern of maternal behaviour, and to ensure the survival of the young.

ULTRASOUND AND AGGRESSIVE BEHAVIOUR

Rats

The emission of ultrasound during aggressive behaviour was first discovered in rats (Sewell, 1967, 1969). Previously, Sewell had found that two litters of rats produced ultrasounds well into adult life when they were handled. This could have been a continuation of juvenile behaviour due to regular handling, but ultrasounds were also detected from 46 out of 91 other adult rats that had not been regularly handled. Two distinct types of pulse were recorded from these adults. Twelve animals produced 'long pulses', up to 700 ms in duration and at a frequency of about 25 kHz, when removed from their cages. 'Short pulses' of 3–60 ms duration and at 45–70 kHz were produced by 38 rats (including two of those that had produced long pulses) when they were rolled onto their backs and restrained. This posture resembled the full submissive posture of a defeated rat as described by Seward (1945) and Grant and Mackintosh (1963). The ultrasounds produced by these adults may therefore have been associated with aggressive or submissive responses, in this case to the experimenter (Sewell, 1967).

Sewell then observed 40 different encounters between pairs of male laboratory rats that had previously been isolated; one rat was placed in the cage of another and the animals were observed for 1 h. Aggressive behaviour occurred in many of these encounters and was often associated with the emission of ultrasound. A typical aggressive bout began with the two animals nosing and then sniffing each other from head to

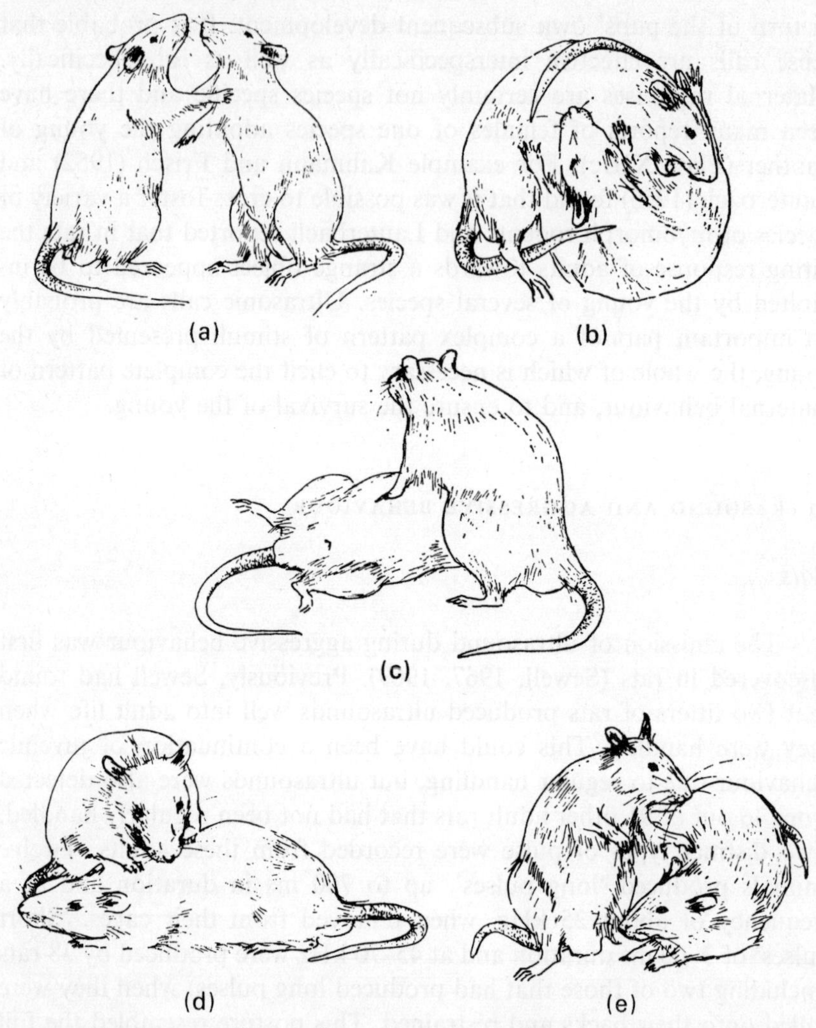

FIGURE 7.9 Some actions and postures seen during aggressive encounters in male laboratory rats. (a) Boxing. (b) Wrestling. (c) Aggressive-submissive postures. (d) Low crouched submissive posture and aggressive groom. (e) Upright submissive posture and sideways threatening posture. (Drawings by A. M. Brown.)

tail. The introduced rat then investigated the cage, followed closely by the resident.

Sooner or later one of the rats, generally the resident, attacked the other and both then stood up on their hind legs, facing each other and 'boxing' with their fore paws (Fig. 7.9a). In very aggressive encounters the combatants would 'wrestle' locked together and rolling around the cage (Fig. 7.9b). The bouts generally ended either with the two rats separating and vigorously sniffing the cage, feeding or cleaning, or else with one of the animals adopting the full submissive posture, lying on its back or side, sometimes with the legs rigidly extended. Occasionally the more aggressive rat forced its opponent into this position and then stood at right-angles over it in the full aggressive posture (Fig. 7.9c). After a time, varying between a few seconds and several minutes, the two rats separated and a short time later the whole procedure might be repeated.

In many encounters one rat adopted the submissive posture more and more frequently and eventually remained in this position for several minutes. This now submissive rat showed characteristic long exhalations separated by short, sharp inhalations. Very little aggression was seen after this and the submissive rat usually spent the rest of the observation period crouched in a corner of the cage (Fig. 7.9d, e). Sometimes it later began to move around and to engage in maintenance activities, feeding, drinking, grooming or sniffing the cage, but still showing irregular respiration. The dominant rat in this situation at first generally showed displacement activities, very abbreviated acts of washing, feeding, drinking or digging which are thought to indicate a high aggressive drive that is somehow inhibited, presumably by the submissive behaviour of the opponent. Later this rat too resumed normal activities. Sometimes the aggressive rat crawled over the submissive one or groomed it, pulling at the fur with its teeth, and on a few occasions one rat attempted to mount the other as in sexual behaviour.

Ultrasounds were detected in every encounter except one. Long pulses, typically 800–1600 ms in duration with a maximum of 3400 ms were detected in 26 of the 40 encounters. They consisted of a single component between 22 kHz and 30 kHz that was either almost constant in frequency or showed one or more slow drifts (Fig. 7.10). These long ultrasonic pulses could always be associated with the characteristic long exhalations of submissive rats so they could easily be associated with some behaviour patterns. They were commonly detected from an animal in a submissive or crouched posture, particularly when the opponent was engaged in maintenance activities. In this situation they were often emitted for long periods of time, either continuously for 5–10 min or intermittently for 15–20 min. Long pulses were sometimes detected later when the defeated rat was feeding or cleaning and some animals continued to produce them for 10–20 min after they had been separated from their opponents.

Short ultrasonic pulses were produced in 39 out of the 40 encounters. They were 3–65 ms in duration and consisted of a single component at 40–70 kHz. Short pulses usually occurred in groups and many pulses showed rapid changes in frequency (Fig. 7.11). Because these pulses could not be associated with an obvious respiratory pattern,

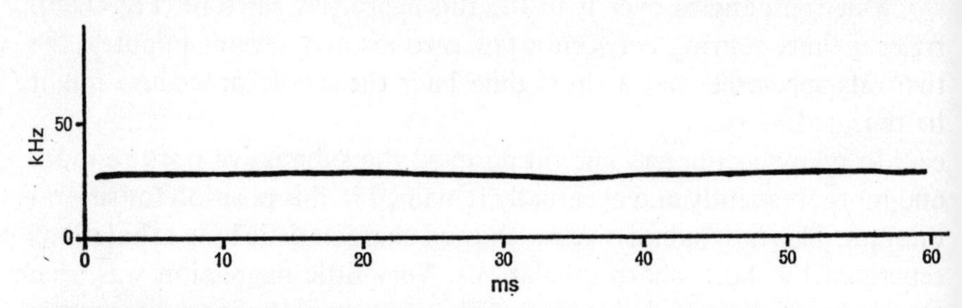

FIGURE 7.10 Sonagram of the first 60 ms of a long low-frequency ultrasonic call produced by a submissive laboratory rat.

it was often difficult to tell which animal was producing them. They often appeared to be synchronous with the movements of the more aggressive rat but sometimes both animals seemed to produce them. The correlation between the emission of short pulses and different behaviour patterns was therefore determined statistically (Sales *née* Sewell, 1972a).

Short pulses were produced during all aggressive acts and they were

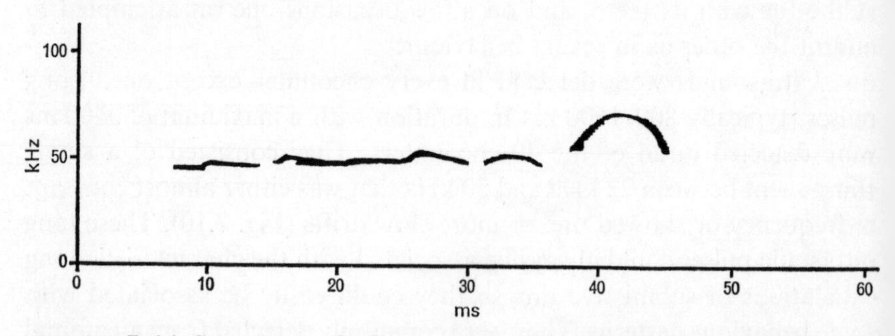

FIGURE 7.11 Sonagram of ultrasounds emitted during an aggressive encounter between male laboratory rats showing rapid frequency changes.

statistically correlated with attacking, boxing and wrestling. They were also produced during the few sexual activities that were observed. But very few short pulses were emitted when one animal was in the full aggressive posture and the other was in the full submissive posture. In this situation the aggressive drive of the dominant rat is probably low as it appears to be reduced by the supine posture of the submissive rat (Grant, 1963). Short pulses were also statistically correlated with the occurrence of displacement activities and with one animal grooming or crawling over the other, all of which are thought to be associated with an aggressive drive (Grant, 1963; Grant and Mackintosh, 1963).

Sales next studied seven encounters between pairs of male wild-trapped rats, *Rattus norvegicus*. These are said to be more aggressive than laboratory rats, but little aggression was seen. Short pulses, similar to those of laboratory rats, were detected when two animals sniffed each other or when one animal kicked the other, and long pulses, 300–600 ms in duration, were detected from rats in submissive postures On one occasion, short pulses were detected when a member of an artificially established colony of wild rats chased and fought a strange male that had been introduced into the colony room.

The emission of short pulses therefore appears to be associated with an aggressive motivation whereas the emission of long pulses appears to indicate a submissive motivation. It is not yet known whether these calls actually play a part in communication although there is some circumstantial evidence that long pulses may have an inhibitory effect on aggressive behaviour. It was mentioned earlier that many aggressive bouts between males ended with one animal adopting the full submissive posture but these were followed by further bouts until long pulses were detected, when little more aggression occurred. This suggests that the full submissive posture may be only a temporary act of submission, ending a particular bout. If, as Grant (1963) suggests, this posture reduces the aggressive drive of the opponent, the effect appears to be only temporary. The production of long pulses, however, may indicate a more long-lasting state of submission and may inhibit the aggression of the dominant rat for longer periods, so allowing the submissive animal to perform 'essential' activities such as feeding and grooming even when in close proximity to the aggressive rat.

The effect, if any, of short pulses on other rats is even less certain. In the wild, rats live in colonies and Barnett (1963) found that in large artificial colonies of wild rats, certain dominance–subordination relationships between rats were initially set up by fighting and then remained fairly stable. It may be that within a colony the emission of short pulses by a dominant animal causes a less aggressive rat to submit with little or no overt conflict. If the submissive rat then produced long pulses, the aggression of the first could be inhibited without either animal having to adopt, and perhaps maintain, aggressive or submissive postures.

Within a rat colony, therefore, ultrasounds may be important not only in establishing dominance–subordination relationships without injury, but also in the maintenance of these with the minimum disruption of normal activities and without resort to conflict. However, all this is merely speculation and has yet to be proved; investigations into the communication value of both long and short pulses are being carried out at present.

Other myomorph rodents

Sales (*née* Sewell, 1972a) observed aggressive encounters in 13 other species of myomorph rodents which are listed in Table 7.2. These species were not studied in such great detail as were the rats and varying degrees of aggression were seen. During these encounters ultrasounds were produced by eight of the 13 species.

Among murid rodents, aggressive behaviour was observed in *Apodemus sylvaticus* when a young male was introduced into an established group of adults. Ultrasonic calls were detected when two adults, a male and a female, postured aggressively towards the stranger and when they chased and fought with him in turn. Ultrasounds were also emitted during chasing and fighting between male *Arvicanthis niloticus*, African grass rats, and when a lactating female *Acomys cahirinus*, chased a male that had been introduced into her cage. However, the situation was complicated as the female's litter had previously been removed and so the emission of ultrasound may have been in response to this rather than to the presence of the intruder. No ultrasounds were detected during two other encounters between males and non-lactating females or during the one aggressive encounter between two males. Very little aggression was seen in *Thamnomys* (species uncertain), the African forest mouse, but a few ultrasonic pulses were detected when one animal postured aggressively towards another. The calls that were emitted during these aggressive encounters in murid rodents were all purely ultrasonic and consisted of a single component only. The three other murid species that were observed, *Mus musculus*, the laboratory mouse, *Praomys natalensis*, the multimammate mouse and *Mus minutoides*, the minute mouse, produced no detectable ultrasound during aggressive behaviour even though intense fighting was sometimes seen, especially in laboratory mice. A few ultrasonic pulses were detected from all three species, however, when one male sniffed the genital region of another.

Both narrow-band and wide-band calls were detected from cricetid rodents during aggressive behaviour. The gerbils, *Meriones shawi* and *Gerbillus sp.* produced both types of call. Narrow-band purely ultrasonic calls were emitted during chasing, boxing and aggressive posturing. They generally consisted of a single frequency component although some of those emitted by *Gerbillus* consisted of several, apparently harmonically unrelated components within a narrow frequency band of 10–15 kHz. In both species, the less aggressive animal produced wide-band 'creaking' sounds, containing audible and ultrasonic components, as it fended off the attacker. Often one of the ultrasonic components was more intense than the others and the pulse ended on this single frequency. Some pulses consisted of several components, 5–12 kHz apart,

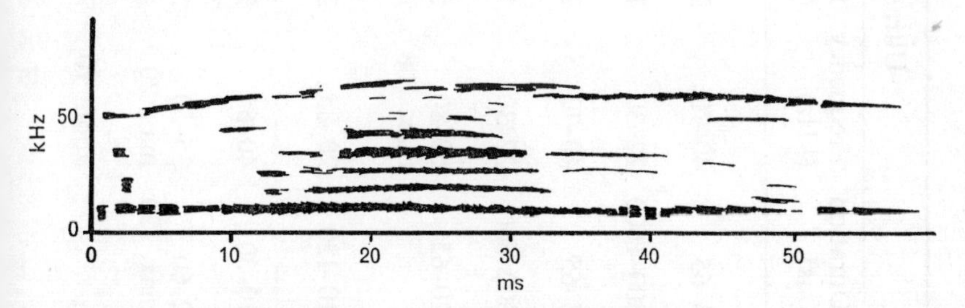

FIGURE 7.12 Sonagram of a broad-band call produced by *Gerbillus* sp. during an aggressive encounter showing possibly harmonically related components extending from the audible to the ultrasonic range.

with possible harmonic relationships (Fig. 7.12). Submissive postures, similar to those of adult rats, were sometimes adopted by *Meriones shawi* but no ultrasounds were detected while these postures were maintained.

Hamsters produced only wide-band calls during aggressive behaviour. These calls consisted of many apparently harmonically unrelated components, both audible and ultrasonic, and they also often ended in a purely ultrasonic note, at 25–35 kHz. Purely ultrasonic calls with a single component were produced during one aggressive encounter between male *Clethrionomys* but these calls were of very low intensity. No ultrasounds were detected during aggressive behaviour in the other two species of cricetids studied, *Microtus agrestis* and *Lagurus lagurus*, the steppe lemming.

Social significance

The emission of ultrasound during aggressive behaviour in rodents appears to be related to the social structure of the different species

TABLE 7.2 Ultrasound emission during aggressive behaviour in rodents

Species	Encounters		Behaviour	Ultrasonic calls			
	Type	No. ultra-sounds/total		Duration (ms)	Frequency (kHz)	Bandwidth (kHz)	Frequency pattern
Family Muridae							
Rattus norvegicus (laboratory)	♂ *v.* ♂	39/40	Fighting	3–65	40–70	2–50	Short pulses, rapid frequency changes
		26/40	Submission	800–1600	22–30	1–5	Long pulses, slow drifts
(wild)	♂ *v.* ♂	3/7	Sniffing, kicking	3–65	40–70	2–25	Short pulses, rapid frequency changes
		4/7	Submission	300–600	22–30	1–5	Long pulses, slow drifts
Apodemus sylvaticus	♂ *v.* group	1/1	Chasing and fighting	20–65	50–80 max. 108	5–20 max. 50	Slow and rapid frequency changes, frequency steps
Arvicanthis niloticus	♂ *v.* ♀ ♀ *v.* ♀	4/4 1/1	Chasing and fighting	10–120	20–85	2–15	Drifts, rapid frequency changes
Acomys cahirinus	♂ *v.* ♂	0/1	(Chasing seen)	—	—	—	—
	♂ *v.* ♀	1/3	Chasing	10–30	40–65	2–10	Rapid frequency changes
Thamnomys sp.	♂ *v.* ♂♂	2/3	Facing and posturing	2–60 max. 130	35–50 max. 70	5–15 max. 27	Rapid frequency changes, frequency steps
Mus musculus (albino)	♂ *v.* ♂	4/10	Genital sniffing, attempted mounting		about 70		

(C_3H)	♂ v. ♂	1/4	Genital sniffing		about 70		
(wild)	♂ v. group	1/1	Genital sniffing (Chasing and fighting seen in all 3 strains)		about 70		
Mus minutoides	♂ v. ♂	0/4	(Attacking and boxing seen)	—	—	—	—
	♂ v. ♀	1/1	Nosing		about 80		
Praomys natalensis	♂ v. ♂	1/5	Genital sniffing, nosing (posturing and chasing seen)		about 70		
	♂ v. ♀	1/4					
Family Cricetidae							
Meriones shawi	♂ v. ♂	4/4	Chasing and fighting (narrow-band)	10–40	35–60	2–25	Drifts, frequency steps
	♂ v. ♀	4/4	Self-defence (wide-band)	10–100	6–60	Up to 60	Many components
Gerbillus sp.	♂ v. ♂	2/2	Chasing and fighting (narrow-band)	30–200	40–65	10–15	One or more components
	♂ v. ♀	2/2	Self-defence (wide-band)	60–120	6–65	Up to 65	Many components
Clethrionomys glareolus	♂ v. ♂	1/1	Chasing and fighting	10–45	17–30	2–5	Slow drifts
Lagurus lagurus	♂ v. ♂	0/4	(Chasing, wrestling, and biting seen)	—	—	—	—
Microtus agrestis	♂ v. ♂	0/1	(Aggressive posturing seen)	—	—	—	—
	♂ v. ♀	0/2					

rather than to their taxonomic, and therefore probably their evolutionary relationships. Species which are said (Walker, 1964) to live in colonies or to show some degree of mutual tolerance, such as rats, *Meriones* and *Gerbillus*, produced pure ultrasounds in aggressive situations. Rats, which have fairly stable dominance–subordination relationships, also emitted ultrasounds in submissive situations. The organization within colonies of *Meriones* and *Gerbillus* does not appear to be as rigid as that of rats, and no ultrasounds were detected during the few submissive postures that were seen in *Meriones*, although the 'creaking' sounds associated with defensive behaviour in both species could have an appeasement function. The social organization of *Apodemus* in the wild is not well known although hierarchies with one dominant animal have been observed in laboratory colonies (Bovet, 1972). Hierarchies possibly exist within communities of *Clethrionomys* (Kikkawa, 1964). Both species emitted ultrasounds during aggressive behaviour. The social structure of *Thamnomys*, which emitted ultrasounds during threat postures, is not known.

No pure ultrasounds were detected during aggressive behaviour in hamsters, laboratory mice, *Mus minutoides*, *Microtus*, *Praomys* or *Lagurus* and they did not appear to be commonly produced during aggression in *Acomys*. Mice and hamsters appear to live solitarily or to set up individual territories if space permits. In mice the apparent absence of ultrasounds during aggressive behaviour may be significant as there appears to be no clear submissive posture in this species and in experimental situations dominant males often do not tolerate subordinate males but attack them whenever they meet (Mackintosh, 1970).

Ultrasonic signals may therefore be important in the evolution of stable social organizations that allow the animals to live in groups without harmful conflict. The possible role of ultrasound in such social organizations has only recently been realized and it should provide a fascinating field for further study.

ULTRASOUND AND MATING BEHAVIOUR

During the comparative survey of aggressive behaviour that has just been described it was found that several species of myomorph rodents emitted ultrasonic calls during investigation of the genital region or attempted mounting of one animal by another. This led Sewell (1969; Sales *née* Sewell, 1972b) to investigate the emission of ultrasound during mating behaviour in rodents. Most of her observations were made on laboratory rats and mice, and the production of ultrasound during mating behaviour in these animals has now also been studied by Barfield and Geyer (1972) and by Whitney, Stockton and Tilson (1971). Sales also observed varying degrees of mating behaviour in nine other species of rodents. Several of the animals were wild-trapped and most were nocturnal and so many of the observations were made by red light.

Mice

Sales (*née* Sewell 1972b) observed 110 heterosexual encounters involving 16 males and 18 females of the E.N. strain of albino mice, and in most of these encounters the males were introduced singly into the cages of the females. The animals generally first approached and nosed each other and then the male sniffed the genital region of the female, often following her round the cage in order to do so. On many occasions the male mounted the female and his attempts to mate were indicated by rapid pelvic thrusting movements. The female usually rebuffed the male, kicking out at him before running away to crouch in a corner of the cage. Sometimes the male retaliated by poking the female with his snout or even biting her. At other times he repeatedly pushed himself between the female and the side of the cage. In twelve encounters the females were receptive to the males and adopted the rigid mating posture when mounted. These females were therefore considered to be in behavioural oestrous. During these encounters the males successfully gained intromission, often for many seconds at a time, but only two males completed the mating pattern and ejaculated during the observation period. Maintenance activities such as feeding, drinking, grooming and sniffing the cage were also seen during the heterosexual encounters.

Ultrasonic calls at about 70 kHz were emitted during 101 of the 110 observations and Sales found that they were most often produced during behaviour patterns that indicated a high level of sexual motivation in the male (Sewell, 1969; Sales *née* Sewell, 1972b). Ultrasound emission was statistically correlated with the behaviour patterns of the male: nosing the female, sniffing her genital region and mounting without gaining intromission. Ultrasounds were also detected during some mounts with intromission, including those leading to ejaculation. However, other intromissions, as well as the two ejaculations, were apparently silent.

The ultrasonic calls produced during mounting appeared to be

TABLE 7.3 Ultrasound emission during heterosexual encounters in rodents

Species	No. of observations	Behaviour	Ultrasonic calls: Duration (ms)	Frequency (kHz)	Bandwidth (kHz)	Frequency pattern	Authors
Family Muridae							
Mus musculus							
E.N.	110	Mounting	50–300	50–110	Up to 50	Sequence of drifts, rapid frequency changes and frequency steps	Sales (1972b)
T.O. Swiss	4	Mounting	50–300	30–112	Up to 60		
	1	Ejaculation					
C_3H	3	Mounting	50–300	30–90	Up to 50		
C57 BL/6J x BALB/cJ	22			About 70			Whitney, Stockton & Tilson (1971)
Rattus norvegicus							
Wistar	26	Sniffing cage and nosing female	3–1000	40–116	2–20	Drifts, frequency steps	Sales (1972b)
		Genital sniffing	100–800	30–56	1–5	Slow drifts, rapid fluctuations	
		Mounting	2–40	34–120	Up to 70		
	4	Ejaculation					
Long Evans	11	Post ejaculation	1000–3000	22–25		Slow drifts	Barfield & Geyer (1972)

Acomys cahirinus	2	Chasing	100–300	28–80	Up to 50	Rapid fluctuations	Sales (1972b)
		Intromission	2–10	50–60	1–5		
Apodemus sylvaticus	2	Nosing, genital sniffing	8–25	65–85	1–10	Drifts	Sales (1972b)
	2	Chasing prior to intromission	2–40 2–20	60–100 70–105	Up to 20	Drifts, steps, rapid fluctuations	Sales & Smith (Unpub.)
Mus minutoides	4	Nosing, genital sniffing	2–30	80–90	1–10	Rapid frequency changes	Sales (1972b)
Praomys natalensis	2	Nosing, genital sniffing	5–30	70–85	2–5	Slow drifts	Sales (1972b)
Family Cricetidae							
Mesocricetus auratus	3	Homosexual nosing	50–120	20–35	1–5	Drifts, amplitude modulation	Sales (1972b)
		Heterosexual nosing	50–200	25–50	1–50	Drifts, amplitude modulation	
Lagarus lagurus	2	Nosing, genital sniffing	15–30	60–100	2–20	Rapid frequency changes	Sales (1972b)
		Mounting	5–60	65–90	10–30	Sweeps or steps	
Calomys callosus	4	Nosing, genital sniffing	2–25	65–90	10–20	Sweeps	Sales (1972b)
	1	Mounting		about 65–70			

TABLE 7.3—*continued*

Species	No. of observations	Behaviour	Ultrasonic calls				Authors
			Duration (ms)	Frequency (kHz)	Bandwidth (kHz)	Frequency pattern	
Family Cricetidae							
Peromyscus maniculatus	3 1	Nosing Genital sniffing, intromission	2–50	35–60	6–20	Rapid frequency changes and drifts	Sales (1972b)
Clethrionomys glareolus	2	Chasing, intromission	2–25	20–55	1–5	Slight drifts	Sales (1972b)
Dicrostonyx groenlandicus	39+	(♂ & ♀) Nose, chase, box, attack, groom (♂) Mounting	70	15–35	3–4	Sweeps	Brooks & Banks (1973)

emitted by the male. Observations with the bat-detector showed that the pulses seemed to be related to the pelvic thrusts of the males and audible cries, obviously emitted by the females, were often produced at the same time as, but were not synchronous with, the ultrasounds. Ultrasounds were sometimes produced when the male approached the female and when, after being rebuffed by the female, he poked her with his nose or pushed past her. These latter actions may represent displacement behaviour in the absence of a receptive outlet for a high sexual drive. Ultrasounds were seldom detected when the female approached or nosed the male or when both animals were engaged in maintenance activities.

Sales also briefly studied ultrasound emission during mating behaviour in the T.O. Swiss and the C_3H strains of mice. Both of these emitted ultrasounds in the same situations as E.N. mice, and again the pulses appeared to be emitted by the males. In all three strains mounting was often accompanied by the emission of sequences of pulses which lasted up to 7 s and consisted of 3 to 25 or more pulses at intervals of up to 200 ms. The physical characteristics of these calls are summarized in Table 7.3. Most of the pulses consisted of a single component although traces of a second harmonic were sometimes present and a few pulses ended in two or more harmonically unrelated components. The pulses at the beginning of a sequence were often shorter than those in the middle and they commonly showed drifts or occasionally rapid changes or steps in frequency. Pulses occurring towards the end of a sequence often contained a step downwards of 2–40 kHz (Fig. 7.13). During intromission and during the early part of an ejaculatory mount (tape-recorded only in T.O. Swiss mice), ultrasound emission continued for longer periods than in other mounts and sequences of up to 80 pulses were found. The frequency patterns of these pulses were similar to those produced during mounting without ejaculation, although a few pulses showed an additional component, for a short time, at half the frequency of the more intense and long-lasting component. Many of the ultrasounds recorded during the ejaculatory mounts contained small, rapid changes in frequency or several successive upwards and downwards frequency drifts. Pulses with a variety of frequency patterns were also produced singly or in small groups and were associated with other behaviour patterns such as the male nosing the female or sniffing the cage.

The ultrasonic pulses that were detected during these heterosexual encounters appeared to be produced by the males, particularly those with a high level of sexual motivation. This conclusion was strengthened by observations on the two males that ejaculated. One was immediately

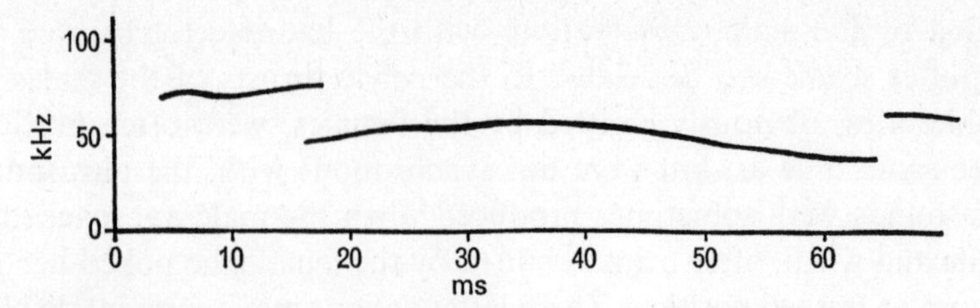

FIGURE 7.13 Call with step-like frequency changes produced during mating in laboratory mice.

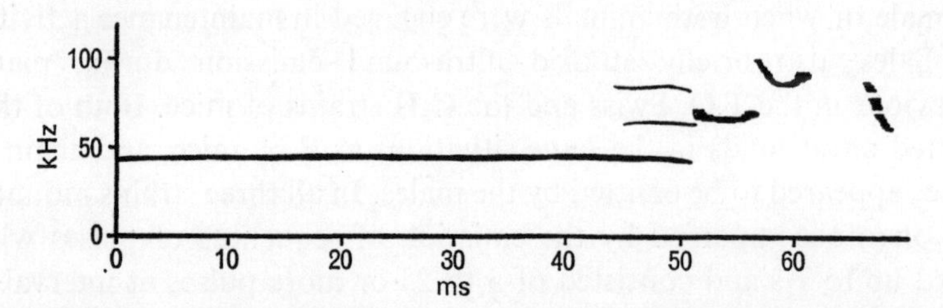

FIGURE 7.14 The last 50 ms of a call with slight changes in frequency, ending in harmonically unrelated components and leading into a rapid frequency fluctuation. Emitted during mating in laboratory rats.

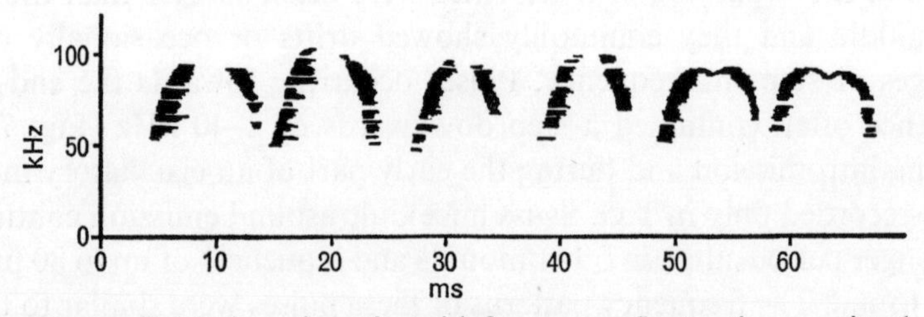

FIGURE 7.15 A series of rapid frequency fluctuations emitted during mating in laboratory rats.

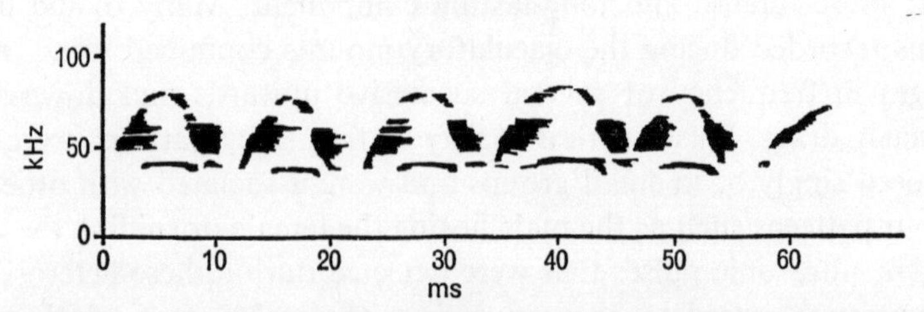

FIGURE 7.16 Calls emitted during mating in *Acomys cahirinus* showing rapid frequency fluctuations, two frequency components and formant structures.

presented with a second receptive female and the other male was presented with three more receptive females in turn. No ultrasounds were detected during these additional encounters in which genital sniffing, but not mounting, occurred. These observations may indicate that ultrasounds are emitted less readily when the sexual drive of the male is low, as it probably is for a time after ejaculation.

Further evidence that ultrasounds are emitted by the males comes from the work of Whitney, Stockton and Tilson (1971). These authors used a bat-detector alone to study ultrasound emission during a variety of encounters between male and female hybrid laboratory mice. They found that ultrasonic calls at about 70 kHz and similar to those of young pups were produced during each of 22 heterosexual encounters although calls were detected from only four out of 11 encounters between pairs of males. No ultrasounds were detected when 12 normal females were 'paired' with anaesthetized males but calls were produced in eight out of 10 'pairings' of normal males with anaesthetized females.

Ultrasound emission in adult mice, however, is not confined to the males. Sewell (1969) reported that pulses at about 70 kHz were produced during two different encounters between females when one animal nosed or sniffed the genital region of the other and Whitney and his colleagues (1971) found that ultrasounds were emitted during 10 out of 15 encounters between females.

The function of the ultrasounds emitted during heterosexual encounters in mice and their effect on other individuals, especially the females, is not clear. Sewell (1969; Sales *née* Sewell, 1972b) suggested that the ultrasonic calls may indicate to a female that an approaching male is sexually motivated rather than aggressively motivated and that they may be important in inhibiting the aggression of the female and inducing the mating posture. Whitney, Stockton and Tilson (1971) also suggested that these calls indicate a lack of aggressive intent and, as the calls 'resembled infant calls', that they represent a ritualized courtship display involving juvenile behaviour. In mice, which are territorial in the wild, ultrasounds may be part of a signal system which is important in overcoming the otherwise intense intraspecific aggression and in allowing mating to occur successfully.

Rats

The ultrasonic calls emitted during mating behaviour in rats have been studied by Sales (*née* Sewell, 1972b) and further calls emitted after ejaculation have been studied and described by Barfield and Geyer (1972).

Sales observed 30 different heterosexual encounters in Wistar laboratory rats and found that, as in mice, the emission of ultrasound was associated with different activities of the male, but these animals produced pulses with several distinct patterns. Short pulses, similar to those recorded during aggressive behaviour, were often emitted when the male entered the female's cage and when he investigated either the female or the cage at intervals throughout the observation period. They were sometimes also detected when the male unsuccessfully attempted to mount the female or when either animal was aggressive towards the other. Some of these short pulses contained slow drifts in frequency but most of them contained rapid frequency changes and sudden steps.

Sequences of pulses were produced during mounts with intromission and ejaculation, which in rats lasts for about a second. These sequences contained three different types of frequency pattern. Some of the pulses were similar to the short ones described above, but most of the ultrasonic emissions were either long pulses with slow drifts in frequency or else trains of very brief pulses with marked, rapid fluctuations in frequency. The long pulses generally consisted of a single component at about 40–50 kHz but occasionally traces of second and third harmonics were also present and sometimes a pulse ended in several, harmonically unrelated components (Fig. 7.14). The very brief pulses occurred 3–15 ms apart in trains of up to 200 ms. Each pulse showed one or more marked fluctuations in frequency of up to 70 kHz and the pulses often appeared to be part of a continuous, frequency-modulated waveform (Fig. 7.15).

Most of the sequences lasted about 2 s. They often began with one or more of the short pulses described earlier, generally followed by two to four, but up to 24, alternating periods of slow frequency drifts and rapid frequency modulation. Intervals of up to 100 ms separated the different types of frequency pattern but the long drift-like pulses often immediately preceded and appeared to lead into trains of frequency fluctuations. Both types of frequency pattern were found in isolated pulses that were associated with the male, or occasionally the female, sniffing the cage or the other animal.

Barfield and Geyer (1972) found that male Long Evans laboratory rats emitted long calls, 1–3 s in duration and at 22–25 kHz, in the post-ejaculatory interval between two successive ejaculations. This interval has two phases, an absolute refractory period, during which the male is incapable of spontaneous copulatory activity, and a relative refractory period, in which sexual behaviour can occur in response to a strong arousing stimulus. During the absolute refractory period, the female tends to stay away from the male and she does not display the hopping, darting and ear-wiggling movements which commonly occur prior to an ejaculatory mount. It was during this absolute refractory period that the 11 males studied by Barfield and Geyer produced the 22 kHz calls, which could be correlated with the long exhalations of the refractory male rat. Barfield and Geyer called them the 'post-ejaculatory song of the male rat'. They suggested that the emission of the 22 kHz signal reflects a state of social withdrawal in the male and that it inhibits the reproductive behaviour of the female during the time when the male is incapable of sexual responses. In this respect the calls may be comparable with the long 'submissive' pulses which are emitted by defeated rats in aggressive encounters and which may inhibit the aggression of the dominant animal.

Other myomorph rodents

Ultrasounds were produced by all of the nine other species of myomorph rodents that Sales (*née* Sewell, 1972b) observed, but these animals were not studied in such great detail as the rats and mice and varying degrees of mating behaviour were observed (Table 7.3).

Mounting with intromission occurred in five species and was accompanied by ultrasound emission in four of these. In *Acomys* ultrasonic calls were detected when a male chased a female prior to mounting and during the early part of mounting with intromission. These pulses were emitted for up to 300 ms at a time. They showed rapid and sometimes very marked frequency modulation and many contained what appeared to be formant structures for short periods (Fig. 7.16). A very few short pulses were also produced. In *Apodemus* chasing and intromission were accompanied by the emission of sequences of pulses containing both long drifts in frequency and some brief frequency fluctuations. The calls were of very low intensity and the frequency fluctuations were not so marked as in rats. *Peromyscus* emitted short pulses with a single frequency component during nosing and genital sniffing of the female by the male as well as during mounts with intromission. These pulses often contained marked frequency drifts downwards, occasionally followed by an upward drift. A few very low intensity pulses were produced during intromission in *Clethrionomys*, which was observed on two different occasions when females were introduced singly into a cage containing several males. In all of these species the emission of ultrasound appeared to be related to the pelvic thrusts of the males as they mounted the females.

Golden hamsters emitted purely ultrasonic calls in encounters that involved females in behavioural oestrous. During two different encounters between an oestrous female and an anoestrous female, low intensity ultrasounds were detected as the oestrous female nosed the other in intervals between adopting the rigid mating posture several times in succession. On each occasion the anoestrous female was then replaced by a male and ultrasonic calls were detected as the female nosed

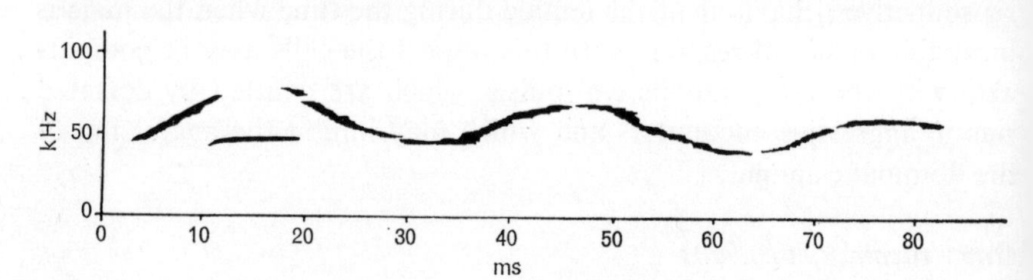

FIGURE 7.17 Sonagram of a call showing frequency steps and frequency modulations produced during mating in *Lagurus lagurus*.

the male, particularly after she had maintained the rigid mating posture without being mounted. No ultrasounds were detected when the male did mount the female. Most of the pulses produced during these and a third heterosexual encounter, consisted of a single component at 25–50 kHz. A few pulses, however, contained several apparently harmonically unrelated components within a total frequency band 10–50 kHz wide and were similar to the calls produced during aggression in this species.

Sequences of pulses were emitted during mounts without intromission in *Lagurus lagurus*. These consisted of five to six pulses emitted 10–20 ms apart. Each pulse showed one or more marked frequency sweeps or large frequency steps upwards and downwards (Fig. 7.17). Very short high frequency calls were produced during heterosexual encounters in *Calomys callosus* when the male nosed or sniffed the female and when he attempted to mount. In both *Mus minutoides* and *Praomys* the males were only seen to sniff the females but ultrasounds were detected from both species during this behaviour.

Recently Brooks and Banks (1973) have recorded high frequency calls from both male and female *Dicrostonyx groenlandicus* during heterosexual encounters. These calls were about 70 ms in duration and they consisted of a single component at 15–35 kHz that showed sweeps and fluctuations in frequency. In both males and females these calls were produced when one animal approached, nosed, chased, boxed, groomed or was attacked by the other, and males also produced them when they attempted to mount the females.

Social significance

So far ultrasounds have been detected during mating behaviour in all species of myomorph rodents in which mating has been observed. Except in hamsters, the ultrasonic calls were associated with activities of the male, such as nosing, genital sniffing and mounting, and they appeared to be emitted by the male, although in *Dicrostonyx* the female apparently also produced high frequency calls. It seems likely that in many other species of myomorph rodents, as in mice, these ultrasounds are important in indicating the sexual motivation of the male and in allowing mating to occur, possibly by inhibiting the aggression or the flight reaction of the female, and by inducing the mating posture. It is interesting that four of the species in which intromission was observed (rats, mice, *Acomys* and *Apodemus*) emitted ultrasounds with similar patterns: frequency modulations or frequency steps occurring either continuously or in sequences. Calls with frequency sweeps or frequency steps were produced during intromission or attempted mounting in *Peromyscus*, *Lagurus* and *Calomys*, while *Dicrostonyx* produced both sweeps and fluctuations in frequency. In *Lagurus* (Fig. 7.17) and *Dicrostonyx* the calls were emitted in sequences.

In hamsters the females appeared to emit the ultrasonic calls rather than the males. These animals are normally solitary and both males and females are generally very aggressive towards other individuals. The ultrasonic signals may therefore be important in indicating the females' temporary lack of aggression and willingness to mate. The emission of ultrasound during mating behaviour thus appears to be a widespread phenomenon in myomorph rodents and this suggests that it probably plays an important part in the reproductive behaviour of these animals.

OTHER SITUATIONS INVOLVING ULTRASOUND EMISSION IN RODENTS

During a study of aggression in rodents, Sewell (1969; Sales *née* Sewell, 1972b) removed the litter of a lactating female *Acomys cahirinus* before introducing a male into the cage (see section on aggressive behaviour). The female chased the male once or twice but most of the time she ran to and fro throughout the cage, passing the male and apparently ignoring him. Many ultrasonic calls were detected as she ran in and out of the nest box and sniffed either the nest box or the rest of the cage. A few days later the litter was again removed but no male was introduced. On this occasion and on two similar occasions when another female was deprived of her litter, many ultrasonic calls were detected as the females repeatedly entered and left the nest box. These calls emitted by adult female *Acomys* were similar to the distress calls of the young of this species. They were 80–120 ms in duration and consisted of a single component at 25–35 kHz which showed slow drifts in frequency. A few shorter pulses of 50–60 kHz with step-like frequency changes were also produced.

Two female *Apodemus sylvaticus* that were deprived of their litters produced calls 25–40 ms in duration and at frequencies of 60–75 kHz. These calls generally consisted of slow drifts in frequency although rapid frequency changes also occurred. Under similar conditions two *Clethrionomys glareolus* females produced very low intensity calls, 20–50 ms long which contained one or two harmonic components with a fundamental between 12 kHz and 25 kHz. Okon (1971b) reported that lactating female albino (E.N.) mice emitted calls 30–110 ms in duration and at 60–80 kHz when their litters were removed. In all of these species the emission of ultrasound by the females was erratic; sometimes the calls were only emitted 30 min or more after the litter had been removed and at other times no calls were detected.

The function of the calls emitted by lactating females is not clear. They may act as warning signals, indicating the possible presence of danger or they may direct other adults to the site of disturbance to help in the retrieval of lost young. But in general the calls were of low intensity and they probably would not carry far. It is unlikely that they stimulate the young to emit ultrasonic calls as, in mice at least, behavioural and physiological responses to sound are absent until 9–14 days after birth (Alford and Ruben, 1963) when the pups are becoming independent.

Apodemus sylvaticus appear to emit ultrasonic calls whenever they 'explore' their surroundings (Sales *née* Sewell, 1972b). Calls at about 70 kHz were often detected when animals began to investigate new surroundings such as a clean cage, particularly when they stood up on their hind legs sniffing, and when established animals first emerged from the nest to feed in the evening. When cages containing *Apodemus* were disturbed by moving or knocking them, ultrasonic calls were detected as the nest material heaved just before the animals emerged, as they poked their heads out of the nest and then as they ran round sniffing the cage. These calls were 10–30 ms long, at 65–90 kHz and generally consisted of slow drifts in frequency; they may have been alarm calls. Other possible alarm calls were detected from a lactating female *Meriones unguiculatus* that was suddenly disturbed when she was out of the nest. She immediately thumped rhythmically on the ground with one hind foot and pulses at about 50 kHz were detected simultaneously.

There are several possible functions for the calls emitted by *Apodemus* while exploring their surroundings. These animals are generally nocturnal and so the signals could serve to keep members of a group, possibly a family group, together while feeding at night or they could have an echo-location function. The social structure of *Apodemus* in the wild is not well known but it is possible that these ultrasounds could act as intraspecific signals, to warn off both subordinates within the territory and members of neighbouring territories, or they could act interspecifically to warn off other species using the same terrain. Andrzejewski and Olszewski (1963) found that although both *Apodemus flavicollis* and *Clethrionomys glareolus* used the same pre-baited feeding site, the *Clethrionomys* seldom entered the area if any *Apodemus* were present and if they did so they were driven off by the *Apodemus*. These authors suggested that the *Clethrionomys* often avoided contact with the *Apodemus* by detecting them at some distance. It is conceivable that this avoidance could be mediated by ultrasonic calls which, in *Apodemus sylvaticus* at least, are apparently emitted whenever the animals are out of the nest.

THE MECHANISM OF ULTRASOUND PRODUCTION IN RODENTS

The mechanism of ultrasound production in rodents is not yet known for certain although there are some very good indications of the kind of mechanism that might be involved. In man, and other mammals including rodents, broad-band audible sounds are produced when air is forced between the vocal cords, causing them to vibrate. These sounds have a low fundamental frequency and a long series of harmonics. The resonant cavities of the mouth and nasal tract selectively enhance certain harmonics in groups called formants. In speech the fundamental frequency remains fairly constant but different vowel sounds are produced by changes in the shape of the cavities which affect their resonant frequencies. Although bats appear to produce ultrasound in this way (Pye, 1967a; Roberts, 1973b), the physical structure of rodent ultrasounds is not easily explained by such a mechanism.

Sewell (1969) suggested that a whistle-like mechanism might be involved in ultrasound emission. She found that some of the calls of baby mice contained large downward frequency jumps, the frequencies on either side of which were often almost in a simple ratio of 2:1 or 3:2 This suggested that there was a sudden change in the mode of vibration similar to that causing the jumps obtained by overblowing a recorder The mechanism of ultrasound production in rodents has since been studied in detail by Roberts (1972b, 1973b).

Roberts found that infant rodents could emit ultrasounds through either the nose or the mouth. The structure of the ultrasonic cries was not affected when either of these were blocked by an acrylic resin although the amplitude of the signals was reduced when the mouth was blocked. This indicated that the calls are probably normally emitted through the mouth but that neither the mouth nor the nose acts as a resonant cavity for these sounds.

The emission of ultrasound in rodents appears to be associated with respiration. The abdominal movements of young pups are more marked during ultrasound production and the long 'submissive' calls of adult rats could always be associated with irregular respiration Roberts (1972b) examined this correlation experimentally using a hot wire anemometer, a wire that is warmed by an electric current and which shows a change in resistance when cooled by an air-flow. The 'hot' wire was placed 1–2 mm in front of one nostril of a rodent pup and the changes in resistance during respiration were recorded on one (f.m.) channel of a P.I. 6100 tape-recorder. The audible and ultrasonic cries produced by the pup were recorded simultaneously on another (direct) channel. Pups of five rodent species were studied: Wistar rats T.O. Swiss mice, *Sigmodon hispidus* (cotton rats), *Meriones unguiculatus* and *Clethrionomys glareolus*. In all of these species only one cry was produced during each exhalation and this was emitted at the onset of exhalation, before the air-flow reached a maximum. The air-flow was therefore low during ultrasound production but, after the cessation of the pulse, it increased to a greater level than in silent respirations. The short, broad-band clicks that are sometimes produced by infant *Clethrionomys* were also produced at the onset of exhalation indicating that they are probably produced by the respiratory air-flow rather than by other structures such as the tongue or the teeth.

The mechanisms for the production of the audible and the ultrasonic cries of rodents appear to be closely linked. Each type of sound can 'break' instantaneously into the other and as they never overlap they may have a common origin. Denervation of the larynx in young rats 2–3 months old prevents the production of either type of call immediately after the operation, although the animals do regain some ability to produce sounds several months later. Roberts (1973b) made a detailed anatomical study of the larynx of infant rodents but he could find no special structure that might be responsible for ultrasound production.

To determine whether there were any major differences in the mechanism of audible and ultrasound production, Roberts studied the animals in a mixture of air and diver's gas (20% oxygen and 80% helium). Helium is an inert gas in which the velocity of sound is greater than in air because of its lower density. This results in an increase in the resonant frequency of tuned cavities, although the frequencies of sounds produced by vibrating structures such as the vocal cords are not affected. Thus when a 'voiced' cry is produced in helium, the fundamental frequency, and therefore the harmonic series, remain the same but the resonant cavities now enhance higher harmonics in the series and so the formants are shifted in proportion to the amount of helium present. Bat echo-location cries and rodent audible cries, including those with ultrasonic harmonics, do not show a change of fundamental frequency in helium and so are assumed to be vocal. But the single component of most rodent ultrasonic cries is increased, although not quite to the extent originally expected. This shows that these sounds are not produced by a vibrating solid structure and it strengthens the argument that a whistle-like mechanism is involved.

Whatever mechanism is proposed for ultrasound production, it must be able to account for all the different frequency patterns, the jumps and the fluctuations as well as the amplitude changes, that rodents themselves produce. Roberts found that all of these could be reproduced by 'bird-whistles'. These consist of two parallel plates a short distance apart, each pierced by a hole in the centre and they are commonly found in 'whistling' kettles. Model bird-whistles made of metal or of soft rubber, could imitate all features of ultrasonic rodent cries (e.g. Fig. 7.18) and moreover behaved in exactly the same way in helium mixtures. Roberts therefore suggested that a mechanism of this sort could be responsible for ultrasound production in rodents. This mechanism appears to be in the larynx but the exact structures involved have yet to be identified.

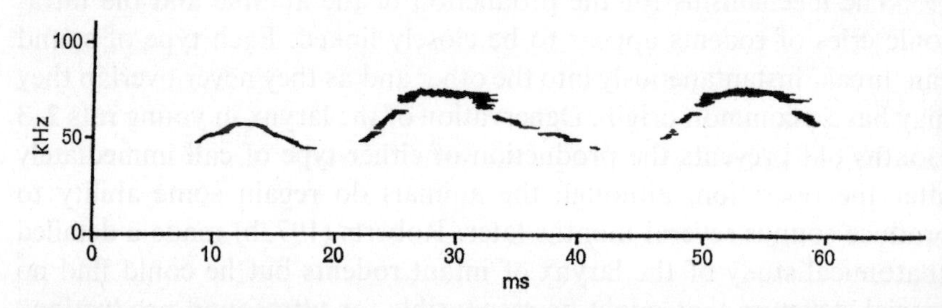

FIGURE 7.18 Sonagram of one of the sounds similar to those of rodents produced by a bird whistle. (Compare with Fig. 7.17) (Redrawn from Roberts, 1973b)

THE ABILITY OF RODENTS TO HEAR HIGH FREQUENCY SOUNDS

The ears of rodents do not show obvious anatomical specializations that might increase their sensitivity to high frequency sounds. There is, however, much evidence to suggest that many species of rodents can hear these sounds very well (reviewed in greater detail by Brown and Pye, 1974).

Several different techniques have been used to study the ability of rodents to hear ultrasounds. Earlier workers studied behavioural responses to a range of acoustic stimuli. Many animals respond to sounds by twitching the ears (the Preyer reflex) or the vibrissae, and Schleidt (1948, 1951, 1952) elicited these responses from a number of rodents to sounds of up to at least 50 kHz; *Mus musculus*, the house mouse, responded to sounds of almost 100 kHz. Certain mutant strains of rodents display epileptic seizures or convulsions when exposed to loud sounds. Dice and Barto (1952) used these epileptic responses as well as the Preyer reflex to show that *Peromyscus* could hear sounds of up to 90 kHz, although their epileptic strains were most sensitive to signals between 5 kHz and 16 kHz.

The conditioned reflex has also been used to determine an animal's auditory sensitivity. Many animals can be conditioned to sweat in response to an acoustic stimulus alone by first presenting this for several trials with a noxious stimulus, such as a mild electric shock. The moisture decreases the electrical resistance between a pair of electrodes on the skin, allowing the response to be recorded on a galvanometer. Such galvanic skin responses have been elicited from laboratory mice by signals up to 40 kHz by Berlin (1963) who found that the mice responded most readily to signals of 10–16 kHz.

The auditory sensitivity of adult rats has been determined by operant conditioning techniques. Animals were trained to respond to acoustic stimuli by pressing a lever (Gourevitch and Hack, 1966) or by moving from one part of the cage to another to avoid a mild electric shock (Gould and Morgan, 1941). Both studies showed that rats were most sensitive to frequencies of 40 kHz. This was the upper limit of Gould and Morgan's apparatus although Gourevitch and Hack obtained responses to signals of 50 kHz but not 60 kHz.

It is possible that the behavioural responses of an animal may be inhibited even though it has 'heard' the sound stimulus. The range of auditory signals that animals are able to detect can be determined using techniques which involve recording the auditory evoked response of the cochlea, or of the acoustic centres of the brain. One response of the cochlea to a sound stimulus is the production of the cochlear microphonic potential which, as its name implies, reproduces the stimulus signal waveform. It can be recorded by a fine wire electrode from the round window of the cochlea in anaesthetized animals. The auditory sensitivity is then determined by measuring the sound pressure level of each stimulus frequency necessary to give a standard response on an oscilloscope screen. Alternatively, but less commonly, a standard signal is used and the magnitude of the response is measured for each frequency studied. Using this technique, Crowley and his co-workers (1965) showed that adult rats were sensitive to signals up to 100 kHz and had peaks of sensitivity at 10 kHz, 20 kHz and 40 kHz. Crowley and Hepp-Raymond (1966) also used the cochlear microphonic response to study the development of hearing in infant rats. Responses were first obtained in 8–9-day-old pups and to only a limited range of frequencies, but as the rats grew older both their sensitivity and the range of frequencies to which they responded increased until adult levels of auditory sensitivity were reached at the age of 16–20 days. Cochlear responses to stimuli up to 30 kHz have been obtained from *Meriones unguiculatus*, the Mongolian gerbil (Finck and Sofouglu, 1966), and up to 100 kHz from

Dipodomys merriami, the kangaroo rat (Vernon *et al.*, 1971). Recently Brown (1970, 1971a, b, 1973a) has studied the cochlear responses of 10 species of rodents to signals up to at least 100 kHz using two different methods.

Brown's first technique was similar to the one described above; a range of short, narrow-band, 'pure-tone' pulses was used to stimulate the cochlea and the responses were observed on an oscilloscope screen. The second technique was completely novel. A high-voltage pulse generator was used to produce a train of sparks across an air gap. These sparks produced broad-band sounds which stimulated the cochlea at all frequencies simultaneously. The cochlear responses were tape-recorded and groups of responses were then analysed using the Kay sonagraph to give an integrated 'frequency section' or spectrum, a graph of amplitude against frequency. After certain corrections had been applied, this gave a clear picture of the response of the cochlea to all frequencies. The cochlear response curves obtained by both the 'tone pulse' and the 'spark' method were similar for each animal. They showed that myomorph rodents have at least two peaks of auditory sensitivity, one in the audible range and a second in the ultrasonic range corresponding to the ultrasonic frequencies that the animals produce.

Although the cochlear microphonic responses are related to the animal's hearing, they cannot be said to represent this ability directly, since they do not necessarily imply the excitation of sensory nerve fibres. A more reliable indication can be obtained from the acoustic information actually reaching the brain. This can most conveniently be recorded from the inferior colliculus of the mid-brain by passing a fine wire electrode through a hole drilled in the skull of anaesthetized animals until its tip just rests on the surface of the inferior colliculus. The sound pressure level necessary to elicit the minimum detectable response is then found for each of the frequencies studied. Ralls (1967) used this technique to study the auditory sensitivity of *Peromyscus* and *Mus musculus*. She found that *P. boylii* and *P. leucopus* responded to signals of up to 70–100 kHz with peaks of sensitivity between 10 kHz and 40 kHz. Wild-type house mice were more sensitive to high frequencies than were the two strains of laboratory mice studied, DBA/2J and BALB/cJ, both of which showed a reduction with age in their sensitivity to high frequencies. Later Brown (1971a, b, 1973b) found that in seven species of rodents the mid-brain response, like that of the cochlea, was bimodal (Fig. 7.19) with peaks of sensitivity occurring at approximately the same frequencies as those of the cochlear response (Table 7.4). The response of the inferior colliculus of each species to high frequencies was often greater than the response to low frequencies. These high frequency responses were found to be dependent on the maintenance of normal body temperature, which may explain why they had not been found by Ralls (1967). As discussed for bats (Chapter 3), large neural responses may indicate a low threshold or a large number of brain cells responding at the frequency involved. These explanations cannot yet be distinguished and may be combined but both would seem to be adaptive.

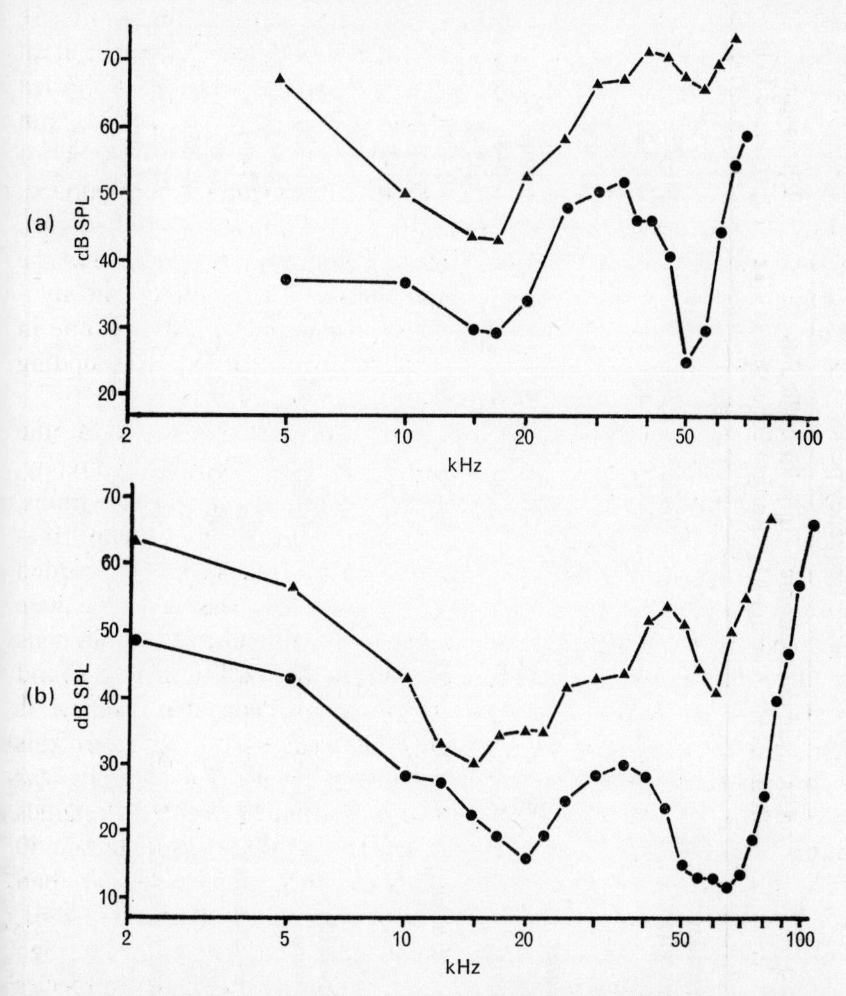

FIGURE 7.19 Auditory response curves of individuals from two species of murid rodents. (a) *Mus musculus*. (b) *Apodemus sylvaticus*. ▲ Cochlear microphonic responses. ● gross evoked responses of the inferior colliculus. ((a) redrawn after Brown, 1971b; (b) redrawn after Brown, 1971a)

TABLE 7.4 The relation between high frequency auditory responses in rodents and the ultrasounds produced*

Species	High frequency peaks				Ultrasound produced kHz	Animals involved
	Cochlear microphonic Range kHz	Cochlear microphonic Average kHz	Inferior colliculus Range kHz	Inferior colliculus Average kHz		
Family Muridae						
Rattus norvegicus			30–32	30	40–75	young
			40–42	40	40–120	adult aggression
					22–30	adult submission
					30–120	adult mating
Mus musculus						
(feral)	50–65	55	50–60	50	about 70	adult
(T.O. Swiss)	45–55	50	45–60	50	(40–88, max. 148)	(young E.N.)
					30–112	adult mating
Apodemus sylvaticus	45–60	50	50–70	60	40–88	young
					50–108	adult aggression
					60–105	adult mating
						adult exploration
Family Cricetidae						
Meriones shawi	30–40	35	30–40	30	45–130	young pure ultrasound
					35–60	adult aggression
Meriones unguiculatus	40–50	45	40	40	37–80	young pure ultrasound
Gerbillus sp.	40–45	45	40–45	40	50–85	young pure ultrasound
					40–65	adult aggression
Clethrionomys glareolus	45–50	50	50–60	50	20–55, max. 110	young
					17–30	adult aggression
					20–55	adult mating
Microtus agrestis	50	50	50	50	25–60	young
Sigmodon hispidus	40–65	40	40–65	50	23–60	young

*Table modified from Brown (1971b).

All of the studies reported here indicate that many species of rodents are extremely sensitive to high frequency sounds. The work of Brown in particular, has shown that the hearing ability of these animals appears to be related to the frequency of the ultrasonic signals that they emit. This suggests in yet another way that ultrasonic signals are of great importance in the lives of rodents.

CHAPTER EIGHT

Other Vertebrate Groups

The ability of bats to determine the distance, speed and often the nature of a target by listening to the echoes of their own cries, whether audible or ultrasonic, was discussed in Chapter 3. This ability is not confined to bats, however. Echo-location using audible cries has now been demonstrated in two genera of birds and in porpoises, while the dolphins use wide-band signals extending into the ultrasonic range. An echo-location function has also been claimed for the sounds produced by several other mammalian groups, although these may eventually prove to be examples of intraspecific communication. All of these cases will be examined in this chapter.

BIRDS

Probably no birds are able to hear ultrasonic frequencies (Payne, 1961, 1971; Schwartzkopff, 1968; Konishi, 1969a, 1970), nevertheless some birds do produce ultrasound. Thorpe and Griffin (1962b, c) recorded and analysed the songs of 20 different species of song-birds and they found that in 10 of these, including *Serinus canarius*, the canary, *Erithacus rubecula*, the robin and *Acrocephalus scirpaceus*, the reed warbler, the songs contained frequencies up to 50 kHz. These ultrasonic frequencies, however, were always associated with audible sounds and were of low intensity.

The acoustic responses of single neurones in the cochlear nucleus of song-birds have been studied by Konishi (1969a, 1970). He found that some neurones had their maximum sensitivity as high as 9–10 kHz, but their thresholds increased rapidly above this. The curve from all the neurones was found to agree with previously published behavioural sensitivity curves. Konishi concluded that the popular belief that song-birds can hear frequencies inaudible to man is not supported by reliable evidence. It therefore seems that, as Thorpe and Griffin suggested, the ultrasonic components present in some bird songs do not play a part in intraspecific communication.

It is interesting to note that, although the flight sounds of some small birds contain ultrasounds, those of owls that prey on small mammals do not (Thorpe and Griffin, 1962a). Owls have three modifications of their feathers: a fringe on the leading edge, another on the trailing edge and a downy upper surface. These were previously thought to eliminate audible flight noise and so allow silent approach to prey. It seems that these same modifications are also effective in the ultra-sonic range. This is obviously advantageous as many of the small mammals that owls prey upon are able to hear ultrasonic frequencies (Chapter 7). These modifications are absent from the fish-eating owls of Asia, and Thorpe and Griffin found that one of these owls produced considerably more ultrasound during flight than other owls of the same size.

Although birds do not appear to be able to hear ultrasound, at least two different groups of birds are able to echo-locate using audible frequencies. These are *Steatornis caripensis*, the oil bird of South America, and various members of the genus *Collocalia*, the cave swiftlets or birds of 'birds' nest soup' that live in South East Asia. The low frequency echo-location of these birds will be discussed briefly as a comparison with the high frequency echo-location of bats and dolphins.

Both *Steatornis* and some species of *Collocalia* live in caves, often nesting where little or no light penetrates. Griffin first suggested in 1954 that *Steatornis* uses echo-location. These birds feed at night and Griffin noted that as they flew around inside the cave they produced continuous and rapid trains of clicks. The clicks were emitted in short bursts of 2–6 with intervals of 1·7–4·4 ms between them. Each click was about 1 ms in duration and the frequency spectrum was 6–10 kHz with the main energy at about 7–8 kHz. When four oil birds were introduced into a completely darkened room they flew around safely, clicking as they did so. Fewer bursts of clicks were emitted when the light was switched on. However, when their ears were plugged, the birds were unable to orientate safely in the dark and they crashed into the walls. Griffin suggested that the clicks were used for echo-location when the birds were in darkness but that in the light the birds relied upon vision for orientation.

This can be supported by one of us (J.D.P.) who recorded a single *Steatornis* flying round an inner chamber of a communal cave. Even a single torch in the large chamber gave enough light for the bird to fly silently, but it clicked continuously when the torch was switched off. Each time the bird landed on its nesting ledge the click-rate increased from about 4–7 bursts per second to a 'buzz' of 11–12 bursts per second (Plate XII a) in much the same way as the pulses of bats. Two birds were also observed by natural light as they fed in the late evening about 3–5 km from their cave. They flew round and round a tall tree occasionally darting in to take something, presumably a nut, from the foliage. There was enough light to be sure of the unique profile of the species and although several observers stood only 10–15 m below the birds, not a single click was heard. This tends to confirm earlier suggestions that *Steatornis* uses echo-location only in order to roost in dark caves and not for feeding purposes.

The ability of various species of *Collocalia* to use echo-location has also been studied (Griffin, 1958; Medway, 1959; 1967; Novick, 1959; Griffin and Suthers, 1970). Like other swifts these birds feed on insects during the day, but during the night they return to caves to roost (Medway, 1960, 1969). Both Medway (1959) and Novick (1959) noted that *Collocalia* click (Plate XII b) when they enter the caves and that the rate of click production increases as they penetrate deeper into the caves. The clicks of *Collocalia* vary somewhat in frequency with the species, ranging from 1·5–4·5 kHz in *C. fuciphaga* (Medway, 1967) to 4·5–7·5 kHz in *C. vanikorensis granti* (Griffin and Suthers, 1970). Although clicks with components up to 16 kHz have been recorded from *C. vanikorensis granti*, the frequency spectra of *Collocalia* have not yet been found to extend into the ultrasonic range (Cranbrook and Medway, 1965; Griffin and Suthers, 1970). The clicks of *C. vanikorensis* have a double structure, with a pulse of moderate amplitude a few milliseconds long, followed after several milliseconds by a second pulse of higher amplitude. These clicks are produced at rates of 3–20 pairs per second.

Novick reported that *C. brevirostris unicolor* could fly if the eyes were covered, but that one bird whose ears were plugged roosted as soon as possible and apparently preferred not to fly at all. By avoidance experiments in darkened rooms Medway (1967) found that *C. fuciphaga* was unable to detect a barrier of wooden rods 1 cm^2 in cross section, whereas Griffin and Suthers (1970) found that *C. vanikorensis* could detect and avoid cylindrical wires as small as 6 mm in diameter. This difference in discrimination is possibly related to the different wave-lengths used by the two species, as it is in bats and, indeed, in all other sonar systems.

Nothing is known as yet of the mechanisms of click production in either *Steatornis* or *Collocalia* or of their auditory acuity. Konishi (1969b) has shown, in five other species of (non-echo-locating) birds, that single auditory neurones in the brain can follow click repetition rates of up to 1000 s^{-1}. This is about ten times the time-resolution that mammals such as the cat are capable of, and the ability would obviously be of advantage to any bird that used rapid bursts of clicks for echo-location.

An echo-location function has also been proposed for the cries of some other birds such as the curlew (Ashton-Freeman, 1953) and the Manx shearwater (Fisher and Lockley, 1954) but these have not yet received experimental support.

CETACEA

After the bats, perhaps the most well known animals in which echo-location has been demonstrated are the Cetacea, the whales, dolphins and porpoises. Only a few groups within this order have been studied in detail and most of these are smaller members of the Odontocetae, or toothed whales, belonging to the family Delphinidae, the dolphins. The Mysticetae, or whalebone whales, have no smaller representatives and have been less studied. They are not known to echo-locate but the humpback whales produce a 'song' containing a very wide variety of sounds extending up to, and perhaps into, the ultrasonic range (Payne, 1970; Payne and McVay, 1971). These whales are also reported to emit narrow-band infrasonic sounds at 20 Hz which may be used for long range communication, possibly up to several hundred kilometres (Payne and Webb, 1971).

The sounds of odontocetes

Toothed whales produce a wide variety of sounds which have been described by many different authors. Only a few review papers will be referred to here although a more complete list of references is given in an addendum to the bibliography. The sounds of odontocetes can be divided into three main groups: pure-tone whistles, complex sounds and impulsive clicks. The pure-tone signals, or whistles, are commonly thought to be used for communication between individuals. Each whistle may last for up to several seconds and, although the fundamental frequency and the major part of the sound energy is generally below 20 kHz (Fig. 8.1), in some signals harmonic components extend well into the ultrasonic range (Lilly, 1962). It appears that some whistles may be individual 'signatures' to identify the emitter while others are associated with different situations, particularly distress and possibly with sexual encounters. The distress calls of *Tursiops truncatus*, the bottlenose dolphin, consist of an upwards sweep from 3 kHz to 20 kHz followed by a downwards sweep back to 5 kHz, and they are thought to elicit assistance from other dolphins (Lilly, 1963).

The complex signals (Plate XII d) are possibly also used for communication and again they may also be characteristic of the emitting animal (Caldwell and Caldwell, 1967). These sounds are emitted as very rapid bursts of clicks, up to 800 per second (Lilly and Miller, 1961a), and they show a wide variety of frequency structures. Most of them are low in frequency with the main energy below 10–16 kHz although some sounds contain components extending into the ultrasonic range (Fish and Mowbray, 1962). They have been recorded after bodily contact between dolphins (Lilly and Miller, 1961b), when the animals were alarmed (Caldwell *et al.*, 1962) and as a result of human intervention (Lilly and Miller, 1961a).

The third type of sound produced by odontocetes is the impulsive click signal which is generally at higher sound frequencies (Plate XII c). It is these sounds that have been implicated in echo-location. They are produced intermittently at rates of 1–2 s^{-1} by awake, captive animals that are gently cruising (Kellogg, 1959a), but when the animal turns towards an object of interest, the rate of click production may increase to several hundred per second. It appears that if the animal approaches very close to an object, as it does in taking a fish, clicks are commonly produced at rates of 400–500 s^{-1}, but if the animal does not approach the object closely, as in discrimination tasks, much lower rates of 50–200 s^{-1} are used (Norris *et al.*, 1967). Bastian (1967) found that these clicks may also have communication value between dolphins. When two animals were separated with only an acoustic link between them and were required to co-operate in a task, their success appeared to depend on the emission of click trains by one animal and only indirectly on whistle signals.

The clicks produced by many dolphins are individually very brief

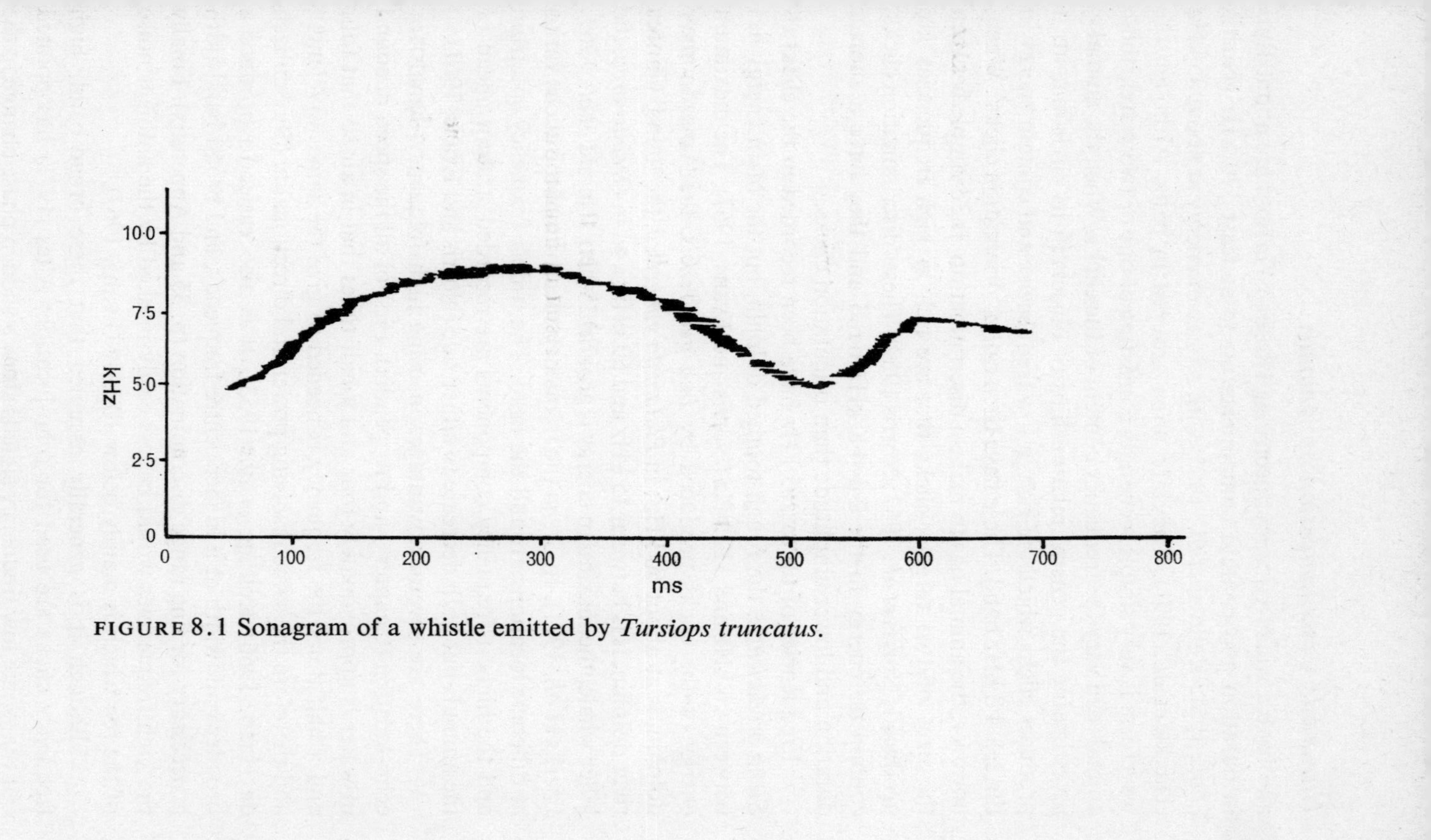

FIGURE 8.1 Sonagram of a whistle emitted by *Tursiops truncatus*.

and have a wide frequency spectrum. Those of *Tursiops* are about 1 ms in duration and contain components up to at least 170 kHz (Norris *et al.*, 1961; Kellogg *et al.*, 1953) with a peak of energy at about 35 kHz (Dierks *et al.*, 1971). They are often grouped in pairs. Although the waveform is very simple, generally consisting of one or two waves, these animals can vary the frequency content of the clicks. When the animal is investigating the general relationships of objects in its environment, it produces clicks which are rich in low frequencies and contain energy in the 2·4–4·8 kHz band. These have been called 'orientation clicks'. When, however, the animal has to make a discrimination, for example between the sizes of two targets, clicks that are rich in high frequencies are produced (Norris *et al.*, 1967; Norris 1969). These 'discrimination clicks' contain no energy in the 2·4–4·8 kHz band and they have a shorter duration and lower amplitude than orientation clicks.

Frequencies of up to 265 kHz have been recorded in the clicks of *Steno bredanensis*, the rough-toothed dolphin, but the main energy lies between 16 kHz and 112 kHz (Norris and Evans, 1967). The maximum energy of the clicks produced by *Inia geoffrensis*, the Amazon river dolphin, is at about 60 kHz; in *Platanista gangetica*, the 'blind' Ganges river dolphin, it is between 15 kHz and 60 kHz; and in *Orcinus orca*, the killer whale, the maximum energy is around 25 kHz (Herald *et al.*, 1969; Dierks *et al.*, 1971). In *Steno* the frequency spectrum of the clicks varies at different locations around the head. The widest frequency spectrum and the highest frequency components are recorded straight ahead of the animals but both decrease to either side (Norris and Evans, 1967).

There are several advantages in using high frequency signals for echo-location (Chapter 3 and Appendix). First of all the speed of sound in water is approximately four and a half times that in air so that four and a half times the frequency is needed to give the same wavelength and therefore the same resolving power and directionality (Reysenbach de Haan, 1966; Kellogg *et al.*, 1953). It is also easier to produce a broader bandwidth at a higher centre frequency, and broad bandwidth is necessary for fine range discrimination (p. 35 and Appendix). Finally the high frequencies would be less easily masked by the ambient noise of the sea which is mainly below 10 kHz (Evans, 1967).

Although it is generally assumed that these broad-band, high frequency clicks are used for echo-location, Altes (1971) has pointed out that some low frequency signals show a sudden phase change in the middle of the pulse and he suggested that these would also form ideal sonar signals.

Some other odontocete whales produce low frequency sounds in which the maximum energy is below about 12 kHz. These signals are often up to 25 ms in duration and therefore are of much longer duration than the broad-band clicks mentioned above. An echo-location function for these signals has been suggested in *Physeter catodon*, the sperm whale (Worthington and Schevill, 1957), *Orcinus orca*, the killer whale (Schevill and Watkins, 1966) and in *Pseudorca crassidens*, the false killer whale (Busnel and Dziedzic, 1968). Echo-location by low frequency signals has been reported in *Phocaena phocaena*, the common porpoise (Busnel *et al.*, 1965a, b; see p. 217).

The site of sound production

There is much controversy over the site of production of the different types of sounds. *Tursiops* can produce whistles and clicks simultaneously (Lilly and Miller, 1961a; Evans and Prescott, 1962) so it is likely that there are at least two different sources of sound in this group. In dolphins, echo-location signals appear to be produced at some internal site. No air bubbles can be seen when the sounds are emitted so it appears that the air must be recycled if it is used in sound production (Norris, 1969). The larynx of dolphins has no vocal cords although noises can be produced by blowing air through it. Evans and Prescott (1962) found that when air was forced through the excised larynx of *Stenella graffmani*, the spotted dolphin, barks or whistles were produced depending on the air pressure. When the air was forced through the nasal sac system which leads to the blowhole, 'echo-location clicks' were produced. It perhaps should be pointed out that in delphinids and most other odontocetes, the larynx is elongated and is held in the palate by a sphincter muscle so that the respiratory and food passages are completely separated as in many bats and other mammals (Negus, 1962). Above the anterior end of the larynx a tube leads to the surface of the skull and opens as the blowhole. Lawrence and Schevill (1956) believed that all of the underwater calls of dolphins are produced by the larynx and this site has also been suggested for the origin of the echo-location calls in particular, by Purves (1967). However Norris (1969) and Norris and Prescott (1961) suggested that the muscular apparatus of the blowhole may be responsible for the production of the echo-location clicks. Other sites of sound production within the nasal–

laryngeal complex have been proposed (see Norris, 1969) but the question has not yet been resolved.

Norris (1969) has suggested that the sounds are directed forwards by the melon, the large cushion of fatty tissue that lies on the forehead. This hypothesis is supported by the fact that the radiated sound fields are sharply directional. In *Tursiops* the most intense sounds are recorded when the head points slightly downwards (Fig. 8.2) so that the melon faces the recording hydrophone (Norris *et al.*, 1961; Norris and Evans, 1967). This could account for the 'scanning' movements of the head

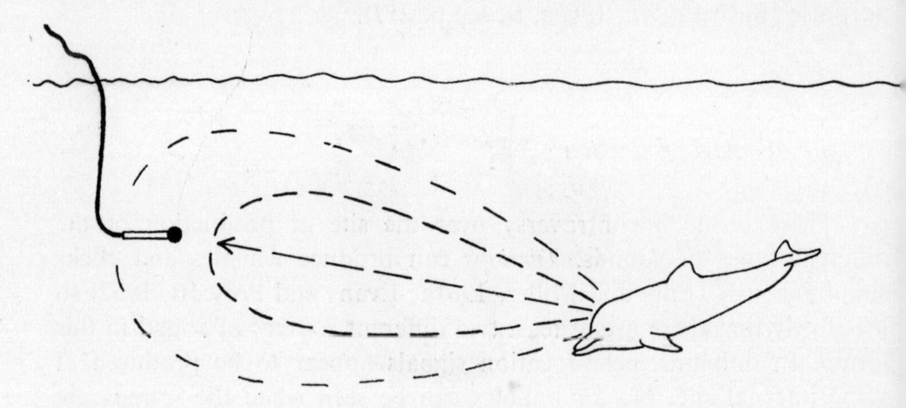

FIGURE 8.2 Diagram of the directional pattern of sound emitted by a dolphin as indicated by hydrophone measurements and by various studies of behaviour

that are often seen when a dolphin approaches a target (Kellogg, 1959b; Norris *et al.*, 1961). It is interesting to note that the skull of the 'blind' river dolphin, *Platanista*, is unique in possessing broad maxillary flanges that extend upwards into the melon. The insides of these flanges are web-like and they may act as sound baffles to direct the sound forward and help in forming the beam that these animals produce (Herald *et al.*, 1969).

The ear of odontocetes

Hearing underwater presents problems different to those of hearing in air, and the mammalian ear is therefore modified in aquatic forms. One of the problems is that the body has approximately the same density, and so has the same acoustic impedance, as water. This means that sound waves can travel freely through the tissues and reach the inner ear from all directions. There are therefore no external ears in whales and the head can be streamlined.

The external auditory meatus or ear canal is very narrow, 0·5–5 mm in diameter, and it leads to a thickened, cone-shaped tympanic membrane, the tympanic conus, which projects into the tympanic cavity. The tympanic bulla and the periotic bones, which surround the middle ear, are relatively massive with thick, very dense walls, but they are

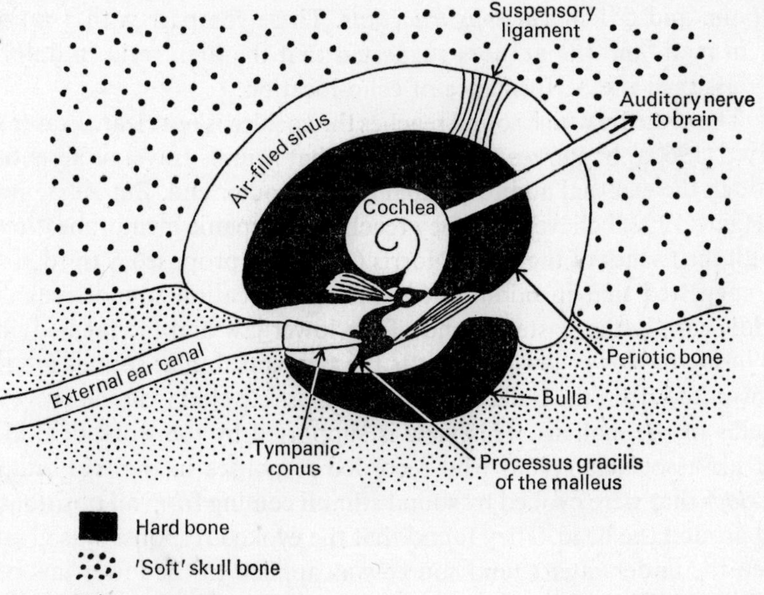

FIGURE 8.3 Stylized diagram of a frontal view of the right ear of an odontocete cetacean. (Somewhat modified after Reysenbach de Haan, 1958)

connected to the rest of the skull only by a thin ligament (Fig. 8.3). This provides some degree of acoustic isolation from the skull. Further isolation is achieved by cavities which surround the middle ear and are filled with an albuminous foam of air bubbles. These air sinuses are extensions of the middle-ear cavity to which they remain connected (Fraser and Purves, 1954, 1960a; Reysenbach de Haan, 1958).

The auditory ossicles of the middle ear, the malleus, the incus and the stapes, are all relatively thick and heavy and their suspension is much more rigid than in any other vertebrate group (Fraser and Purves, 1960a, b; Reysenbach de Haan, 1958; Giraud–Sauveur, 1969). The malleus is attached to the bulla by a stiff bony strip, as it is in bats and mice, and at the other end of the chain of ossicles, the stapes is very firmly set in the oval window of the inner ear.

The cochlea consists of 1·75–2 turns with the basal turn almost completely enclosing the following one. Reysenbach de Haan (1958) has studied the histology of the cochlea in detail and he found that in certain features it resembles that of bats. Wever and his colleagues (1972) have shown that the ratio of the number of ganglion cells to sensory hair cells is 4:1 in *Lagenorhyncus obliquidens*, the Pacific white-sided dolphin, and 5:1 in *Tursiops truncatus*. These compare with a ratio of 2:1 in man, and the authors suggested that the high ratio in dolphins may assist in the performance of echo-location.

The route by which sound reaches the cochlea is not clear. Fraser and Purves (1960a, b; Purves, 1966) believe that sounds travel preferentially through the external auditory canal from its outer end. But Reysenbach de Haan (1958) believes that they reach the tympanic membrane directly via all the tissues of the head. Norris (1969) has proposed a third route. He suggested that in odontocetes, the echo-location sounds reach the middle ear via the posterior end of the lower jaw and the fat body that lies inside the lower jaw and abutts the tympanic bulla; sounds are then transmitted from the bulla directly to the ossicle chain via the processus gracilis of the malleus. This view is supported by the work of Bullock and his associates (1968) who recorded responses in the mid-brain of *Tursiops* that were evoked by sound stimuli coming from all positions on and around the head. They found that the evoked response was greatest when the underwater sound source was applied to the lower jaw or to the melon. When a distant sound source was used, the evoked response decreased if an acoustic shield was placed over the lower jaw. McCormick and his colleagues (1970) also supported a bone conduction theory and they found that, under experimental conditions, cochlear responses to sound were not affected by removal of the tympanic membrane, which they therefore suggested was unnecessary for sound transmission to the inner ear. But removal of the malleus itself only reduced sound transmission slightly in these experiments.

If the head is transparent to sound, the only clue to the directionality of a signal is the difference in the time of arrival at the effective sound receivers on the two sides, and for a given direction this depends on the distance between the receivers. Dudok van Heel (1959, 1962) used an operant training technique to determine the minimum angle between two signal sources that could be detected by *Phocaena phocaena*, the common porpoise. He found that for tones of 3·5 kHz the minimum detectable angle was 22°, while for 6 kHz the angle was reduced to 16°. The performance of man at similar wavelengths in air, was about twice as good as this, suggesting that the effective interaural distance was half that of man, or 10 cm. This corresponds to the actual distance between the lateral sides of the bullae in *Phocaena*. Dudok van Heel therefore supported Reysenbach de Haan's view that the effective distance between the ears is the distance between the bullae, rather than that between the two external auditory openings as suggested by Fraser and Purves. It must be admitted, however, that the position is complex and Mills (1963) for example, has published rather different figures for man.

Hearing in odontocetes

The ability of odontocetes to hear high frequency sounds has been established by experimental studies, most of which have used behavioural responses to determine the animals' ability to hear. Kellogg and his associates noted that two species of dolphins, *Tursiops truncatus* and *Stenella plagiodon*, responded to tones up to 80 kHz by a suddenly accelerated swimming movement or by jumping from the water (Kellogg, 1953; Kellogg and Kohler, 1952; Morris *et al.*, 1953). Schevill and Lawrence (1953a, b) trained a female *Tursiops* to respond consistently to signals from 0·15 kHz to 120 kHz by collecting a food reward. The number of correct responses decreased with higher frequencies up to 153 kHz, the upper limit of the apparatus, but the authors suggested that this probably does not represent the upper limit of hearing.

A detailed study of the acoustic sensitivity of *Tursiops* was carried out by Johnson (1966, 1967) who trained a male dolphin to respond to a pure tone by pressing a lever to obtain a reward. By varying the sound intensity, Johnson attempted to determine the auditory response threshold at different frequencies. As Johnson pointed out, this experiment only determined the sound level at which the animal was willing to respond. Nevertheless the results were remarkably constant over several days and are probably near to the true threshold. The animal responded over the frequency range 0·075–150 kHz and was most sensitive to frequencies around 50 kHz (Fig 8.4). Johnson concluded that the effective upper limit of hearing in dolphins is 150 kHz.

Jacobs (1972), using a similar operant training technique, showed that *Tursiops* is most sensitive to differences in frequency within the range 2–20 kHz. Dolphins therefore differ from man in which the region of maximum frequency discrimination coincides with that of maximum sensitivity. Jacobs pointed out that the region of maximum frequency discrimination in dolphins is the same as that of the whistle signals that are presumed to be used for intra-group communication. The maximum sensitivity at higher frequencies is presumably associated with the reception of high frequency echoes.

Operant training techniques have also been used to determine the auditory thresholds in other species of odontocetes. Jacobs and Hall

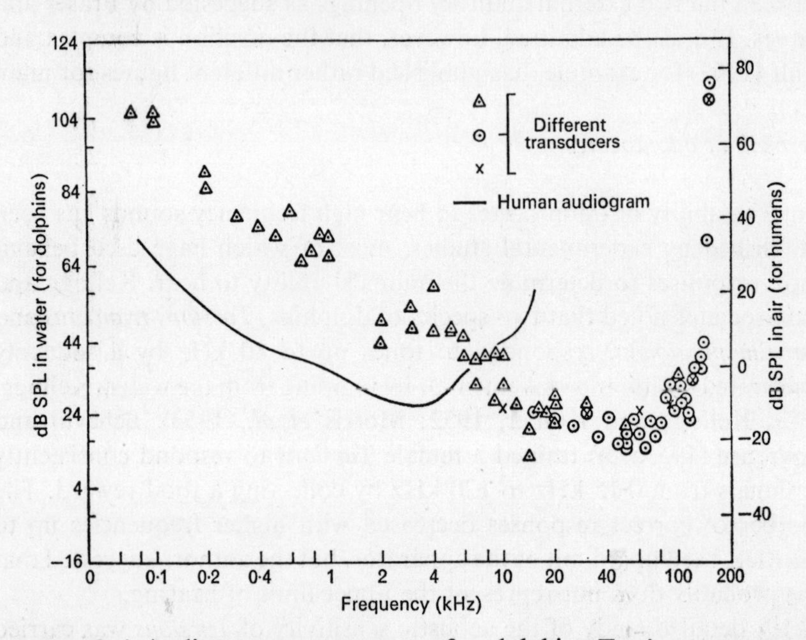

FIGURE 8.4 Auditory response curve of a dolphin, *Tursiops truncatus* as established by behavioural methods. The three symbols represent stimuli delivered by three different transducers. The line shows a human auditory response curve in air for comparison. (After Johnson, 1966)

(1972) found that *Inia geoffrensis* responded to tones of 1–105 kHz and was most sensitive to tones of 79–90 kHz. But *Orcinus orca* responded only to tones up to 31 kHz and showed maximum sensitivity to tones of around 15 kHz (Hall and Johnson, 1972). It has been claimed that *Delphinus delphis*, the common dolphin, responded to tones up to 280 kHz (Bel'kovitch and Solntseva, 1970).

There have been very few physiological studies of hearing in odontocetes, but those that have been carried out confirm the behavioural evidence that these animals are sensitive to high frequencies. Bullock and his associates (1968) recorded the responses of the inferior colliculus of four species of dolphins to pure tones up to 150 kHz in air and in water. They found that the underwater auditory sensitivity curves of all four species, *Stenella caerulea*, *Stenella attenuata*, *Steno bredanensis* and *Tursiops gilli*, were in close agreement with that obtained behaviourally for *Tursiops truncatus* by Johnson (1966). The acoustic sensitivity of these species was high to frequencies between 20 kHz and 70 kHz and was at a maximum around 60 kHz. The highest frequency that elicited a response was between 120 kHz and 140 kHz. After one stimulus, the acoustic responsiveness recovered very rapidly, within 1–5 ms, so that the animals could obviously detect and resolve very rapid trains of clicks.

Bullock and Ridgway (1972) have shown that the inferior colliculus of *Tursiops truncatus* is specialized for analysing very brief ultrasonic sounds that are closely spaced and have very fast rise times, like the echo-location clicks. By implanting chronic recording electrodes into the inferior colliculus, Bullock and Ridgway were able to study the auditory evoked responses to the animals' own echo-location sounds. Surprisingly perhaps, the highest intensity clicks often resulted in fairly modest potentials, while weaker clicks often gave maximum potentials. The authors suggested that differences in the form of the clicks may be important here. With artificial stimuli, Bullock and Ridgway found that no potentials could be evoked in the inferior colliculus if the amplitude of the stimulus signal increased slowly over more than 5 ms or if the frequency was below 5 kHz. In the cerebral cortex, however, long duration, slowly recovering potentials were evoked by signals from less than 1 kHz up to 10–15 kHz. This region did not respond to echo-location clicks and the most effective frequency for eliciting cerebral potentials was 5 kHz. Thus on both physiological and behavioural evidence it seems that dolphins have 'two different auditory systems', one for intraspecific communication and one for echo-location.

Evidence for echo-location in odontocetes

Perhaps the first person to suspect that dolphins and porpoises use sound for navigation was McBride, a curator at the Marine-land Aquarium in Florida. In July 1947 McBride wrote in his notebook: 'In view of the enormous development of the cerebral cortex of porpoises and of the obvious importance of the acoustic sense, might we not suspect . . . a specialised mechanism of information by sound?' But McBride never published his idea and it was only made known after his death by Schevill (McBride, 1956). The first experimental evidence of echo-location was produced in 1956 by Schevill and Lawrence. They trained a dolphin, that was blind in one eye, to collect a fish from an experimenter at a variable location when a tone was sounded. The animal was tested in turbid water and in darkness. Schevill and Lawrence reported that it could find the fish successfully when it produced clicks, but that it also relied to some extent on vision as well as echo-location: often the fish was unnoticed if it was placed on the side of the animal's blind eye.

Kellogg (1958, 1959a, b) produced much more detailed evidence for echo-location which is summarized in his book 'Porpoises and Sonar' (1961). He found that even in turbid water at night, *Tursiops* could distinguish, at a distance, between a 30-cm-long mullet, a fish that the animals always rejected, and a 15-cm-long spot, a fish that they would eat. Kellogg showed that this discrimination was based on the size of the fish; when both fish were held in the water so that only 15 cm of the mullet projected beneath a screen, the dolphin could not discriminate between them so well.

As a further test, Kellogg suspended 36 metal poles in rows 2·4 m apart in the dolphins' pool. The poles made a 'ringing' sound when they were hit. Two animals swam through this 'maze' for 20 min in darkness on six different occasions. Four poles were touched during the first session, three in the second and none in the last four sessions. The animals produced click sounds continuously while swimming and they were able to avoid the poles even when tape-recordings of their own sounds were replayed to them in an attempt to 'mask' the real echoes. On another occasion Kellogg placed a wire-mesh barrier across the dolphins' pool. There were two openings in the barrier, one of which was covered by perspex (plexiglass). The perspex was moved to the centre of the barrier between trials and then over either opening at random. On each trial the animals were induced to swim through the open gap in the barrier from one end of the pool to the other. The animals chose the 'closed' gap in only two out of 100 trials.

Indirect evidence for the use of echo-location by dolphins comes from the observations of Herald and his associates (1969) on *Platanista gangetica*, the 'blind' river dolphin which lives in muddy rivers in India and Pakistan. This dolphin has very degenerate eyes and correspondingly poor vision. Nevertheless it is very adept at avoiding obstacles and produces trains of clicks all the time it is swimming. This contrasts strongly with *Inia geoffrensis*, a South American freshwater dolphin, and with the marine dolphins studied, all of which produce click trains only sporadically unless their 'attention' is focused on a particular object. The continuous sound production of *Platanista*, together with the degeneration of the eyes, the ability to avoid obstacles and life in a murky habitat, all strongly suggest the use of echo-location (Herald *et al.*, 1969).

The first experiments that excluded all possibility of vision being used in echo-location tests were carried out by Norris and his co-workers in 1961. They trained a female *Tursiops truncatus* to wear rubber suction-cups over its eyes. Even when blindfolded in this way, the dolphin could avoid obstacles and could distinguish between small pieces of fish and water-filled capsules of the same size. It could also locate a variably placed target and press this with its snout to obtain a reward. Busnel and his associates (1963, 1965a, b; Busnel and Dziedzic, 1967) also used this method of blindfolding to demonstrate the echo-location ability of *Phocaena phocaena*, the common porpoise, which apparently produces low frequency clicks of about 2 kHz. A blindfolded animal was made to negotiate a maze of nylon, iron, copper or steel wires. It was able to avoid nylon threads of 1 mm diameter and metal wires as small as 0·2 mm diameter.

Norris and his colleagues (1961) attempted to interfere with the sound-emitting and receiving mechanisms of the same dolphin that they had blindfolded, but covering the external auditory openings with rubber suction-cups (Plate XIIId) had no obvious effects on the animal's behaviour. It was suggested that either the cups might not have blocked the openings effectively or that sound was conducted by adjacent blubber and muscle, as discussed earlier. A second attempt was made by using a mask which covered the upper jaw and forehead region. Even after two months of intensive training, however, the animal refused to wear the mask for more than a few seconds.

Later Evans, Norris and Turner tested the discrimination ability of another blindfolded dolphin (Turner and Norris, 1966; Norris *et al.*, 1967). Two steel spheres 12·7 cm apart and two levers corresponding in position to the two spheres were placed in the animal's pool. One of the spheres was a 'standard' one of 6·3 cm diameter, the other one was of variable diameter. The dolphin was rewarded only for pressing the lever corresponding to the standard sphere and it was able to distinguish, at a distance, between a 5-cm sphere and the standard one (6·3 cm) with 100% accuracy. The number of errors increased markedly when the difference in diameter between the spheres was reduced to 9 mm. Evans and Powell (1967) demonstrated that a blindfolded *Tursiops* could distinguish between targets of the same area but made of different metals. A copper plate 0·22 cm thick was used as the standard target and the dolphin was able to differentiate perfectly between this and targets of aluminium 0·32 cm thick and of brass 0·64 cm thick. These results showed that at least under certain circumstances the animal can detect differences between various materials and so should be able to obtain detailed information about the nature of surrounding objects and of potential food.

Evans and Powell (1967) also recorded the echo-location clicks of the dolphin as it approached the response levers and they found that, as in bats, the wanted echoes almost always fell within the interclick interval. They suggested that the dolphin was using the time interval between the emitted click and the reflected echo to determine the distance and the rate of approach to the object. This would anyway be expected, by radar theory, from the broad bandwidth nature of the emitted signals (p. 35 and Appendix). Whatever the mechanism used for echo-location, dolphins are obviously capable of accurate navigation and of making very fine discriminations. It has been suggested (Chapman, 1968) that dolphins may be able to construct three dimensional images of the objects that they inspect by echo-location and they thus may be able to 'see' objects acoustically, although their acuity is limited by the wavelength to about 1 cm.

OTHER MARINE MAMMALS

An echo-location function has been claimed for the low frequency clicks produced underwater by *Zalophus californianus*, the California sea-lion, a member of a group of carnivores called the Pinnipedia (Poulter, 1963, 1966, 1967; Poulter and Jennings, 1969; Shaver and Poulter, 1967, 1968). Poulter and his associates recorded these signals using equipment that had a flat response to 100 kHz. They reported that the clicks were double pulses up to 1 ms in duration and with the main energy between 2 kHz and 13 kHz. The sea-lions were able to locate food readily in the dark. They could find a dead fish that was wrapped in an odour-proof bag, and distinguish between horse meat, beef and fish at a distance. The animals produced clicks during these tests and Poulter suggested that they were actively echo-locating. He has also suggested that *Callorhinus ursinus*, fur seals, and *Spheniscus humboldti*, Humboldt's penguins, use echo-location (Poulter, 1969).

Other workers, however, have been unable to confirm these claims. Schevill, Watkins and Ray (1963) recorded the clicks of *Zalophus* with equipment that had a flat response to 10 kHz but was 'usable' up to 30 kHz. These authors could not find the double pulse structure and they reported that the main energy of the clicks was between 0·6 kHz and 2 kHz. The possibility of active echo-location in the California sea-lion and also the role of vision in discrimination have been studied in some detail (Evans and Haugen, 1963; Schusterman, 1966, 1967; Schusterman *et al.*, 1965; Schusterman and Feinstein, 1965). Schusterman (1967) in particular has made some very extensive investigations. None of the results of these other workers support Poulter's conclusions. They show that sea-lions have good visual acuity underwater and rely mainly on vision for discrimination purposes. Although the clicks may be produced when the animal is roused (Schusterman, 1966), they are probably not used for echo-location (Evans, 1967; Schevill, 1968).

Schevill, Watkins and Ray (1963) recorded the sounds produced by five species of pinnipedes including *Phoca vitulina*, the harbour seal, and *Phoca* (*Pagophilus*) *groenlandia*, the harp seal, while the animals searched for food underwater. The main energy produced by these five species was near or below 12 kHz and only one of them, *Halichoerus grypus*, the grey seal, produced dominant frequencies in the ultrasonic range, up to 30 kHz. These observations have apparently not been repeated at higher frequencies even though it has now been shown that some of these species are sensitive to sounds above 30 kHz (see below).

Schusterman and Feinstein (1965) reported that *Zalophus* could readily be trained to click in response to a conditioning stimulus. This fact was later used by Schusterman, Balliet and Nixon (1972) to obtain a behavioural audiogram for this species. A sea-lion was trained to produce clicks when a light and a tone were presented simultaneously. Thirteen different tones between 250 Hz and 64 kHz were used and the animal responded to signals up to 60 kHz, but it was most sensitive to frequencies between 1 kHz and 28 kHz and had a peak of sensitivity at 16 kHz.

Møhl (1968) also used an operant conditioning technique to determine the auditory threshold of *Phoca vitulina*, the common seal, in both air and in water. In air, tones of up to 22·5 kHz were used and the animal showed two peaks of sensitivity, one at 2 kHz and a second at 12 kHz which corresponds to the peak energy of the calls of this species as reported by Schevill, Watkins and Ray (1963). Above 12 kHz the response thresholds increased sharply. In water, the auditory sensitivity of the common seal was much greater. It responded to signals up to 180 kHz and had a peak of sensitivity at 32 kHz. Terhune and Ronald (1971) reported that in air *Pagophilus groenlandicus*, the harp seal, responded to signals up to 32 kHz and its greatest sensitivity was at 4 kHz. In water, there were indications that this species could hear signals up to 100 kHz.

Bullock, Ridgway and Suga (1971) have studied the auditory evoked responses of the mid-brain in *Zalophus californianus* and in *Phoca vitulina*, both in air and in water. Both species responded to signals up to 30–35 kHz and were most sensitive to frequencies of 4–6 kHz. The authors found that, unlike that of porpoises, the inferior colliculus did respond well to sounds of long duration and with a slow rise time, and that the directionality of the response was weak. They concluded that the mid-brain of pinnipedes is apparently not specialized for the analysis of brief, sharp, high frequency clicks and echoes. Thus, although pinnipedes are capable of hearing and to some extent of producing ultrasonic frequencies, at present it appears that they do not use them for echo-location, but that for orientation and discrimination purposes they rely primarily on vision.

INSECTIVORA

Another group of mammals in which echo-location has been reported is the order Insectivora which includes shrews (Plate XIV), hedgehogs, moles and tenrecs.

Purely ultrasonic signals have been recorded from three species of shrews of the genus *Sorex* by Gould, Negus and Novick (1964) who suggested that they are used for echo-location. The authors trained shrews to jump from a circular platform to a smaller one placed at variable positions around its periphery and 10 cm or more below. From here the shrews could obtain food and water. Very low intensity sounds, 0·02–0·14 μbar in 'amplitude', 4·9–33 ms in duration and at frequencies of 30–60 kHz, were detected as the animals searched the rim of the disc in complete darkness. The shrews were able to find the lower platform successfully except when their ears were plugged; then only two out of seven animals were successful at a single trial. Ultrasonic pulses 1·5–20 ms in duration and at 29–55 kHz were also detected from another species *Blarina brevicauda* when it searched unfamiliar surroundings. It was suggested that these findings support the hypothesis that shrews use echo-location. The suggestion has recently received support from Buchler (personal communication) who has made a similar but more detailed study of echo-location in *Sorex vagrans* and obtained similar results.

Grünwald (1969), however, studying a different genus of shrews, *Crocidura*, came to a different conclusion. He investigated the respective roles of vision, olfaction, audition and tactile senses in the orientation of two species, *Crocidura russula* and *C. oliviera*. Grünwald recorded some purely ultrasonic signals of 32–34 kHz from *C. russula* but he concluded that although the auditory sense is well developed in both of the species he studied, neither of them used echo-location.

Click-like pulses have been recorded from *Crocidura russula judaica*, Israeli shrews (Pye, unpublished). These clicks had a frequency spread from 10 kHz to 70–110 kHz and they were produced, either singly or in bursts of 8–25 with interclick intervals as short as 1 ms, when the shrews first investigated the recording microphone. A variety of sounds at frequencies between 5 kHz and at least 55 kHz were recorded (Sales, unpublished) from an Israeli shrew and from *Neomys fodiens*, a water shrew, when these animals jumped from a platform 5–10 cm above the floor of their cage. These signals were produced before and after the animals landed and they included purely ultrasonic signals with a single component at 30–55 kHz. The significance of these calls is not known, but no ultrasounds were detected when the animals were on the platform so that an echo-location function seems unlikely.

Hedgehogs are a group of insectivores that are known to respond to high frequency sounds by jerking their bodies or by turning their heads (Chang, 1936; Poduschka, 1968a). Chang used this response to study the auditory sensitivity of hedgehogs to frequencies of up to 84 kHz produced by a Galton high frequency whistle. He found that the animals responded to tones between 7 kHz and 84 kHz but that the optimal stimulus for eliciting a response was 20 kHz (Chang, 1936).

Poduschka (1968a) reported that sounds such as clicks normally elicited the turning response in *Erinaceus europaeus roumanicus*, a European hedgehog, but failed to do so when tape-recordings of the same sounds were replayed to the animals. As the recording equipment had an upper frequency limit of 24 kHz, Poduschka concluded that ultrasonic components above this limit might be responsible for eliciting the response. He also suggested (1968b) that European hedgehogs can produce sounds with ultrasonic components.

The anatomy of the middle ear of some insectivores, including

Erinaceus europaeus, has been studied by Henson (1961) who concluded that, although hedgehogs may be able to hear sounds approaching the ultrasonic range, it was unlikely that they would have any special acuity or use for such sounds. The auditory sensitivity of a different species, *Hemiechinus auritus*, has been investigated using a conditioned response technique (Ravizza *et al.*, 1969b). The three animals studied responded to tones up to 40 kHz but they were most sensitive to frequencies of 8 kHz.

Echo-location using audible frequencies and communication using ultrasonic frequencies have been described in another group of insectivores, the tenrecs. Tenrecs are closely related to both hedgehogs and shrews and they are found on the island of Madagascar. Many species are tail-less and are covered with quills or a mixture of hair and quills. The animals produce a wide variety of sounds: squeals, sniffs, twitters and also tongue-clicks which Novick and Gould (1964; Gould, 1965) suggested are used for echo-location. The tongue clicks are 0·1–2 ms in duration and the main energy is between 9 kHz and 17 kHz.

Novick and Gould trained tenrecs to jump from a platform to a smaller one 11 cm or 14 cm below to obtain a food reward. Clicks were detected as the animals searched the rim of the platform. The animals were able to find the lower platform successfully at night, or when blindfolded or when open plastic tubes were placed in their ears. But when the tubes were plugged with cotton, only two out of six animals landed on the lower platform at a single trial.

One of the tenrecs studied by Novick and Gould, *Hemicentetes*, also produced sounds by a special stridulatory organ. This consists of three rows of modified dorsal spines which vibrate against each other when the animal is excited and produce wide-band sounds with a frequency spectrum of 10–70 kHz, and with most of the energy in the ultrasonic range. Novick and Gould suggested that these sounds are not used for echo-location but rather for communication within a group. *Hemicentetes* live exclusively on earthworms and they hunt in groups of about 20 animals. When one animal finds a worm, the others rush towards it and although the tenrec with the worm flees, the others remain in the same general area. The disadvantage of a highly specialized diet may therefore be offset by the development and use of this stridulatory organ to communicate with other members of a hunting group (Novick and Gould, 1964; Gould, 1965).

The auditory sensitivity of *Hemicentetes* was studied by Wever and Herman (1968). They measured electrical responses from the round window of the cochlea to tones of 2–50 kHz. Cochlear potentials were obtained over the whole range but maximum sensitivity was at 10–16 kHz. Thus these tenrecs appear to be most sensitive to the frequencies of their own tongue clicks and the sounds produced by the stridulatory organ are at the upper end of their audible range.

Broad-band clicks have also been recorded from *Solenodon paradoxus*, a close relative of the tenrec that is found only on the island of Hispaniola (Eisenberg and Gould, 1966). These clicks were 0·1–3·6 ms in duration and the main energy was at 9–31 kHz. They were produced in bursts of 1–6 when the animals explored a new area or encountered a strange animal.

CHAPTER NINE

Review and Speculations

Previous chapters have shown that ultrasounds are produced by many different groups of animals and in a variety of situations. These high frequency sounds are used for auto-communication or echo-location in cetacea and bats and possibly other small mammals, and they are apparently important in social behaviour in rodents and insects. They are also used in interspecific communication between bats and those moths that are able to detect the ultrasonic cries of their predators and to take appropriate avoiding action or even to discourage them.

The different mechanisms of ultrasound production are almost as varied as the animals producing the sounds. In both bats and rodents the larynx is involved, but ultrasound production is achieved in different ways: by the vibration of vocal folds in bats but probably by a whistle mechanism in rodents. The larynx may also be important in cetacea, but here other sites of sound production have also been implicated and the the picture is by no means clear. Insects use very different mechanisms of sound production called stridulation, which is the mechanical vibration of parts of the exoskeleton. This may either be by strigilation, the friction of two hard parts one against the other as in grasshoppers and crickets, or by the alternate buckling and relaxation of blister-like tymbal organs as in some moths and cicadas. In many insects, the production of ultrasound is coincidental with the production of lower frequency sounds.

The ability to hear ultrasounds is also widespread. In mammals this has been considered to be a primitive feature (Masterton *et al.*, 1969). Apart from the groups discussed earlier, auditory sensitivity to high frequencies has been reported in a wide variety of other mammals, including a marsupial (Ravizza *et al.*, 1969a), three edentates (Suga,

1967b), hystricomorph rodents (Pestalozza and Davis, 1956; Peterson, 1968; Miller, 1970; Heffner *et al.*, 1971), a variety of primates (e.g. Wever and Vernon, 1961c; Stebbins, *et al.*, 1966; Heffner *et al.*, 1969a, b; Heffner and Masterton, 1970; Mitchell *et al.*, 1970, 1971) and a variety of fissipede carnivores (e.g. Wever *et al.*, 1958; Peterson *et al.*, 1969). The extensive literature has been reviewed by Brown and Pye (1974). Indeed man appears to be unusual in having such a low upper frequency limit (Manley, 1971) (Fig. 9.1). Some structural adaptations of the

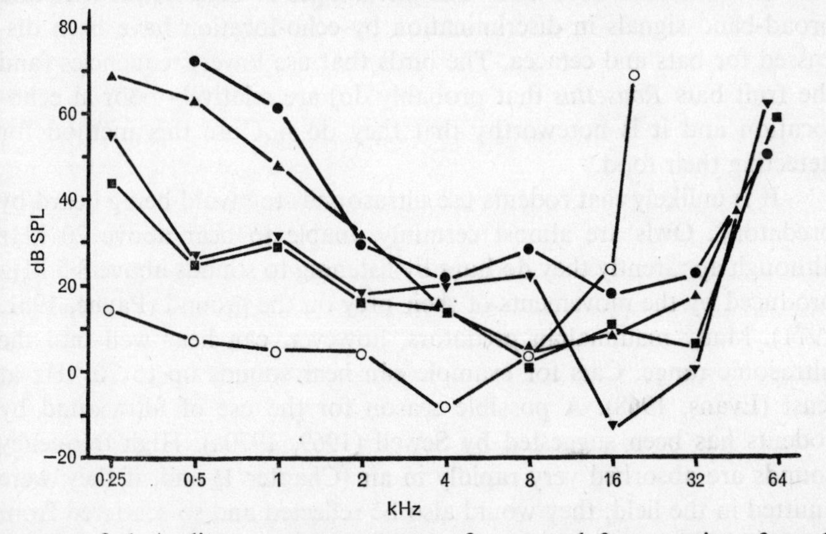

FIGURE 9.1 Auditory response curves of man and four species of small mammals obtained by behavioural methods. ○ man, *Homo sapiens*. ■ bush baby, *Galago senegalensis*. ▼ tree shrew, *Tupaia glis*. ▲ hedgehog, *Hemiechinus auritus*. ● opossum, *Didelphis virginiana*. (Redrawn after Masterton *et al.*, 1969)

middle and inner ears are seen in bats and in cetacea, but their functional significance is mostly obscure; other mammals sensitive to high frequencies appear to have no special modifications.

In the insects a wide range of hearing organs has developed in different parts of the body. This variation is clearly seen in moths and it indicates that rather than hearing being a primitive phenomenon in these insects, auditory sensitivity to high frequencies probably arose relatively late in their evolution, when predatory bats appeared.

The widespread occurrence of ultrasound among animals inevitably raises the problem of why so many animals use these high frequencies rather than lower ones. There is no simple answer to this question. In bats, the production of ultrasound could be a function of their size. Man can produce fundamental frequencies from about 66 Hz to 1056 Hz; bats are much smaller and a smaller vocal apparatus would be expected to produce higher frequencies. Cetacea, which are generally much larger than man, can produce very high frequencies, but like the small rodents their mechanism of sound production is probably different from that of man. Bats, rodents and cetacea also produce lower frequency sounds and therefore they have a much wider bandwidth available to them than does man. The advantages of high frequencies and broad-band signals in discrimination by echo-location have been discussed for bats and cetacea. The birds that use lower frequencies (and the fruit bats *Rousettus* that probably do) are relatively poor at echo-location and it is noteworthy that they do not use this method for detecting their food.

It is unlikely that rodents use ultrasounds to 'avoid being heard by predators'. Owls are almost certainly unable to hear above 20 kHz although apparently they do hunt by listening to sounds above 8·5 kHz produced by the movements of their prey on the ground (Payne, 1961, 1971). Many mammalian predators, however, can hear well into the ultrasonic range. Cats for example can hear sounds up to 70 kHz at least (Evans, 1968). A possible reason for the use of ultrasound by rodents has been suggested by Sewell (1969, 1970a). High frequency sounds are absorbed very rapidly in air (Chapter 1) and, if they were emitted in the field, they would also be reflected and so scattered from small objects such as blades of grass because of their short wavelengths (pp. 7–8 and Appendix). So although ultrasonic signals could be located close to their source, further away they would be difficult both to detect and to locate. Ultrasounds could therefore be used during social en-encounters between animals in close proximity without unduly advertising their presence or position to predators.

It is also not clear why insects use ultrasounds. It is not known whether the ultrasonic clicks produced by moths represent an aposematic, warning signal or whether they are used to 'jam' the bats' 'radar' system. If the latter is true, the moths would obviously have to use the frequencies of the bats' echoes. For aposematic signals the agreement is not so important but the sounds would be more effective within the range of greatest sensitivity of the bats' ears.

In crickets and grasshoppers the use of high frequency sounds may help to increase the directionality of the signals but they may also help to prevent 'congestion' of the frequency bands used for communication.

Most insects produce broad-band signals and are unable to analyse the frequency of sound to any great degree. By using widely separated frequency bands and having auditory organs 'tuned' to their own band, different groups of insects might distinguish their own songs without being confused by those of other groups. Within a group of insects having a characteristic frequency spectrum, further specific recognition then appears to depend on the spike, syllable and possibly chirp repetition rates.

From the foregoing discussion on the production, detection and use of ultrasound in animals, it is possible to suggest other instances where ultrasounds may be produced and where they may be important in communication. For example it was pointed out that many species of small mammals can hear high frequencies. These animals have a similar respiratory mechanism to bats and rodents and it is possible that they too produce ultrasounds. Some marsupials are convergent with rodents in both size and habits, and their acoustic behaviour may well prove interesting. The marsupial desert mouse has an enlarged auditory bulla like that of all the diverse and unrelated placental desert rats, and similar convergence involving ultrasound may occur in other forms.

Within the rodents themselves, pure ultrasounds have so far been found only in two families of the one sub-order Myomorpha. Ultrasonic components have been reported for the audible sounds of the hystricomorph guinea-pigs (Anderson, 1954; Berryman, *in litt.*) but apart from this there appear to have been few studies using ultrasonic equipment on the sounds of other hystricomorph rodents, of sciuromorph rodents or indeed of other families of myomorph rodents. Here, as in the insectivores and the smaller primates, further investigations may well be rewarding.

Within the bats, two small and geographically restricted families of the Microchiroptera, the Furipteridae and the Myzopodidae, have not yet been examined for echo-location. Representatives of every subfamily within all the other families have been examined but the signals of one family, the Thyropteridae, have been observed briefly with only an oscilloscope (Griffin, 1971), and have not yet been tape-recorded. Even within the groups already recorded, there may be more surprises in the variety of frequency patterns of the signals, and far more cases of flexibility in signal structure are now suspected (Chapter 3).

The acoustic behaviour of megachiropteran bats should also repay further study. So far only a few genera have been examined and only *Rousettus* has been found to use echo-location. Many other fruit bats live in caves and one genus, *Notopteris*, a nectar-feeding bat from New Guinea, is said to roost in complete darkness. These bats are not known to produce clicks but their means of navigation is worth investigating critically. Two other genera, *Nyctimene* and *Paranyctimene*, both tubular nosed bats, have been reported to whistle in flight. Walker (1964) has suggested that this behaviour might represent echo-location but so far this has not been substantiated.

The discussion of stridulation by insects in Chapter 5 suggested that the production of ultrasound during audible stridulation was not surprising. Members of many other arthropod groups are known to produce sound by rubbing two parts of the exoskeleton together (Dumortier, 1963a) and one may therefore expect to find ultrasounds produced by groups such as the crustacea and myriapods. Ultrasound up to at least 35 kHz has been recorded from a variety of marine animals (Hashimoto and Maniwa, 1959) and in particular from the snapping shrimps *Crangon* and *Synalpheus* (Johnson *et al.*, 1947; Everest *et al.*, 1948). The noises of crustacea are well known to interfere with the sonar (man-made underwater echo-location) systems of ships and it has been suggested that the snapping shrimps may be responsible for several curious and unexplained sonar detections, including in 1942 a 'new-fangled gadget of the Japanese'!

Blisters with rows of striae resembling microtymbals have been reported in various positions on the wings of some moths of the families Agaristidae and Noctuidae (Dumortier (1963a) cites a number of papers by Edwards, 1889; Hampson, 1892; Jordan, 1921; Scharrer, 1931; Bourgogne, 1951; Viette, 1955; Hanneman, 1956). It has sometimes been suggested that these moths strigilate in flight by rubbing the ridged tarsus of the hind leg over the ribbed organ of the wing. But these 'blisters' are curious structures and since strigilation in this way during flight would seem to be very difficult, it is possible that they act instead like a tymbal mechanism. Conceivably changes in the angle of aerodynamic attack of the wings could initiate sound production by forcing the blisters to click in and out during each wing beat.

A ribbed crest has been described on the abdomen of the noctuid moth, *Arcte caerulea*, by Gravely (1915, cited by Dumortier, (1963a)), and this 'may have some connection with clicking noises made in flight' (Dumortier, 1963a). *Hylophila prasinana*, the scarce silver lines moth, has a tymbal organ at the back of the thorax and in 1872 White suggested that the squeaking sound of this moth is produced by 'rumpling', bending and unbending the membrane of this organ. There appears to have been no further study of sound production in either of these moths. Many insects have other file-like structures, apparently without being able to bring a plectrum to bear on them. The possibility that these represent tymbal arrays and sources of ultrasound should not perhaps be overlooked.

Some female grasshoppers are known to make strigilatory movements although no sounds are apparently produced (Ragge, 1955, 1965); the production of ultrasound is again a possibility. Ultrasounds of 20–22 kHz have been reported from domestic bees while in flight (Rose *et al.*, 1948) although the mechanism of sound production is not known.

This book has shown that ultrasound is now known to be produced by a wide variety of different animals in different situations and that the occurrence of ultrasound is probably more widespread than is realized at present. We are just beginning to appreciate the extent to which sound production and communication have been used by animals for millions of years. With the advances in techniques for detecting, recording and analysing ultrasonic signals, the field of ultrasonic research in animals is widening rapidly. But there are still very many unanswered questions about the groups of animals already being investigated and there are many more groups of animals that deserve to be studied. It is hoped that the curiosity of many readers will be stimulated by at least some of these problems and that they will join in this field of research, even if only in a small way. Naturalists, biologists, engineers, physicists and even mathematicians can all contribute to the subject and help to broaden the present knowledge and understanding of communication by animals at ultrasonic frequencies.

APPENDIX

Some formulae summarizing the rules of echo-location for a single emitted pulse

1. The radar equation

The strength of an echo from small targets varies inversely as the fourth power of the range. If we assume that the transmitter radiates equally in all directions, that the (small) target scatters its received energy equally in all directions and that the receiver is equally sensitive in all directions, then:

$$P_r = \frac{P_t \,.\, \sigma}{(4\pi)^2 \,.\, r^4}$$

where P_t = power transmitted; P_r = power received in the echo; r = range of the target; σ = effective cross-section area of the target.

If the transmitter concentrates the same energy, giving a gain of G_t in one direction, and the receiver is similarly sensitive in the same direction with gain G_r, then:

$$P_r = \frac{P_t \,.\, \sigma \,.\, G_t \,.\, G_r \,.\, \lambda^2}{(4\pi)^3 \,.\, r^4}$$

where λ = wavelength used.

Thus double the range needs sixteen times the transmitted power, or +12 dB. Note, however, that the minimum P_r and σ are not under the control of the echo-location system. The maximum range at which a given target can be detected depends on the echo-power (see *5* below) and on the sensitivity of the receiver. The latter depends in turn on the level of spurious 'noise' generated within the system. If the receiver bandwidth is 'matched' to that of the received signal, then the ratio of the energies of signal and noise is $\frac{2E}{N_0}$ where E is the total signal energy received and N_0 is the noise power per unit bandwidth (having the dimensions of energy). This dimensionless quantity $\frac{2E}{N_0}$ is a fundamental one, for noise also causes uncertainties about the nature of the echo, that is its time of arrival, its frequency and its direction.

This is especially true for weak echoes and these uncertainties limit the resolution of the system in measuring the range, velocity and direction of targets, as shown in the next three sections.

2. Uncertainty about time

The time of arrival of the echo defines the range of the target since:

$$r = \frac{c \, . \, t}{2}$$

where c = velocity of pulse propagation and t = time from transmission of pulse to reception of the echo.

Clearly t must be shorter than the time between transmitted pulses, otherwise it will not be clear which pulse produced a given echo. For shorter ranges the rate of pulse transmission can be increased.

But in noise it is difficult to measure the exact time of arrival of the echo and it has been shown that *time resolution and therefore range resolution depends on signal bandwidth:*

$$\delta_t = \frac{1}{\beta \, . \sqrt{\left(\frac{2E}{N_0}\right)}}$$

where δ_t = echo-delay increment that can just be discriminated and β = effective signal bandwidth.

or

$$\delta_r = \frac{c}{2\beta \, . \sqrt{\left(\frac{2E}{N_0}\right)}}$$

where δ_r = target range increment that can just be discriminated.

The minimum usable value of $\frac{2E}{N_0}$ is 1 and for a 'square' pulse of constant carrier frequency this relation then simplifies to:

$$\delta_r = \frac{c \, . \, \tau}{2} = \frac{\text{pulse length}}{2}$$

where τ = pulse duration, $\left(\text{since } \frac{1}{\tau} = \text{effective bandwidth}\right)$

A shorter pulse, having a wider bandwidth, gives better range resolution and a 'click' of very few waves satisfies this condition. But as the pulse duration is reduced, the pulse energy is also reduced unless the amplitude can be increased accordingly. An alternative method is to obtain the wide bandwidth by sweeping the frequency of a rather longer pulse of more moderate amplitude or of greater total energy. The range resolution then becomes:

$$\delta_r = \frac{c}{2\,.\,\Delta f}$$

where Δf = the extent of the frequency sweep.

Strictly Δf is the effective bandwidth only for rather slow sweeps where the product of total frequency variation and sweep duration is high. This appears to be assumed for radar theory and is effectively true for most bat signals, even the rapid sweeps of cruising *Myotis*. But the shorter interception pulses of *Myotis* may have a bandwidth that is wider than the extent of their sweep suggests and Cahlander (1967) has shown that they actually have an improved range discrimination.

3. Uncertainty about frequency

Any difference in frequency between the transmitted pulse and the received echo defines the relative velocity of the target since the Doppler shift is:

$$\Delta f' = f_t\,.\,\frac{2v}{c}$$

where $\Delta f'$ = difference in frequency between transmitted pulse and its echo; v = approach velocity of the target; f_t = transmitted frequency.

But in noise it is difficult to measure the echo-frequency accurately and it has been shown that *frequency resolution and therefore velocity resolution depends on pulse duration;*

$$\delta_f = \frac{1}{\tau\,.\,\sqrt{\left(\frac{2E}{N_0}\right)}}$$

where δ_f = frequency increment that can just be discriminated

or:

$$\delta_v = \frac{c}{2f_t\,.\,\tau\,.\,\sqrt{\left(\frac{2E}{N_0}\right)}}$$

where δ_v = velocity increment that can just be discriminated.

The velocity discrimination is therefore improved by increasing $f_t.\tau$ which is the number of cycles in each pulse. Half the frequency needs twice the duration for the same resolution. Clearly the best signal-to-noise ratio will be obtained if the signal has a narrow bandwidth so that the total noise can be restricted by filtering. Thus the optimum bandwidth requirements for velocity discrimination are directly opposed to those for range discrimination and a pulse of given energy can be designed to achieve one or the other or to effect a compromise between them.

4. Uncertainty about direction

The directionality of any echo-location system depends on the effective beam-widths of the transmitter and/or receiver and these depend on their apertures expressed in terms of the wavelength used. *Any given system has a greater directionality at higher frequencies* and echo-location is therefore more accurate at ultrasonic than at lower, audible frequencies. But again noise obscures the direction of maximum response and it has been shown that:

$$\delta_\theta = \frac{1}{l \cdot \sqrt{\left(\frac{2E}{N_0}\right)}}$$

where δ_θ = angular increment that can just be discriminated and l = effective aperture in wavelengths.

Clearly large ears and perhaps large nose-leaves are an advantage although their irregular shapes make the response more complicated. Also two ears are better than one but the situation is not as simple as for a single aperture.

5. Reflective properties of targets

Higher frequencies and therefore shorter wavelengths not only improve the directionality of an echo-location system, they also give stronger echoes from small targets. Fig. A.1 shows that the echo returned from a small spherical target becomes stronger as the sphere becomes larger in relation to the wavelength (Rayleigh scattering region). The maximum echo is produced when the circumference of the sphere is exactly one wavelength $\left(\text{when } \frac{\pi d}{\lambda} = 1\right)$. After oscillating somewhat in the resonance region, the echo becomes independent of the size when the sphere is many wavelengths in circumference. The sphere then appears to have its true or 'optical' diameter. Targets of other shapes have more complex responses to their size in relation to the wavelength used (Griffin, 1958; King and Wu, 1959) but the spherical case gives an adequate approximation for many targets important to both radar and echo-location systems.

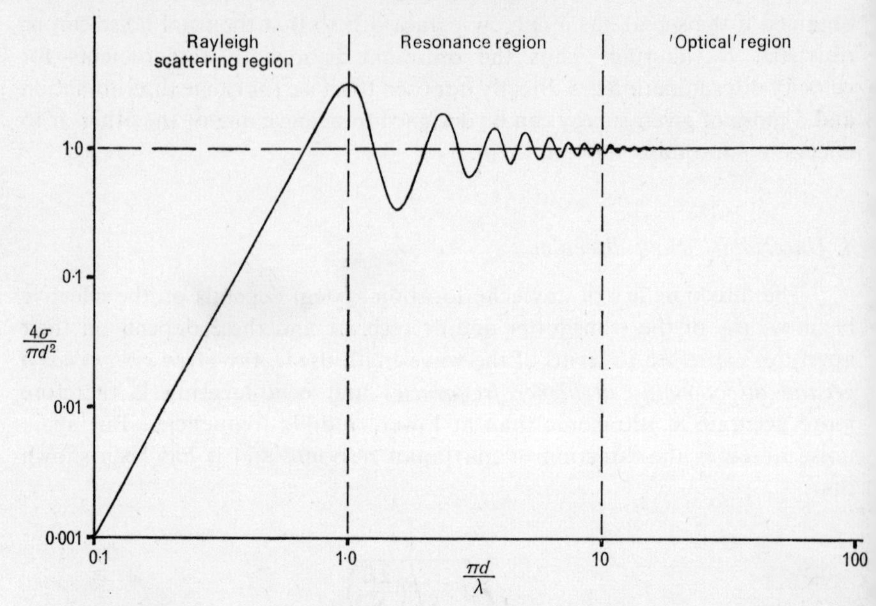

FIGURE A.1 The relationship between the echo-reflecting power of a sphere and the wavelength of the transmitted signal. Both axes have been 'normalized' for greater generality. The ordinate shows the apparent cross-section area (σ) of the sphere expressed in terms of its actual cross-section area $\left(\frac{\pi d^2}{4}\right)$. The abscissa shows the circumference of the sphere (πd) in terms of the wavelength (λ).

Some references on radar theory are: Woodward (1953), Skolnik (1962), Barton and Ward (1969). An excellent reference on telecommunication and modulation theory is Brown and Glazier (1964).

Barton, D. K. and Ward, H. R. (1969), *Handbook of Radar Measurement*, Prentice-Hall, Englewood Cliffs, New Jersey.

Brown, J. and Glazier, E. V. D. (1964), *Telecommunications: the General Principles*, Chapman and Hall, London.

King, R. W. P. and Wu, T. T. (1959), *The Scattering and Diffraction of Waves*, Harvard University Press, Cambridge, Mass.

Skolnik, M. I. (1962), *Introduction to Radar Systems*, McGraw Hill, New York.

Woodward, P. M. (1953), *Probability and Information Theory with Applications to Radar*, Pergamon Press, Oxford.

References

Adam, L. J. and Schwartzkopff, J. (1967), 'Getrennte nervöse Representation für verschiedene Tonbereiche im Protocerebrum von *Locusta migratoria*', *Zeitschrift für vergleichende Physiologie*, **54**, 246–255.

Agee, H. R. (1967), 'Response of acoustic sense cell of the bollworm and tobacco budworm to ultrasound', *Journal of Economic Entomology*, **60**, 366–369.

Agee, H. R. (1969a), 'Response of flying bollworm moths and other tympanate moths to pulsed ultrasound', *Annals of the Entomological Society of America*, **62**, 801–807.

Agee, H. R. (1969b), 'Response of *Heliothis* spp. (Lepidoptera: Noctuidae) to ultrasound when resting, feeding, courting, mating or ovipositing', *Annals of the Entomological Society of America*, **62**, 1122–1128.

Agee, H. R. (1971), 'Ultrasound produced by wings of adults of *Heliothis zea*', *Journal of Insect Physiology*, **17**, 1267–1273.

Alexander, R. D. (1956), *A Comparative Study of Sound Production in Insects with Special Reference to the Singing Orthoptera and Cicadidae of the Eastern United States*. Doctoral dissertation, Ohio State University.

Alexander, R. D. (1957), 'Sound production and associated behaviour in insects', *Ohio Journal of Science*, **57**, 101–113.

Alexander, R. D. (1960), 'Sound communication in Orthoptera and Cicadidae'. In *Animal Sounds and Communication*, eds. Lanyon, W. E. and Tavolga, W. N., pp. 38–92. American Institute of Biological Sciences, Washington.

Alexander, R. D. (1961), 'Aggressiveness, territoriality and sexual behaviour in field crickets (Orthoptera: Gryllidae)', *Behaviour*, **17**, 130–223.

Alexander, R. D. (1967), 'Acoustical communication in Arthropods', *Annual Review of Entomology*, **12**, 496–526.

Alford, B. R. and Ruben, R. J. (1963), Physiological, behavioural and anatomical correlates of the development of hearing in the mouse', *Annals of Otology, Rhinology and Laryngology*, **72,** 237–247.

Allin, J. T. and Banks, E. M. (1971), 'Effects of temperature on ultrasound production by infant albino rats', *Developmental Psychobiology*, **4,** 149–156.

Allin, J. T. and Banks, E. M. (1972), 'Functional aspects of ultrasound production by infant albino rats (*Rattus norvegicus*)', *Animal Behaviour*, **20,** 175–185.

Altes, R. A. (1971), 'Computer derivation of some dolphin echolocation signals', *Science, New York*, **173,** 912–914.

Altes, R. A. and Titlebaum, E. L. (1970), 'Bat signals as optimally doppler tolerant waveforms', *Journal of the Acoustical Society of America*, **48,** 1014–1020.

Anderson, J. W. (1954), 'The production of ultrasonic sounds by laboratory rats and other mammals', *Science, New York*, **119,** 808.

Andrejewski, R. and Olszewski, J. (1963), 'Social behaviour and interspecific relations in *Apodemus flavicollis* (Melchior, 1834) and *Clethrionomys glareolus* (Schreber, 1780)', *Acta Theriologica*, **7,** 155–168.

Anstee, J. H. (1971), 'The stridulatory apparatus of two species of tettigoniid', *Tissue and Cell*, **3,** 71–76.

Arntz, B. (1972), 'The hearing capacity of water bugs', *Journal of Comparative Physiology*, **80,** 304–311.

Ashton Freeman, J. (1953), 'The curlew's echo system', *The Field*, October 15th, pp. 646–647.

Auger, D. and Fessard, A. (1920), 'Observations sur l'excitabilité de l'organe tympanique du criquet', *Comptes rendus des séances de la Société de Biologie*, **99,** 400–401.

Autrum, H. (1936), 'Über lautaüsserungen und schallwahrnehmung bei Arthropoden. I. Untersuchungen an Ameisen. Eine allgemeine theorie der Schallwahrnehmung bei Arthropoden', *Zeitschrift für vergleichende Physiologie*, **23,** 332–373.

Autrum, H. (1940), 'Über lautaüsserungen und schallwahrnehmung bei Arthropoden. II. Das richtungshören von *Locusta* und versuch einer hörtheorie für Tympanalorgane vom Locustidentyp', *Zeitschrift für vergleichende Physiologie*, **28,** 326–352.

Autrum, H. (1941), 'Über gehör und Erschutterungssinn bei Locustiden', *Zeitschrift für vergleichende Physiologie*, **28,** 580–637.

Autrum, H. (1960), 'Phasische und tonische Antworten vom Tympanalorgan von *Tettigonia viridissima*', *Acustica*, **10,** 340–348.

Autrum, H. (1963), 'Anatomy and physiology of sound receptors in invertebrates'. In *Acoustic Behaviour of Animals*, ed. Busnel, R.-G., Ch. 15. Elsevier Publishing Co., Amsterdam.

Bailey, W. J. (1967), 'Further investigations into the function of the "mirror" in Tettigonioidea (Orthoptera)', *Nature, London*, **215,** 762–763.

Bailey, W. J. (1970), 'The mechanics of stridulation in bush crickets (Tettigonioidea, Orthoptera). I. The tegminal generator', *Journal of Experimental Biology*, **52**, 495–505.

Bailey, W. J. and Broughton, W. B. (1970), 'The mechanics of stridulation in bush crickets (Tettigonioidea, Orthoptera). II. Conditions for resonance in the tegminal generator', *Journal of Experimental Biology*, **52**, 505–517.

Bailey, W. J. and Robinson, D. (1971), 'Song as a possible isolating mechanism in the genus *Homorocoryphus* (Tettigonioidea, Orthoptera)', *Animal Behaviour*, **19**, 390–397.

Barfield, R. J. and Geyer, L. A. (1972), 'Sexual behaviour: ultrasonic post-ejaculatory song of the male rat', *Science, New York*, **176**, 1349–1350.

Barnett, S. A. (1963), *A Study in Behaviour*, Methuen, London.

Bastian, J. (1967), 'The transmission of arbitrary environmental information between bottlenose dolphins'. In *Animal Sonar Systems, Biology and Bionics*, ed. Busnel, R.-G. Vol. I, pp. 803–873, Laboratoire de Physiologie Acoustique, INRA–CNRZ, Jouy-en-Josas.

Beier, M. (1967), 'Neuer Beiträge zur Kenntnis der Gattungsgrubbe *Amytta* Karsch (Ort. Mecanematinae)', *Eos*, **42**, 305–310.

Bel'Kovich, U. M. and Solntseva, G. N. (1970), 'Morfo-funktsional' l'nye osobennosti organa slukha del'finou' (Morpho-functional features of the organ of hearing in dolphins. In Russian). *Zoologicheskiĭ zhurnal*, **49**, 275–282.

Bell, R. W., Nitschke, W., Gorry, T. H. and Zachman, T. A. (1971), 'Infantile stimulation and ultrasonic signalling. A possible mediator of early handling phenomena', *Developmental Psychobiology*, **4**, 181–191.

Bell, R. W., Nitschke, W. and Zachman, T. A. (1972), 'Ultra-sounds in three inbred strains of young mice', *Behavioural Biology*, **7**, 805–814.

Belton, P. (1962a), 'Responses to sound in pyralid moths', *Nature, London*, **196**, 1188–1189.

Belton, P. (1962b), 'Effects of sound on insect behaviour', *Proceedings of the Entomological Society of Manitoba*, **18**, 1–9.

Belton, P. and Kempster, R. H. (1962), 'A field test on the use of sound to repel the European corn borer', *Entomologia experimentalis et applicata*, **5**, 281–288.

Benedetti, E. (1950a), 'Potenziali neuroacustici prodotti da ultrasuoni in alcuni ortotteri', *Bollettino della Società italiana di biologia sperimentale*, **26**, 741–743.

Benedetti, E. (1950b), 'La percezione degli ultrasuoni in alcuni ortotteri rilevata mediante la registrazione oscillografica della correnti bioelettriche', *Ateneo parmense*, **22**, 105–116.

Beniest-Noirot, E. (1957), 'Le comportement dit "maternal" de la souris', *Extrait de l'ouvrage de l'Union Internationale des Sciences Biologiques* (Section de Psychologie Expérimentale et Comportement Animal), pp. 139–146. Bruxelles.

Beniest-Noirot, E. (1958), 'Analyse du comportement dit "maternal" chez la souris', *Monographies Françaises de Psychologie*, **1,** 1–114. Centre National de la Recherche Scientifique.

Bennet-Clark, H. C. (1970), 'The mechanism and efficiency of sound production in mole crickets', *Journal of Experimental Biology*, **52,** 619–652.

Bentley, D. R. and Kutsch, W. (1966), 'The neuromuscular mechanism of stridulation in crickets (Orthoptera: Gryllidae)', *Journal of Experimental Biology*, **45,** 151–164.

Berlin, C. I. (1963), 'Hearing in mice via GSR audiometry', *Journal of Speech and Hearing Research*, **6,** 359–368.

Blest, A. D. (1964), 'Protective display and sound production in some New World arctiid and ctenuchid moths', *Zoologica, New York*, **49,** 161–181.

Blest, A. D., Collett, T. S. and Pye, J. D. (1963), 'The generation of ultrasonic signals by a New World arctiid moth', *Proceedings of the Royal Society, Series B*, **158,** 196–207.

Borror, D. J. (1954), 'Audio-spectrographic analysis of the song of the cone-headed grasshopper, *Neoconocephalus ensiger* (Orthoptera: Tettigoniidae)', *Ohio Journal of Science*, **54,** 297–303.

Bovet, J. (1972), 'On the social behaviour in a stable group of long-tailed field mice (*Apodemus sylvaticus*). II. Its relations with distribution of daily activity', *Behaviour*, **41,** 55–67.

Bradbury, J. W. (1970), 'Target discrimination by the echolocating bat *Vampyrum spectrum*', *Journal of Experimental Zoology*, **173,** 23–46.

Brooks, R. J. and Banks, E. M. (1973), 'Behavioural biology of the collared lemming [*Dicrostonyx groenlandicus* (Traill)]: An analysis of acoustic communication', *Animal Behaviour Monographs*, **6,** (1) 1–83.

Broughton, W. B. (1963), 'Method in bioacoustic terminology'. In *Acoustic Behaviour of Animals*, ed. Busnel, R.-G., Ch. 1. Elsevier Publishing Co, Amsterdam.

Broughton, W. B. (1964), 'Function of the "mirror" in tettigonioid Orthoptera', *Nature, London*, **201,** 949–950.

Brown, A. M. (1970), 'Bimodal cochlear response curves in rodents', *Nature, London*, **228,** 576–577.

Brown, A. M. (1971a), 'High frequency responsiveness in rodents at the level of the inferior colliculus', *Nature, London*, **232,** 223–224.

Brown, A. M. (1971b), *Auditory Responses to High Frequencies in Small Mammals*. Unpublished Ph.D thesis, University of London.

Brown, A. M. (1973a), High frequency peaks in the cochlear microphonic response of rodents', *Journal of Comparative Physiology*, **83,** 377–392.

Brown, A. M. (1973b), 'High levels of responsiveness from the inferior colliculus of rodents at ultrasonic frequencies', *Journal of Comparative Physiology*, **83,** 393–406.

Brown, A. M. (1973c), 'An investigation of the cochlear microphonic response of two species of echolocating bats, *Rousettus aegyptiacus* (Geoffroy) and *Pipistrellus pipistrellus* (Schreber)', *Journal of Comparative Physiology*, **83**, 407–413.

Brown, A. M. and Pye, J. D. (1974), 'Auditory sensitivity at high frequencies in mammals', *Advances in Comparative Physiology and Biochemistry*, ed. Lowenstein, O. E., **5**, Academic Press, New York and London.

Brown, E. S. (1955), 'Méchanismes du comportement dans les émissions sonores chez les orthoptères'. In *Colloque sur l'acoustique des Orthoptères*, ed. Busnel, R.-G., pp. 168–170. Fascicule hors série des *Annales des épiphyties*. Institut National de la Recherche Agronomique, Paris.

Brown, J. and Glazier, E. V. D. (1964), *Telecommunication: the General Principles*. Chapman Hall, London.

Bullock, T. H., Grinnell, A. D., Ikezono, E., Kameda, K., Katsuki, Y., Nomoto, M., Sato, O., Suga, N. and Yanagisawa, K. (1968), 'Electrophysiological studies of central auditory mechanisms in cetaceans', *Zeitschrift für vergleichende Physiologie*, **59**, 117–156.

Bullock, T. H. and Ridgway, S. H. (1972), 'Evoked potentials in the central auditory system of alert porpoises to their own and artificial sounds', *Journal of Neurobiology*, **3**, 79–99.

Bullock, T. H., Ridgway, S. H. and Suga, N. (1971), 'Acoustically evoked potentials in midbrain auditory structures in sea lions (Pinnipedia)', *Zeitschrift für vergleichende Physiologie*, **74**, 372–387.

Busnel, M. C. (1953), 'Contribution à l'étude des émissions acoustiques des Orthoptères 1er Memoire', *Annales des épiphyties*, **3**, 333–421.

Busnel, R.-G. (ed.) (1955), *Colloque sur l'acoustique des Orthoptères*. Fascicule hors serie des *Annales des épiphyties*. Institut National de la Recherche Agronomique, Paris.

Busnel, R.-G. (1956), 'Some new aspects of acoustical animal behaviour', *Journal of Scientific and Industrial Research*, **15A**, 306–310.

Busnel, R.-G. (ed.) (1963a), *Acoustic Behaviour of Animals*, Elsevier Publishing Co., Amsterdam.

Busnel, R.-G. (1963b), 'On certain aspects of animal acoustic signals'. In *Acoustic Behaviour of Animals*, ed. Busnel, R.–G., Ch. 5. Elsevier Publishing Co., Amsterdam.

Busnel, R.-G. and Chavasse, P. (1951), 'Recherche sur les émissions sonores et ultrasonores d'Orthoptères nuisibles à l'agriculture'. Supplemento al Vol. VII ser. IX del *Nuovo Cimento* (Bologna), **1**, 470–486.

Busnel, R.-G. and Dumortier, B. (1959), 'Vérification par des methodes d'analyse acoustique des hypothèses sur o'origine du cri du Sphinx *Acherontia atropos* (Linné)', *Bulletin de la Société entomologique de France*, **64**, 44–58.

Busnel, R.-G. and Dziedzic, A. (1967), 'Résultats métrologiques experimentaux de l'echolocation chez le *Phocaena phocaena*, et leur comparaison avec ceux de certaines chauves-souris'. In *Animal Sonar Systems, Biology and Bionics*, ed. Busnel, R.-G., Vol. I, pp. 306–335. Laboratoire de Physiologie Acoustique, INRA–CNRZ, Jouy-en-Josas.

Busnel, R.-G. and Dziedzic, A. (1968), 'Caracteristiques physiques des signaux acoustiques de *Pseudorca crassidens* (Cetace odontocete)', *Mammalia*, **32,** 1–5.

Busnel, R.-G., Dziedzic, A. and Andersen, S. (1963), 'Sur certaines caracteristiques des signaux acoustiques du Marsouin, *Phocaena phocaena* L.', *Comptes rendus hebdomadaire des Séances de l'Académie des Sciences*, **257**, 2545–2548.

Busnel, R.-G., Dziedzic, A. and Andersen, S. (1965a), 'Seuils de perception du système sonar du Marsouin *Phocaena phocaena* L., un fonction du diamètre d'un obstacle filiforme', *Comptes rendus hebdomadaire des Séances de l'Académie des Sciences*, **260,** 295–297.

Busnel, R.-G., Dziedzic. A, and Andersen, S. (1965b), 'Rôle de l'impédance d'une cible dans le seuil de sa détection par le système sonar du Marsouin *Phocaena phocaena*', *Comptes rendus des Séances de la Société de Biologie*, **159,** 69–74.

Busnel, R.-G. and Loher, W. (1955), 'Recherches sur les actions de signaux acoustiques artificiels sur le comportement de divers Acrididae males'. In *Colloque sur l'acoustique des Orthoptères*, ed. Busnel, R.-G., pp. 365–394. Fascicule hors serie des *Annales des épiphyties*. Institut National de la Recherche Agronomique, Paris.

Cahlander, D. (1967), 'Theories of sonar systems and their application to biological organisms: discussion'. In *Animal Sonar Systems, Biology and Bionics*, ed. Busnel, R.-G., Vol. II, pp. 1052–1081. Laboratoire de Physiologie Acoustique, INRA–CNRZ, Jouy-en-Josas.

Caldwell, M. C. and Caldwell, D. K. (1967), 'Intraspecific transfer of information via the pulsed sound in captive odontocete cetaceans'. In *Animal Sonar Systems, Biology and Bionics*, ed. Busnel, R.-G., Vol. II, pp. 879–936. Laboratoire de Physiologie Acoustique, INRA–CNRZ. Jouy-en-Josas.

Caldwell, M. C., Haugen, R. M. and Caldwell, D. K. (1962), 'High-energy sound associated with fright in the dolphin', *Science, New York*, **138,** 907–908.

Carpenter, G. D. H. (1938), 'Audible emission of defensive froth by insects. With an appendix on the anatomical structures concerned in a moth by H. Eltringham', *Proceedings of the Zoological Society of London, Series A*, **108,** 243–252.

Chang, H.-T. (1936), 'An auditory reflex of the hedgehog', *Chinese Journal of Physiology*, **10,** 119–124.

Chapman, S. (1968), 'Dolphins and multifrequency, multiangular images', *Science, New York*, **160,** 208–209.

Corso, J. F. (1963), 'Bone conduction thresholds for sonic and ultrasonic frequencies', *Journal of the Acoustical Society of America*, **35,** 1138–1143.

Corso, J. F. and Levine, M. (1963), 'Pitch discrimination by air conduction and by bone conduction for sonic and ultrasonic frequencies', *Journal of the Acoustical Society of America*, **35,** 804.

Cranbrook, Earl and Medway, Lord (1965), 'Lack of ultrasonic frequencies in the calls of swiftlets', *Ibis*, **107,** 258.

Crowley, D. E. and Hepp-Reymond, M.-C. (1966), Development of cochlear function in the ear of the infant rat', *Journal of Comparative and Physiological Psychology*, **62,** 427–432.

Crowley, D. E., Hepp-Reymond, M.-C., Tabowitz, D. and Palin, J. (1965), 'Cochlear potentials in the albino rat', *Journal of Auditory Research*, **5,** 307.

Dalland, J. I. (1965a), 'Hearing sensitivity in bats', *Science, New York*, **150,** 1185–1186.

Dalland, J. I. (1965b), 'Auditory thresholds in the bat: a behavioural technique', *Journal of Auditory Research*, **5,** 95–108.

Dalland, J. I., Vernon, J. A. and Peterson, E. A. (1967), 'Hearing and cochlear microphonic potentials in the bat, *Eptesicus fuscus*', *Journal of Neurophysiology*, **30,** 697–709.

Deatherage, B. H., Jeffress, L. A. and Blodgett, H. C. (1954), 'A note on the audibility of intense ultrasonic sound', *Journal of the Acoustical Society of America*, **26,** 582.

Dice, L. R. and Barto, E. (1952), 'Ability of mice of the genus *Peromyscus* to hear ultrasonic sounds', *Science, New York*, **116,** 110–111.

Diercks, K. J., Trochta, R. T., Greenlaw, C. F. and Evans, W. E. (1971), 'Recording and analysis of dolphin echolocation signals', *Journal of the Acoustical Society of America*, **49,** 1729–1732.

Dijkgraaf, S. (1946), 'Die Sinneswelt der Fledermäuse, *Experientia*, **2,** 438–448.

Dijkgraaf, S. (1957), 'Sinnesphysiologische Beobachtungen an Fledermäusen', *Acta physiologica et pharmalogica néerlandica*, **6,** 675–684.

Dijkgraaf, S. (1960), 'Spallanzani's unpublished experiments on the sensory basis of object perception in bats', *Isis*, **51,** 9–20.

Dudok van Heel, W. H. (1959), 'Auto-direction finding in the porpoise (*Phocaena phocaena*)', *Nature, London*, **183,** 1063.

Dudok van Heel, W. H. (1962), 'Sound and Cetacea', *Netherlands Journal of Sea Research*, **1,** 407–507.

Dunning, D. C. (1968), 'Warning sounds of moths', *Zeitschrift für Tierpsychologie*, **25,** 129–138.

Dunning, D. C. and Roeder, K. D. (1965), 'Moth sounds and the insect-catching behaviour of bats', *Science, New York*, **147,** 173–174.

Dumortier, B. (1963a), Morphology of sound emission apparatus in Arthropoda'. In *Acoustic Behaviour of Animals*, ed. Busnel, R.-G., Ch. 11. Elsevier Publishing Co., Amsterdam.

Dumortier, B. (1963b), 'The physical characteristics of sound emissions in

Arthropoda'. In *Acoustic Behaviour of Animals*, ed. Busnel, R.-G., Ch. 12. Elsevier Publishing Co., Amsterdam.

Dumortier, B. (1963c), 'Ethological and physiological study of sound emissions in Arthropoda'. In *Acoustic Behaviour of Animals*, ed. Busnel, R.-G., Ch. 21. Elsevier Publishing Co., Amsterdam.

Eggers, F. (1919), 'Das thoracale bitympanale Organ einer Gruppe der Lepidoptera Heterocera', *Zoologische Jahrbücher* (*Anatomie*), **41,** 273–376.

Eisenberg, J. F. and Gould, E. (1966), 'The behaviour of *Solenodon paradoxus* in captivity with comments on the behaviour of other Insectivora', *Zoologica, New York*, **51,** 49–58.

Elder, H. Y. (1971), 'High frequency muscles in sound production by a katydid. II. Ultrastructure of the singing muscles', *Biological Bulletin*, **141,** 434–448.

Elias, H. (1907), 'Zur Anatomie des Kehlkopfes der Microchiropteren', *Morphologisches Jarhbuch*, **37,** 70–118.

Evans, E. F. (1968), 'Cortical representation.' In *Hearing Mechanisms in Vertebrates*, Ciba Foundation Symposium, eds. de Reuck, A.V.S. & Knight, J., pp. 272–287. J. and A. Churchill, London.

Evans, W. E. (1967a), 'Discussion of A. D. Grinnell's paper Mechanisms of overcoming interference in echolocating animals'. In *Animal Sonar Systems, Biology and Bionics*, ed. Busnel, R.-G., Vol. I, pp. 495–503. Laboratoire de Physiologie Acoustique, INRA–CNRZ, Jouy-en-Josas.

Evans, W. E. (1967b), 'Vocalization among marine mammals'. In *Marine Bio-acoustics*, Vol II, ed. Tavolga, W. N., pp. 159–186. Pergamon Press, Oxford.

Evans, W. E. and Haugen, R. M. (1963), 'An experimental study of the echolocation ability of a California sea lion, *Zalophus californianus* (Lesson)', *Bulletin of the Southern California Academy of Science*, **62,** 165–175.

Evans, W. E. and Powell, B. A. (1967), 'Discrimination of different metallic plates by an echolocating delphinid'. In *Animal Sonar Systems, Biology and Bionics*, ed. Busnel, R.-G., Vol. I, pp. 363–383. Laboratoire de Physiologie Acoustique, INRA–CNRZ, Jouy-en-Josas.

Evans, W. E. and Prescott, J. H. (1962), 'Observations on the sound production capabilities of the bottlenosed porpoise: A study of whistles and clicks', *Zoologica, New York*, **47,** 121–128.

Everest, F. D., Young, R. W. and Johnson, M. W. (1948), 'Acoustical characteristics of noise produced by snapping shrimp', *Journal of the Acoustical Society of America*, **20,** 137–142.

Fant, C. G. M. (1960) *Acoustic Theory of Speech Production*. Mouten, The Hague.

Finck, A. and Sofouglu, M. (1966), 'Auditory sensitivity of the Mongolian gerbil', *Journal of Auditory Research*, **6,** 313–319.

Fisher, J. and Lockley, R. M. (1954), *Sea Birds*. Collins, London.

Flanagan, J. L. (1972), 'The synthesis of speech', *Scientific American*, **226,** 48–58.

Flieger, E. and Schnitzler, H.-U. (1973), 'Ortungsleistungen der Fledermaus *Rhinolophus ferrumequinum* bei ein- und beitseitiger Ohrverstopfung', *Journal of Comparative Physiology*, **82,** 93–102.

Fraser, F. C. and Purves, P. E. (1954), 'Hearing in cetaceans', *Bulletin of the British Museum (Natural History), Zoology*, **2,** 103–116.

Fraser, F. C. and Purves, P. E. (1960a), 'Hearing in cetaceans', *Bulletin of the British Museum (Natural History), Zoology*, **7,** 1–140.

Fraser, F. C. and Purves, P. E. (1960b) 'Anatomy and function of the cetacean ear', *Proceedings of the Royal Society, Series B*, **152,** 62–77.

Friedman, M. H. (1972a), 'A light and electron microscopic study of the sensory organs and associated structures in the foreleg tibia of the cricket, *Gryllus assimilis*', *Journal of Morphology*, **138,** 263–328.

Friedman, M. H. (1972b), 'An electron microscopic study of the tympanal organ and associated structures in the foreleg tibia of the cricket, *Gryllus assimilis*', *Journal of Morphology*, **138,** 329–348.

Friend, J. H., Suga, N. and Suthers, R. A. (1966), 'Neural responses in the inferior colliculus of echolocating bats to artificial orientation sounds and echoes', *Journal of Cellular Physiology*, **67,** 319–332.

Frischkopf, L. S. (1964), 'Excitation and inhibition of primary auditory neurons in the little brown bat', *Journal of the Acoustical Society of America*, **36,** 1016 (A).

Fulton, B. B. (1933), 'Stridulating organs of female Tettigoniidae (Orthoptera)', *Entomological News*, **44,** 270–275.

Galambos, R. (1941), 'Cochlear potentials elicited from bats by supersonics', *American Journal of Physiology*, **133,** 285–286.

Galambos, R. (1942a), 'Cochlear potentials elicited from bats by supersonic sounds', *Journal of the Acoustical Society of America*, **13,** 41–49.

Galambos, R. (1942b), 'The avoidance of obstacles by flying bats. Spallanzani's ideas (1794) and later theories', *Isis*, **34,** 132–140.

Galambos, R. and Griffin, D. R. (1942), 'Obstacle avoidance by flying bats: the cries of bats', *Journal of Experimental Zoology*, **89,** 475–490.

Gandelman, R., Zarrow, M. X. and Denenberg, V. H. (1971a), 'Olfactory bulb removal eliminates maternal behavior in the mouse', *Science, New York*, **171,** 210–211.

Gandelman, R., Zarrow, M. X. and Denenberg, V. H. (1971b), 'Stimulus control of cannibalism and maternal behavior in anosmic mice', *Physiology and Behaviour*, **7,** 583–586.

Ghiradella, H. (1971), 'Fine structure of the noctuid moth ear. I. The transducer area and connections to the tympanic membrane in *Feltia subgothica*, Haworth', *Journal of Morphology*, **134,** 21–46.

Giraud-Sauveur, D. (1969), 'Recherches biophysiques sur les osselets des cetaces', *Mammalia*, **33,** 285–340.

Gould, E. (1965), 'Evidence for echolocation in the Tenrecidae of Madagascar', *Proceedings of the American Philosophical Society*, **109,** 352–360.

Gould, E. (1970), 'Echolocation and communication in bats'. In *About Bats*, eds. Slaughter, B. A. and Walton, D. W., pp. 144–161. Southern Methodist University Press, Dallas.

Gould, E. (1971), 'Studies of maternal–infant communication and development of vocalizations in the bats *Myotis* and *Eptesicus*', *Communications in Behavioural Biology*, **5,** 263–313.

Gould, E., Negus, N. C. and Novick, A. (1964), 'Evidence for echolocation in shrews', *Journal of Experimental Zoology*, **156,** 19–38.

Gould, J. and Morgan, C. (1941), 'Hearing in the rat at high frequencies', *Science, New York*, **41,** 168.

Goureau, M. (1837, 'Sur la stridulation des insectes', *Annales de la Société Entomologique de France*, **6,** 31–75.

Gourevitch, G. and Hack, M. H. (1966), 'Audibility in the rat', *Journal of Comparative and Physiological Psychology*, **62,** 289–291.

Grant, E. C. (1963), 'An analysis of the social behaviour of the male laboratory rat', *Behaviour*, **21,** 260–281.

Grant, E. C. and Mackintosh, J. H. (1963), 'A comparison of the social postures of some common laboratory rodents', *Behaviour*, **21,** 246–259.

Gray, E. G. (1960), 'The fine structure of the insect ear', *Philosophical Transactions of the Royal Society*, **243,** 75–94.

Gray, E. G. and Pumphrey, R. J. (1958), 'Ultrastructure of the insect ear', *Nature, London*, **181,** 618.

Griffin, D. R. (1950), 'Measurements of the ultrasonic cries of bats', *Journal of the Acoustical Society of America*, **22,** 247–255.

Griffin, D. R. (1954), 'Acoustic orientation in the oil bird, *Steatornis*', *Proceedings of the National Academy of Sciences of the United States of America*, **39,** 884–893.

Griffin, D. R. (1958), *Listening in the Dark*. Yale University Press, New Haven.

Griffin, D. R. (1971), 'The importance of atmospheric attenuation for the echolocation of bats (Chiroptera)', *Animal Behaviour*, **19,** 55–61.

Griffin, D. R., Dunning, D. C., Cahlander, D. A. and Webster, F. A. (1962), 'Correlated orientation sounds and ear movements of horseshoe bats', *Nature, London*, **196,** 1185–1186.

Griffin, D. R. and Galambos, R. (1941), 'The sensory basis of obstacle avoidance by flying bats', *Journal of Experimental Zoology*, **86,** 481–506.

Griffin, D. R. and Grinnell, A. D. (1958), 'Ability of bats to discriminate echoes from louder noise', *Science, New York*, **128,** 145–147.

Griffin, D. R., McCue, J. J. G. and Grinnell, A. D. (1963), 'The resistance of bats to jamming', *Journal of Experimental Zoology*, **152,** 229–250.

Griffin, D. R. and Novick, A. (1955), 'Acoustic orientation of neotropical bats', *Journal of Experimental Zoology*, **130,** 251–300.

Griffin, D. R., Novick A. and Kornfield, M. (1958), 'The sensitivity of echo-location in the fruit bat *Rousettus*', *Biological Bulletin*, **115,** 107–113.

Griffin, D. R. and Suthers, R. A. (1970), 'Sensitivity of echo-location in cave swiftlets', *Biological Bulletin*, **139,** 495–501.

Griffin, D. R., Webster, F. A. and Michael, C. R. (1960), 'The echolocation of flying insects by bats', *Animal Behaviour*, **8,** 141–154.

Grinnell, A. D. (1963a), 'The neurophysiology of audition in bats: intensity and frequency parameters', *Journal of Physiology, London*, **167,** 38–66.

Grinnell, A. D. (1963b), 'The neurophysiology of audition in bats: temporal parameters', *Journal of Physiology, London*, **167,** 67–96.

Grinnell, A. D. (1963c), 'The neurophysiology of audition in bats: directional localization and binaural interaction', *Journal of Physiology, London*, **167,** 97–113.

Grinnell, A. D. (1963d), 'The neurophysiology of audition in bats: resistance to interference', *Journal of Physiology, London*, **167,** 114–127.

Grinnell, A. D. (1967), 'Mechanisms of overcoming interference in echolocating animals'. In *Animal Sonar Systems, Biology and Bionics*, ed. Busnel, R.-G., Vol. I, pp. 451–481. Laboratoire de Physiologie Acoustique, INRA–CNRZ, Jouy-en-Josas.

Grinnell, A. D. (1970), 'Comparative auditory neurophysiology of neotropical bats employing different echolocation signals', *Zeitschrift für vergleichende Physiologie*, **68,** 117–153.

Grinnell, A. D. (1973), 'Rebound excitation (off-responses) following non-neural suppression in the cochleas of echolocating bats', *Journal of Comparative Physiology*, **82,** 179–194.

Grinnell, A. D. and Griffin, D. R. (1958), 'The sensitivity of echolocation in *bats*', *Biological Bulletin*, **114,** 10–22.

Grinnell, A. D. and Grinnell, V. S. (1965), 'Neural correlates of vertical localization by echo-locating bats', *Journal of Physiology, London*, **181,** 830–851.

Grinnell, A. D. and Hagiwara, S. (1972a), 'Adaptations of the auditory nervous system for echolocation: studies of New Guinea bats', *Zeitschrift für vergleichende Physiologie*, **76,** 41–81.

Grinnell, A. D. and Hagiwara, S. (1972b), 'Studies of auditory neurophysiology in non-echolocating bats, and adaptations for echolocation in one genus, *Rousettus*', *Zeitschrift für vergleichende Physiologie*, **76,** 82–96.

Grinnell, A. D. and McCue, J. J. G. (1963), 'Neurophysiological investigations of the bat, *Myotis lucifugus*, stimulated by frequency modulated acoustical pulse', *Nature, London*, **198,** 453–455.

Grumman, R. A. and Novick, A. (1963), 'Obstacle avoidance in the bat *Macrotus mexicanus*', *Physiological Zoölogy*, **36,** 361–369.

Grünwald, A. (1969), 'Untersuchungen zur Orientierung der Weißzahnspitzmäuse (Soricidae–Crocidurinae)', *Zeitschrift für vergleichende Physiologie*, **65,** 191–217.

Hahn, W. L. (1908), 'Some habits and sensory adaptations of cave-inhabiting bats', *Biological Bulletin*, **15,** 135–193.

Hall, J. D. and Johnson, C. S. (1972), 'Auditory thresholds of a killer whale, *Orcinus orca*, Linnaeus', *Journal of the Acoustical Society of America*, **51,** 515–517.

Harrison, J. B. (1965), 'Temperature effects on responses in the auditory system of the little brown bat, *Myotis l. lucifugus*', *Physiological Zoölogy*, **38,** 34–48.

Hart, F. H. and King, J. A. (1966), 'Distress vocalisations of young in two subspecies of *Peromyscus maniculatus*', *Journal of Mammalogy*, **47,** 287–293.

Hartridge, H. (1920), 'The avoidance of objects by bats in their flight', *Journal of Physiology, London*, **54,** 54–57.

Hartridge, H. (1945), 'Avoidance of obstacles by bats', *Nature, London*, **156,** 55.

Hashimoto, T. and Maniwa, Y. (1959), 'Noises of creatures in the sea in region of ultrasound'. (Article in Japanese, English summary.) *Technical Report of Fishing Boat Laboratory.* (*Tokyo*) *Japan Fisheries Agency*, **12,** 99–135.

Haskell, P. T. (1957), 'Stridulation and associated behaviour in certain Orthoptera. I. Analysis of the stridulation and behaviour between males', *British Journal of Animal Behaviour*, **5,** 139–148.

Haskell, P. T. and Belton, P. (1956), 'Electrical responses of certain lepidopterous tympanal organs', *Nature, London*, **177,** 139–140.

Hawker, P. (1972a), 'Synchronous detection in radio reception. I', *Wireless World*, **76,** 419–422.

Hawker, P. (1972b), 'Synchronous detection in radio reception. II. Phase-locking and the "bi-aural" detector', *Wireless World*, **78,** 525–528.

Heath, J. E. and Josephson, R. K. (1970), 'Body temperature and singing in the katydid *Neoconocephalus robustus* (Orthoptera, Tettigoniidae)', *Biological Bulletin*, **138,** 272–285.

Heffner, H. E. and Masterton, B. (1970), 'Hearing in primitive primates: slow loris (*Nycticebus coucong*) and potto (*Perodicticus potto*)', *Journal of Comparative and Physiological Psychology*, **71,** 175–182.

Heffner, H. E., Ravizza, R. J. and Masterton, B. (1969a), 'Hearing in primitive mammals. III. Tree shrews (*Tupaia glis*)', *Journal of Auditory Research*, **9,** 12–18.

Heffner, H. E., Ravizza, R. J. and Masterton, B. (1969b), 'Hearing in primitive mammals. IV. Bush baby (*Galago senegalensis*)', *Journal of Auditory Research*, **9,** 19–23.

Heffner, R., Heffner, H. E. and Masterton, B. (1971), 'Behavioural measurements of absolute frequency difference thresholds in guinea pig', *Journal of the Acoustical Society of America*, **49,** 1888–1895.

Henson, O. W. Jr (1961), 'Some morphological and functional aspects of certain structures of the middle ear in bats and insectivores', *University of Kansas Science Bulletin*, **42,** 151–255.

Henson, O. W. Jr (1965), 'The activity and function of the middle-ear muscles in echo-locating bats', *Journal of Physiology, London*, **180,** 871–887.

Henson, O. W. Jr (1967a), 'The perception and analysis of biosonar signals by bats'. In *Animal Sonar Systems, Biology and Bionics*, ed. Busnel, R.-G., Vol. II, pp. 949–1003. Laboratoire de Physiologie Acoustique, INRA–CNRZ, Jouy-en-Josas.

Henson, O. W. Jr (1967b), 'Auditory sensitivity in Molossidae (Chiroptera)', *Anatomical Record*, **157,** 363–364.

Henson, O. W. Jr (1970), 'The ear and audition'. In *Biology of Bats*, ed. Wimsatt, W. A., Vol. II, pp. 181–263. Academic Press, New York and London.

Henson, O. W. Jr. and Henson, M. M. (1972), 'Middle ear muscle contractions and their relation to pulse- and echo-evoked potentials in the bat, *Chilonycteris parnellii*'. In *Animal Orientation and Navigation*, eds. Galler, S. R., Schmidt–Koenig, K., Jacobs, G. J. and Bellville, R. E., pp. 355–363. Scientific and Technical Information Office, National Aeronautics and Space Administration, Washington.

Henson, O. W. Jr and Pollak, G. (1972), 'A technique for chronic implantation of electrodes in the cochleae of bats', *Physiology and Behaviour*, **8,** 1185–1187.

Herald, E. S., Brownell, R. L., Frye, F. L., Morris, E. J., Evans, W. E. and Scott, A. B. (1969), 'Blind river dolphin: first side swimming cetacean', *Science, New York*, **166,** 1408–1410.

Hinchcliffe, R. and Pye, A. (1968), 'The cochlea in Chiroptera: a quantitative approach', *International Audiology*, **7,** 259–266.

Hinchcliffe, R. and Pye, A. (1969), 'Variations in the middle ear of the Mammalia', *Journal of Zoology, London*, **157,** 277–288.

Hirsch, I. J. (1952), *The Measurement of Hearing*. McGraw-Hill, New York.

Hissa, R. and Lagerspetz, K. (1964), 'The postnatal development of homoiothermy in the golden hamster', *Annales medicinae experimentalis et biologiae Fenniae*, **42,** 43–45.

Hooper, J. H. D. (1969), 'Potential use of a portable receiver for the field identification of flying bats', *Ultrasonics*, **7,** 177–181.

Horridge, G. A. (1960), 'Pitch discrimination in Orthoptera (Insecta) demonstrated by responses of central auditory neurones', *Nature, London*, **185,** 623–624.

Horridge, G. A. (1961), 'Pitch discrimination in locusts', *Proceedings of the Royal Society, Series B*, **155,** 218–231.

Howse, P. E., Lewis, D. B. and Pye, J. D. (1971), 'Adequate stimulus of the insect tympanic organ', *Experentia*, **27,** 598–600.

Huber, F. (1963), 'The role of the central nervous system in Orthoptera during the co-ordination and control of stridulation.' In *Acoustic Behaviour of Animals*, ed. Busnel, R.-G., Ch. 17. Elsevier Publishing Co., Amsterdam.

Huber, F. (1970), 'Nervöse Grundlagen der Akustischen Kommunikation bei Insekten', *Rheinisch–Westfälische Akademie der Wissenschaften*, **205,** 41–91.

Imms, A. D. (1957), *A General Textbook of Entomology*. 9th edition, revised by Richards, O. W. and Davies, R. G. Methuen, London.

Jacobs, D. W. (1972), 'Auditory frequency discrimination in the Atlantic bottlenose dolphin *Tursiops truncatus.* A preliminary report', *Journal of the Acoustical Society of America*, **52,** 696–698.

Jacobs, D. W. and Hall, J. D. (1972), 'Auditory thresholds of a freshwater dolphin, *Inia geoffrensis.* Blainville', *Journal of the Acoustical Society of America*, **51,** 530–533.

Jepsen, G. L. (1966), 'Early Eocene bat from Wyoming', *Science, New York*, **154,** 1333–1339.

Jepsen, G. L. (1970), 'Bat origins and evolution'. In *Biology of Bats*, ed. Wimsatt, W. A., Vol. I, pp. 1–64. Academic Press, New York and London.

Johnson, C. S. (1966), 'Auditory thresholds of bottlenose porpoise (*Tursiops truncatus*, Montagu)', U.S. Naval Ordnance Test Station Technical Publication No. 4178, pp. 1–28.

Johnson, C. S. (1967), 'Sound detection thresholds in marine mammals'. In *Marine Bio-acoustics*, Vol. II, ed. Tavolga, W. N., pp. 247–255. Pergamon Press, Oxford.

Johnson. M., Everest, F. and Young, R. W. (1947), 'The role of the snapping shrimp (*Crangon* and *Synalpheus*) in the production of underwater noise in the sea', *Biological Bulletin*, **93,** 122–138.

Johnstone, B. M., Saunders, J. C. and Johnstone, J. R. (1970), 'Tympanic membrane response in the cricket', *Nature, London*, **227,** 625–626.

Jones, C. (1972), 'Comparative ecology of three pteropid bats in Rio Muni, West Africa', *Journal of Zoology, London*, **167,** 353–370.

Jones, M. D. R. (1964), 'Inhibition and excitation in the acoustic behaviour of *Pholidoptera*', *Nature, London*, **203,** 322–323.

Jones, M. D. R. (1966a), 'The acoustic behaviour of the bush cricket *Pholidoptera griseoaptera.* I. Alternation, synchronism and rivalry between males', *Journal of Experimental Biology*, **45,** 15–30.

Jones, M. D. R. (1966b), 'The acoustic behaviour of the bush cricket *Pholid-*

optera griseoaptera. II. Interaction with artificial sound signals', *Journal of Experimental Biology*, **45**, 30–44.

Josephson, R. K. and Elder, H. Y. (1968), 'Rapidly contracting muscles used in sound production by a katydid', *Biological Bulletin*, **135**, 409.

Josephson, R. K. and Halverson, R. C. (1971), 'High frequency muscles in sound production by a katydid. I. Organisation of the motor system', *Biological Bulletin*, **141**, 411–433.

Kahmann, H. and v. Frisch, O. (1952), 'Uber die Beziehungen von Muttertier und Nestling bei kleinen Säugetieren', *Experentia*, **6**, 221–223.

Kahmann, H. and Ostermann, K. (1951), 'Wahrnehmen und Hervorbringen hoher Töne bei kleinen Säugetieren', *Experentia*, **7**, 268–269.

Kalmring, K. (1971), 'Akustische Neuronen im Unterschlundganglion der Wanderheuschrecke *Locusta migratoria*', *Zeitschrift für vergleichende Physiologie*, **72**, 95–110.

Kalmring, K., Rheinlaender, J. and Romer, H. (1972), 'Akustische Neuronen im Bauchmark von *Locusta migratoria*: Der Einfluβ der Schallrichtung auf die Autswortmusten', *Journal of Comparative Physiology*, **80**, 325–352.

Katsuki, Y. and Suga, N. (1958), 'Electrophysiological studies on hearing in common insects in Japan', *Proceedings of the Japan Academy*, **34**, 633–638.

Katsuki, Y. and Suga, N. (1960), 'Neural mechanisms of hearing in insects', *Journal of Experimental Biology*, **37**, 279–290.

Kay, L. and Pickvance, T. J. (1963), 'Ultrasonic emissions of the lesser horseshoe bat, *Rhinolophus hipposideros* (Bech)', *Proceedings of the Zoological Society of London*, **141**, 163–171.

Kay, R. E. (1969), 'Acoustic signalling and its possible relationship to assembling and navigation in the moth *Heliothis zea*', *Journal of Insect Physiology*, **15**, 989–1001.

Kellogg, W. N. (1953), 'Ultrasonic hearing in the porpoise, *Tursiops truncatus*', *Journal of Comparative and Physiological Psychology*, **46**, 446–450.

Kellogg, W. N. (1958), 'Echo ranging in the porpoise', *Science, New York*, **128**, 982–988.

Kellogg, W. N. (1959a), 'Size discrimination by reflected sound in a bottlenose porpoise', *Journal of Comparative and Physiological Psychology*, **52**, 509–514.

Kellogg, W. N. (1959b), 'Auditory perception of submerged objects by porpoises', *Journal of the Acoustical Society of America*, **31**, 1–6.

Kellogg, W. N. (1961), *Porpoises and Sonar*. University of Chicago Press, Chicago.

Kellogg, W. N. and Kohler, R. (1952), 'Reactions of the porpoise to ultrasonic frequencies', *Science, New York*, **116**, 250–252.

Kellogg, W. N., Kohler, R. and Morris, H. N. (1953), 'Porpoise sounds as sonar signals', *Science, New York*, **117**, 239–243..

Kikkawa, J. (1964), 'The movement, activity and distribution of small rodents, *Clethrionomys glareolus* and *Apodemus sylvaticus* in woodland', *Journal of Animal Ecology*, **33,** 259–299.

Kingdon, J. (1974), *East African Mammals; an Atlas of Evolution in Africa.* Vol. II. Academic Press, London and New York.

Klauder, J. R., Price, A. C., Darlington, S. and Albersheim, W. J. (1960), 'The theory and design of chirp radars', *Bell System Technical Journal,* **39,** 745–808.

Konishi, M. (1969a), 'Hearing, single unit analysis and vocalizations in song-birds', *Science, New York*, **166,** 1178–1181.

Konishi, M. (1969b), 'Time resolution by single auditory neurones in birds', *Nature, London*, **222,** 566–567.

Konishi, M. (1970), 'Comparative neurophysiological studies of hearing and vocalizations in song birds', *Zeitschrift für vergleichende Physiologie*, **66,** 257–272.

Kuhl, W., Schodder, G. R. and Schröder, F. K. (1954), 'Condenser transmitters and microphones with solid dielectric for airborne ultrasonics', *Acustica*, **4,** 519–532.

Kulzer, E. (1958), 'Untersuchungen über die Biologie von Flughunden der Gattung *Rousettus* Gray', *Zeitschrift für Morphologie und Ökologie der Tiere*, **47,** 374–402.

Kulzer, E. (1960), 'Physiologische und Morphologische Untersuchungen über die Erzeugung der Orientierungslaute von Flughunden der Gattung *Rousettus*', *Zeitschrift für vergleichende Physiologie*, **43,** 231–268.

Lagerspetz, K. (1962), 'The post-natal development of homoiothermy and cold resistance in mice', *Experientia*, **18,** 282–284.

Lauterbach, G. (1956), 'Zum Eltern-Kind-Verhältnis bei Nagetieren', *Zoologisches Beiträge, Neue Folge*, **2,** 51–61.

Lawrence, B. and Novick, A. (1963), 'Behavior as a taxonomic clue: relationships of *Lissonycteris* (Chiroptera)', *Breviora*, **184,** 1–16.

Lawrence, B. and Schevill, W. E. (1956), 'The functional anatomy of the delphinid nose', *Bulletin of the Museum of Comparative Zoology at Harvard College*, **114,** 103–151.

Lechtenberg, R. (1971), 'Acoustic response of the B cell in noctuid moths', *Journal of Insect Physiology*, **17,** 2395–2408.

Leston, D. and Pringle, J. W. S. (1963), 'Acoustic behaviour of hemiptera'. In *Acoustic Behaviour of Animals*, ed. Busnel, R.-G., Ch. 14. Elsevier Publishing Co., Amsterdam.

Levine, S. (1962), 'Psychophysiological effects of infantile stimulation'. In *Roots of Behaviour*, ed. Bliss, E. L., Harper, New York.

Lewis, D. B., Pye, J. D. and Howse, P. E. (1971), 'Sound reception in the bush cricket, *Metrioptera brachyptera* (L.) (Orthoptera, Tettigonioidea)', *Journal of Experimental Biology*, **55,** 241–251.

Lilly, J. C. (1962), 'Vocal behaviour of the bottlenose dolphin', *Proceedings of the American Philosophical Society*, **106**, 520–529.

Lilly, J. C. (1963), 'Distress call of the bottlenose dolphin: stimuli and evoked behavioural response', *Science, New York*, **139**, 116–118.

Lilly, J. C. and Miller, A. M. (1961a), 'Sounds emitted by the bottlenose dolphin', *Science, New York*, **133**, 1689–1693.

Lilly, J. C. and Miller, A. M. (1961b), 'Vocal exchanges between dolphins', *Science, New York*, **134**, 1873–1876.

Loftus-Hills, J. J., Littlejohn, M. J. and Hill, K. G. (1971), 'Auditory sensitivity of the crickets *Teleogryllus commodus* and *T. oceanicus*', *Nature New Biology*, **233**, 184–185.

Lottermoser, W. (1952), 'Aufnahme und Analyse von Insektenlauten', *Acustica, Akustische beihefte*. **2**, 66–71.

Mackintosh, J. H. (1970), 'Territory formation by laboratory mice', *Animal Behaviour*, **18**, 177–183.

Manley, G. A. (1971), 'Some aspects of the evolution of hearing in vertebrates', *Nature, London*, **230**, 506–509.

Manley, G. A., Irvine, D. R. F. and Johnstone, B. M. (1972), Frequency response of bat tympanic membrane', *Nature, London*, **237**, 112–113.

Masterton, B., Heffner, H. E. and Ravizza, R. (1969), 'The evolution of human hearing', *Journal of the Acoustical Society of America*, **45**, 966–985.

Matsuzawa, K. (1958), 'Condenser microphones with plastic diaphragms for airborne ultrasonics, I', *Journal of the Physical Society of Japan*, **13**, 1533–1543.

McBride, A. F. (1956), 'Evidence for echolocation by cetaceans', *Deep Sea Research*, **3**, 153–154.

McCormick, J. G., Wever, E. G. and Palin, J. (1970), 'Sound conduction in the dolphin ear', *Journal of the Acoustical Society of America*, **48**, 1418–1428.

McCue, J. J. G. (1961), 'How bats hunt with sound', *National Geographic Magazine*, **119**, 570–578.

McCue, J. J. G. (1969), 'Signal processing by the bat, *Myotis lucifugus*', *Journal of Auditory Research*, **9**, 100–107.

McCue, J. J. G. and Bertolini, A. (1964), 'A portable receiver for ultrasonic waves in air', *Institute of Electrical and Electronics Engineers Transactions, Sonics and Ultrasonics*, **SU–11**, 41–49.

McKay, J. M. (1969), 'The auditory system of *Homorocoryphus* (Tettigonioidea, Orthoptera)', *Journal of Experimental Biology*, **51**, 787–802.

McKay, J. M. (1970), 'Central control of an insect auditory interneurone', *Journal of Experimental Biology*, **53**, 137–146.

Medway, Lord (1959), 'Echo-location among *Collocalia*', *Nature, London*, **184**, 1352–1353.

Medway, Lord (1960), 'Cave Swiftlets', a contribution in *The Birds of Borneo* by Smythies, B. E., Ch. 5, pp. 62–70. Oliver and Boyd, Edinburgh.

Medway, Lord (1967), 'The function of echonavigation among swiftlets', *Animal Behaviour*, **15,** 416–420.

Medway, Lord (1969), 'Studies on the biology of the edible-nest swiftlets of South-East Asia', *Malayan Nature Journal*, **22,** 57–63.

Michelsen, A. (1966), 'Pitch discrimination in the locust ear: observations on single sense cells', *Journal of Insect Physiology*, **12,** 1119–1131.

Michelsen, A. (1968), 'Frequency discrimination in the locust ear by means of four groups of receptor cells', *Nature, London*, **220,** 585–586.

Michelsen, A. (1971a), 'The physiology of the locust ear. I. Frequency sensitivity of single cells in the isolated ear', *Zeitschrift für vergleichende Physiologie*, **71,** 49–62.

Michelsen, A. (1971b), 'The physiology of the locust ear. II. Frequency discrimination based upon resonances in the tympanum', *Zeitschrift für vergleichende Physiologie*, **71,** 63–101.

Miller, J. D. (1970), 'Audibility curve in the chinchilla', *Journal of the Acoustical Society of America*, **48,** 513–523.

Miller, L. A. (1970), 'Structure of the green lacewing tympanal organ (*Chrysopa carnea*, Neuroptera)', *Journal of Morphology*, **131,** 359–382.

Miller, L. A. (1971), 'Physiological responses of green lacewings (*Chrysopa*, Neuroptera) to ultrasound', *Journal of Insect Physiology*, **17,** 491–506.

Miller, L. A. and MacLeod, E. G. (1966), 'Ultrasonic sensitivity: a tympanal receptor in the green lacewing *Chrysopa carnea*', *Science, New York*, **154,** 891–893.

Mills, A. W. (1963), 'Auditory perception of spatial relations'. In *Proceedings of the International Congress on Technology and Blindness*, ed. Clark, L. L., Vol. II, pp. 111–139. The American Foundation for the Blind, Inc., New York.

Mitchell, C., Gillette, L., Vernon, J. and Herman, P. (1970), 'Pure-tone auditory behavioural thresholds in three species of lemurs', *Journal of the Acoustical Society of America*, **48,** 531–535.

Mitchell, C., Vernon, J. and Herman, P. (1971), 'What does the lemur really hear?' *Journal of the Acoustical Society of America*, **50,** 710–711.

Møhl, B. (1968), 'Auditory sensitivity of the common seal in air and water', *Journal of Auditory Research*, **8,** 27–38.

Möhres, F. P. (1950), 'Zur Funktion der Nasenaufsätze bei Fledermäusen', *Naturwissenschaften*, **22,** 526.

Möhres, F. P. (1953), 'Über die Ultraschallorientierung der Hufeisennasen (Chiroptera: Rhinolophinae)', *Zeitschrift für vergleichende Physiologie*, **34,** 547–588.

Möhres, F. P. (1967a), 'Ultrasonic orientation in megadermatid bats'. In *Animal Sonar Systems, Biology and Bionics*, ed. Busnel, R.-G. Vol. I, pp. 115–127. Laboratoire de Physiologie Acoustique, INRA–CNRZ, Jouy-en-Josas.

Möhres, F. P. (1967b), 'Communicative character of sonar signals in bats'. In *Animal Sonar Systems, Biology and Bionics*, ed. Busnel, R.-G., Vol. II, pp. 939–945. Laboratoire de Physiologie Acoustique, INRA–CNRZ, Jouy-en-Josas.

Möhres, F. P. and Kulzer, E. (1955), 'Ein neuer, kombinierter Typ der Ultraschallorientierung bei Fledermäusen', *Naturwissenschaften*, **42**, 131–132.

Möhres, F. P. and Kulzer, E. (1956a), 'Über die Orientierung der Flughunde (Chiroptera, Pteropodidae)', *Zeitschrift für vergleichende Physiologie*, **38**, 1–29.

Möhres, F. P. and Kulzer, E. (1956b), 'Untersuchungen über die Ultraschallorientierung von vier afrikanischen Fledermaus-familien', *Verhandlungen der Deutschen zoologischen Gesellschaft*, 1955, 59–65.

Möhres, F. P. and Neuweiler, G. (1966), 'Die Ultraschallorientierung der Grossblatt-Fledermäuse (Chiroptera-Megadermatidae)', *Zeitschrift für vergleichende Physiologie*, **53**, 195–227.

Morris, G. K. (1970), 'Sound analysis of *Metrioptera sphagnorum* (Orthoptera: Tettigoniidae)', *Canadian Entomologist*, **102**, 363–368.

Morris, G. K. and Pipher, R. E. (1967), 'Tegminal amplifiers and spectrum consistencies in *Conocephalus nigropleurum* (Bruner) Tettigoniidae', *Journal of Insect Physiology*, **13**, 1075–1085.

Morris, H. N., Kohler, R. and Kellogg, W. N. (1953), 'Ultrasonic porpoise communication', *Electronics*, August 1953, 208–214.

Negus, V. E. (1962), *The Comparative Anatomy and Physiology of the Larynx*. Hafner Publishing Co., London.

Neuweiler, G. (1967), 'Interaction of other sensory systems with the sonar system'. In *Animal Sonar Systems, Biology and Bionics*, ed. Busnel, R.-G., Vol. I, pp. 509–533. Laboratoire de Physiologie Acoustique, INRA–CNRZ, Jouy-en-Josas.

Neuweiler, G. (1970a), 'Neurophysiologische Untersuchungen zum Echoortungssystem der Grossen Hufeisennase, *Rhinolophus ferrumequinum* Schreber, 1774', *Zeitschrift für vergleichende Physiologie*, **67**, 273–306.

Neuweiler, G. (1970b), 'Neurophysiological investigations in the colliculus inferior of *Rhinolophus ferrumequinuum*', *Bijdragen tot de Dierkunde*, **40**, 59–61.

Neuweiler, G. and Möhres, F. P. (1967), 'The role of spatial memory in orientation'. In *Animal Sonar Systems, Biology and Bionics*, ed. Busnel, R.-G., Vol. I, pp. 129–140. Laboratoire de Physiologie Acoustique, INRA–CNRZ, Jouy-en-Josas.

Neuweiler, G., Schuller, G. and Schnitzler, H.-U. (1971), 'On- and off-responses in the inferior colliculus of the greater horseshoe bat to pure tones', *Zeitschrift für vergleichende Physiologie*, **74**, 57–63.

Nitschke, W., Bell, R. W. and Zachman, T. (1972), 'Distress vocalizations of young in three inbred strains of mice', *Developmental Psychobiology*, **5**, 363–370.

Nocke, H. (1971), 'Biophysik der Schallerzeugung durch die Vorderflügel der Grillen', *Zeitschrift für vergleichende Physiologie*, **74**, 272–314.

Nocke, H. (1972), 'Physiological aspects of sound communication in crickets (*Gryllus campestris L*)', *Journal of Comparative Physiology*, **80**, 141–162.

Noirot, E. (1964a), 'Changes in responsiveness to young in the adult mouse. I. The problematical effect of hormones', *Animal Behaviour*, **12**, 52–58.

Noirot, E. (1964b), 'Changes in responsiveness to young in the adult mouse. II. The effect of external stimuli', *Journal of Comparative and Physiological Psychology*, **57**, 97–99.

Noirot, E. (1964c), 'Changes in responsiveness to young in the adult mouse. III. The effect of an initial contact with a strong stimulus', *Animal Behaviour*, **12**, 442–445.

Noirot, E. (1965), 'Changes in responsiveness to young in the adult mouse. IV. The effect of immediately preceding performances', *Behaviour*, **24**, 318–325.

Noirot, E. (1966a), 'Ultrasons et comportements maternels chez les petits rongeurs', *Annales de la Société royale Zoologique de Belgique*, **95**, 47–56.

Noirot, E. (1966b), 'Ultrasounds in young rodents. I. Changes with age in albino mice', *Animal Behaviour*, **14**, 459–462.

Noirot, E. (1968), 'Ultrasounds in young rodents. II. Changes with age in albino rats', *Animal Behaviour*, **16**, 129–134.

Noirot, E. (1969a), 'Changes in responsiveness to young in the adult mouse. V. Priming', *Animal Behaviour*, **17**, 542–546.

Noirot, E. (1969b), 'Interactions between reproductive and territorial behaviour in female mice', *International Mental Health Research Newsletter*, **11**, 10–11.

Noirot, E. (1972), 'The onset of maternal behaviour in rats, hamsters and mice', *Advances in the Study of Behaviour*, **4**, 107–145.

Noirot, E. and Pye, J. D. (1969), 'Sound analysis of ultrasonic distress calls of mouse pups as a function of their age', *Animal Behaviour*, **17**, 340–349.

Norris, K. S. (1969), 'The echolocation of marine mammals'. In *The Biology of Marine Mammals*, ed. Andersen, H. T., Ch. 10. Academic Press, New York and London.

Norris, K. S. and Evans, W. E. (1967), 'Directionality of echolocation clicks in the rough-tooth porpoise *Steno bredanensis* (Lesson)'. In *Marine Bioacoustics*, ed. Tavolga, W. N., pp. 305–316. Pergamon Press, Oxford.

Norris, K. S., Evans, W. E. and Turner, R. N. (1967), 'Echolocation in an

Atlantic bottlenose porpoise during discrimination'. In *Animal Sonar Systems, Biology and Bionics*, ed. Busnel, R.-G., Vol. I, pp. 409–437. Laboratoire de Physiologie Acoustique, INRA–CNRZ, Jouy-en-Josas.

Norris, K. S. and Prescott, J. H. (1961), 'Observations on Pacific cetaceans of California and Mexican waters', *University of California Publications in Zoology*, **63,** 291–402.

Norris, K. S., Prescott, J. H., Asa-Dorian, P. V. and Perkins, P. (1961), 'An experimental demonstration of echo-location behaviour in the porpoise, *Tursiops truncatus* (Montagu)', *Biological Bulletin*, **120,** 163–176.

Novick, A. (1958a), 'Orientation in paleotropical bats. I. Microchiroptera', *Journal of Experimental Zoology*, **138,** 81–154.

Novick, A. (1958b), 'Orientation in paleotropical bats. II. Megachiroptera', *Journal of Experimental Zoology*, **137,** 443–462.

Novick, A. (1959), 'Acoustic orientation in the cave swiftlet', *Biological Bulletin*, **117,** 497–503.

Novick, A. (1962), 'Orientation in neotropical bats. I. Natalidae and Emballonuridae', *Journal of Mammalogy*, **43,** 449–455.

Novick A. (1963a), 'Orientation in neotropical bats. II. Phyllostomatidae and Desmodontidae', *Journal of Mammalogy*, **44,** 44–56.

Novick, A. (1963b), 'Pulse duration in the echolocation of insects by the bat, *Pteronotus*', *Ergebnisse der Biologie*, **26,** 21–26.

Novick, A. (1965), 'Echolocation of flying insects by the bat, *Chilonycteris psilotis*', *Biological Bulletin*, **128,** 297–314.

Novick, A. and Gould, E. (1964), 'Comparative study of echolocation in the Tenrecidae of Madagascar and other Old World insectivores'. Final research report No. AFOSR. 64–0245 Yale University.

Novick, A. and Griffin, D. R. (1961), 'Laryngeal mechanisms in bats for the production of orientation sounds', *Journal of Experimental Zoology*, **148,** 125–146.

Novick, A. and Vaisnys, J. R. (1964), 'Echolocation of flying insects by the bat, *Chilonycteris parnellii*', *Biological Bulletin*, **127,** 478–488.

Noyes, A. and Pierce, G. W. (1938), 'Apparatus for acoustic research in the supersonic frequency range', *Journal of the Acoustical Society of America*, **9,** 205–211.

Okon, E. E. (1970a), 'The effect of environmental temperature on the production of ultrasound by isolated, non-handled, albino mouse pups', *Journal of Zoology, London*, **162,** 71–83.

Okon, E. E. (1970b), 'The ultrasonic responses of albino mouse pups to tactile stimuli', *Journal of Zoology, London*, **162,** 485–492.

Okon, E. E. (1971a), 'Temperature relations of vocalisation in infant golden hamsters and wistar rats', *Journal of Zoology, London*, **164,** 227–237.

Okon, E. E. (1971b), *Motivation for the Production of Ultrasound in Infant Rodents*. Unpublished Ph.D thesis, University of London.

Okon, E. E. (1972), 'Factors affecting ultrasound production in infant rodents', *Journal of Zoology, London*, **168,** 139–148.

Payne, R. S. (1961), 'Acoustic orientation of prey by the Barn Owl, *Tyto alba*'. Technical report 1960. Division of Engineering and Applied Physics, Harvard University, Cambridge, Massachusetts. 61 pp.

Payne, R. S. (1970), *Songs of the humpback whale*. Long playing record, no. SWR–11. Communications Research Machines Inc., Del Mar, California.

Payne, R. S. (1971), 'Acoustic location of prey by barn owls (*Tyto alba*)', *Journal of Experimental Biology*, **54,** 535–573.

Payne, R. S. and McVay, S. (1971), 'Songs of humpback whales', *Science, New York*, **173,** 587–597.

Payne, R. S., Roeder, K. D. and Wallman, J. (1966), 'Directional sensitivity of the ears of noctuid moths', *Journal of Experimental Biology*, **44,** 17–31.

Payne, R. S. and Webb, B. (1971), 'Orientation by means of long range acoustic signalling in baleen whales'. In *Orientation: Sensory Basis*, ed. Adler, H. E., pp. 110–141. *Annals of the New York Academy of Sciences*, **188.**

Pestalozza, A. G. and Davis, H. (1956), 'Electric responses of the guinea pig ear to high audio frequencies', *American Journal of Physiology*, **185,** 599–608.

Peterson, E. A. (1968), 'High frequency and ultrasonic overstimulation of the guinea pig ear', *Journal of Auditory Research*, **8,** 43–61.

Peterson, E. A., Heaton, W. C. and Wruble, S. (1969), 'Levels of auditory response in fissipede carnivores', *Journal of Mammalogy*, **50,** 566–578.

Phillips, L. H. and Konishi, M. (1973), 'Control of aggression by singing in crickets', *Nature, London*, **241,** 64.

Pielemeier, W. H. (1946), 'Supersonic insects', *Journal of the Acoustical Society of America*, **17,** 337–338.

Pierce, G. W. (1948), *The Songs of Insects*. Harvard University Press, Cambridge, Massachusetts.

Pierce, G. W. and Griffin, D. R. (1938), 'Experimental determination of supersonic notes emitted by bats', *Journal of Mammalogy*, **19,** 454–455.

Poduschka, W. (1968a), 'Über die Wahrnehmung von Ultraschall beim Igel, *Erinaceus europaeus roumanicus*', *Zeitschrift für vergleichende Physiologie*, **61,** 420–426.

Poduschka, W. (1968b), 'Ergänzungen zum Wissen über *Erinaceus e. roumanicus* und kritische Überlegungen zur bisherigen Literatur über europäische Igel', *Zeitschrift für Tierpsychologie*, **26,** 761–804.

Pollak, G., Henson, O. W. Jr and Novick, A. (1972), 'Cochlear microphonic audiograms in the "pure tone" bat *Chilonycteris parnellii parnellii*', *Science, New York*, **176,** 66–68.

Popov, A. V. (1971), 'Synaptic transformations in the auditory system of insects'. In *Sensory Processes at the Neuronal and Behavioural Levels*, ed. Gersuni, G. V., Ch. 17, Academic Press, London and New York.

Poulter, T. C. (1963), 'Sonar signals of the sea lion', *Science, New York*, **139,** 753–755.

Poulter, T. C. (1966), 'The active use of sonar by the California sea lion, *Zalophus californianus* (Lesson)', *Journal of Auditory Research*, **6,** 165–173.

Poulter, T. C. (1967), 'Systems of echolocation'. In *Animal Sonar Systems, Biology and Bionics*, ed. Busnel, R.-G., Vol. I, pp. 157–186. Laboratoire de Physiologie Acoustique, INRA–CNRZ, Jouy-en-Josas.

Poulter, T. C. (1969), 'Sonar of penguins and fur seals', *Proceedings of the California Academy of Sciences*, **34,** 363–380.

Poulter, T. C. and Jennings, R. A. (1969), 'Sonar discrimination ability of the California sea lion, *Zalophus californianus*', *Proceedings of the California Academy of Sciences*, **34,** 381–389.

Pringle, J. W. S. (1954), 'A physiological analysis of cicada song', *Journal of Experimental Biology*, **31,** 525–560.

Pumphrey, R. J. (1940), 'Hearing in insects', *Biological Reviews*, **15,** 107–132.

Pumphrey, R. J. (1950), 'Upper limit of frequency for human hearing', *Nature, London*, **166,** 571.

Purves, P. E. (1966), 'Anatomy and physiology of the outer and middle ear in cetaceans'. In *Whales, Dolphins and Porpoises*, ed. Norris, K. S., Ch. 16. University of California Press, Berkeley, California.

Purves, P. E. (1967), 'Anatomical and experimental observations on the cetacean sonar system'. In *Animal Sonar Systems, Biology and Bionics*, ed. Busnel, R.-G., Vol. I, pp. 197–270. Laboratoire de Physiologie Acoustique, INRA–CNRZ, Jouy-en-Josas.

Pye, A. (1966a), 'The structure of the cochlea in Chiroptera. I. Microchiroptera: Emballonuroidea and Rhinolophoidea', *Journal of Morphology*, **118,** 495–510.

Pye, A. (1966b), 'The Megachiroptera and Vespertilionoidea of the Microchiroptera', *Journal of Morphology*, **119,** 101–120.

Pye, A. (1967), 'The structure of the cochlea in Chiroptera. III. Microchiroptera: Phyllostomoidea', *Journal of Morphology*, **121,** 241–254.

Pye, A. (1970a), 'The aural anatomy of bats', *Bijdragen tot de Dierkunde*, **40,** 67–70.

Pye, A. (1970b), 'The structure of the cochlea in Chiroptera. A selection of Microchiroptera from Africa', *Journal of Zoology*, **162,** 335–342.

Pye, A. (1971), 'The effect of exposure to intense pure tones on the hearing organ of animals', *Revista de Acustica*, **2,** 199–203.

Pye, A. (1972), 'Variations in the structure of the ear in the different mammalian species', *Sound*, **6,** 14–18.

Pye, A. (1973), 'The structure of the cochlea in Microchiroptera from Africa', *Periodicum biologorum*, **75,** 83–87.

Pye, A. and Hinchcliffe, R. (1968), 'Structural variations in the mammalian middle ear', *Medical and Biological Illustration*, **18,** 122–127.

Pye, J. D. (1960), 'A theory of echolocation by bats', *Journal of Laryngology and Otology*, **74,** 718–729.

Pye, J. D. (1963), 'Mechanisms of echolocation', *Ergebnisse der Biologie*, **26,** 12–20.

Pye, J. D. (1967a), 'Synthesizing the waveforms of bats' pulses'. In *Animal Sonar Systems, Biology and Bionics*, ed. Busnel, R.-G., Vol I, pp. 43–64. Laboratoire de Physiologie Acoustique, INRA–CNRZ, Jouy-en-Josas.

Pye, J. D. (1967b), 'Theories of sonar systems and their application to biological organisms: discussion.' In *Animal Sonar Systems, Biology and Bionics*, ed. Busnel, R.-G., Vol II, pp. 1121–1136. Laboratoire de Physiologie Acoustique, INRA–CNRZ, Jouy-en-Josas.

Pye, J. D. (1968a), 'Animal sonar in air', *Ultrasonics*, **6,** 32–38.

Pye, J. D. (1968b), 'Hearing in bats'. In *Hearing Mechanisms in Vertebrates*, Ciba Foundation Symposium, eds. de Reuck, A. V. S. and Knight, J., pp. 66–84. J. and A. Churchill, London.

Pye, J. D. (1969), 'The diversity of bats', *Science Journal*, **5,** 47–52.

Pye, J. D. (1971), 'Bats and fog', *Nature, London*, **229,** 572–574.

Pye, J. D. (1972), 'Bimodal distribution of constant frequencies in some hipposiderid bats (Mammalia: Hipposideridae)', *Journal of Zoology, London*, **166,** 323–335.

Pye, J. D. (1973), 'Echolocation by constant frequency in bats', *Periodicum biologorum*, **75,** 21–26.

Pye, J. D. (in preparation) *Bat Sonar Signals*. English Universities Press, London.

Pye, J. D. and Flinn, M. (1964), 'Equipment for detecting animal ultrasound', *Ultrasonics*, **2,** 23–28.

Pye, J. D., Flinn, M. and Pye, A. (1962), 'Correlated orientation sounds and ear movements of horseshoe bats', *Nature, London*, **196,** 1186–1188.

Pye, J. D., Godson, S. and Au, Y. M. (in preparation). (A new wideband bat detector.)

Pye, J. D. and Roberts, L. H. (1970), 'Ear movements in a hipposiderid bat', *Nature, London*, **225,** 285–286.

Ragge, D. R. (1955), 'Le problème de la stridulation des femelles acridinae (Orthoptera, Acrididae)'. In *Colloque sur l'acoustique des Orthoptères*, ed. Busnel, R.-G., pp. 171–174. Fascicule hors serie des *Annales des épiphyties*. Institute National de la Recherche Agronomique, Paris.

Ragge, D. R. (1965), *Grasshoppers, Crickets and Cockroaches of the British Isles* (Wayside and Woodland series). Frederick Warne and Co. Ltd., London.

Ralls, K. (1967), 'Auditory sensitivity in mice: *Peromyscus* and *Mus musculus*', *Animal Behaviour*, **15,** 123–128.

Ravizza, R. J., Heffner, H. E. and Masterton, B. (1969a), 'Hearing in primitive mammals. I. Opossum (*Didelphis virginianus*)', *Journal of Auditory Research*, **9,** 1–7.

Ravizza, R. J., Heffner, H. E. and Masterton, B. (1969b), 'Hearing in primitive mammals. II. Hedgehog (*Hemiechinus auritus*)', *Journal of Auditory Research*, **9,** 8–11.

Rayleigh, Lord (1894), *The Theory of Sound*, 2nd edition. Macmillan, London (Dover Paperbacks, 1945).

Regen, J. (1913), 'Über die Anlockung des Weibchens von *Gryllus campestris* L. durch telephonisch übertragene Stridulationslaute des Mannchens', *Archiv für Physiologische Menchen und Tiere*, **155,** 193–200.

Regen, J. (1926), 'Über die Beeinflussung der Stridulation von *Thamnotrizon apterus*. Fab. durch Künstlich erzeugte Töne und verschiedenartige Geräusche', *Sitzungberichte der Akademie der Wissenschaften in Wien*, **135,** 329–368.

Reid, K. H. (1971), 'Periodical cicada: mechanism of sound production', *Science, New York*, **172,** 949–951.

Reysenbach de Haan, F. W. (1958), 'Hearing in whales', *Acta Oto-Laryngologica Supplementum*, **134,** 1–114.

Reysenbach de Haan, F. W. (1966), 'Listening underwater: thoughts on sound and cetacean hearing'. In *Whales, Dolphins and Porpoises*, ed. Norris, K. S., Ch. 26. University of California Press, Berkeley, California.

Rheinlaender, J., Kalmring, K. and Römer, H. (1972), 'Akustische Neuronen mit T-Struktur im Bauchmark von Tettigoniiden', *Journal of Comparative Physiology*, **77,** 208–224.

Richardson, E. G. (1952), *Ultrasonic Physics*. Elsevier, Amsterdam.

Riley, D. A. and Rosenzweig, M. R. (1957), 'Echolocation in rats', *Journal of Comparative and Physiological Psychology*, **50,** 323.

Roberts, L. H. (1972a), 'Variable resonance in constant frequency bats', *Journal of Zoology, London*, **166,** 337–348.

Roberts, L. H. (1972b), 'Correlation of respiration and ultrasound production in rodents and bats', *Journal of Zoology*, **168,** 439–449.

Roberts, L. H. (1973a), 'Cavity resonances in the production of orientation cries', *Periodicum biologorum*, **75,** 27–32.

Roberts, L. H. (1973b), *Comparative Studies of Sound Production in Small Mammals with Special Reference to Ultrasound.* Unpublished Ph.D thesis, University of London.

Robin, H. A. (1881), 'Recherches anatomiques sur les mammifères de l'ordre des chiroptères'. *Annales des sciences naturelles*, **12,** 1–180.

Roeder, K. D. (1959), 'A physiological approach to the relation between prey and predator', *Smithsonian Institution Miscellaneous Collections*, **137,** 287–308.

Roeder, K. D. (1962), 'The behaviour of free flying moths in the presence of artificial ultrasonic pulses', *Animal Behaviour*, **10,** 300–304.

Roeder, K. D. (1963), (revised 1967) *Nerve Cells and Insect Behavior*. Harvard University Press, Cambridge, Massachusetts.

Roeder, K. D. (1964), 'Aspects of the noctuid tympanic nerve response having significance in the avoidance of bats', *Journal of Insect Physiology*, **10,** 529–546.

Roeder, K. D. (1965), 'Moths and ultrasound', *Scientific American*, **212,** 94–102.

Roeder, K. D. (1966a), 'Auditory system of noctuid moths', *Science, New York*, **154,** 1515–1521.

Roeder, K. D. (1966b), 'Acoustic sensitivity of the noctuid tympanic organ and its range for the cries of bats', *Journal of Insect Physiology*, **12,** 843–859.

Roeder, K. D. (1966c), 'A differential anemometer for measuring the turning tendency of insects in stationary flight', *Science, New York*, **153,** 1634–1636.

Roeder, K D. (1966d), 'Interneurones of the thoracic nerve cord activated by tympanic nerve fibres in noctuid moths', *Journal of Insect Physiology*, **12,** 1227–1244.

Roeder, K. D. (1967a), 'Prey and predator', *Bulletin of the Entomological Society of America*, **13,** 6–9.

Roeder, K. D. (1967b), 'Turning tendency of moths exposed to ultrasound while in stationary flight', *Journal of Insect Physiology*, **13,** 873–888.

Roeder, K. D. (1969a), 'Acoustic interneurons in the brain of noctuid moths', *Journal of Insect Physiology*, **15,** 825–838.

Roeder, K. D. (1969b), 'Brain interneurons in noctuid moths: differential suppression by high sound intensities', *Journal of Insect Physiology*, **15,** 1713–1718.

Roeder, K. D. (1969c), 'Acoustic interneurones in the brain of noctuid moths', *Journal of Experimental Zoology*, **134**, 127–157.

Roeder, K. D. (1970), 'Episodes in insect brains', *American Scientist*, **58,** 378–389.

Roeder, K. D. (1971), 'Acoustic alerting mechanisms in insects'. In *Orientation: Sensory Basis*, ed. Adler, H. E., pp. 63–79. *Annals of the New York Academy of Sciences*, **188.**

Roeder, K. D. (1972), 'Acoustic and mechanical sensitivity of the distal lobe of the pilifer in choerocampine hawkmoths', *Journal of Insect Physiology*, **18,** 1249–1264.

Roeder, K. D. and Payne, R. S. (1966), 'Acoustic orientation of a moth in flight by means of two sense cells', *Symposium of the Society for Experimental Biology*, **20,** 251–272.

Roeder, K. D. and Treat, A. E. (1957), 'Ultrasonic reception by the tympanic organ of noctuid moths', *Journal of Experimental Zoology*, **134,** 127–158.

Roeder, K. D. and Treat, A. E. (1961a), 'The reception of bat cries by the tympanic organ of noctuid moths'. In *Sensory Communication*, ed. Rosenblith, W., Ch. 28. Massachusetts Institute of Technology Press, Cambridge, Massachusetts.

Roeder, K. D. and Treat, A. E. (1961b), 'The detection and evasion of bats by moths', *American Scientist*, **49,** 135–148.

Roeder, K. D. and Treat, A. E. (1970), 'An acoustic sense in some hawkmoths (Choerocampinae)', *Journal of Insect Physiology*, **16,** 1069–1086.

Roeder, K. D., Treat, A. E. and Vande Berg, J. S. (1968), 'Auditory sense in certain sphingid moths', *Science, New York*, **159,** 331–333.

Roeder, K. D., Treat, A. E. and Vande Berg, J. S. (1970), 'Distal lobe of the pilifer: an ultrasonic receptor in choerocampine hawkmoths', *Science, New York*, **170,** 1098–1099.

Rollinat, R. and Trouessart, E. (1900), 'Sur le sens de la direction chez les chauves-souris', *Comptes rendus des séances de la Société de Biologie*, **52,** 604–607.

Rose, M., Savornin, J. and Casanova, J. (1948), 'Sur l'emission d'ondes ultra-sonores par les Abeilles domestiques', *Comptes rendus hebdomadaire des séances de l'Académie des Sciences*, **227,** 912–913.

Rosenzweig, M., Riley, D. and Krech, D. (1955), 'Evidence for echolocation in the rat', *Science, New York*, **121,** 600.

Rowell, C. F. H. and McKay, J. M. (1969a), 'An acridid interneurone. I. Functional connexions and response to single sounds', *Journal of Experimental Biology*, **51,** 231–245.

Rowell, C. H. F. and McKay, J. M. (1969b), 'An acridid interneurone. II. Habituation, variation in response level and central control', *Journal of Experimental Biology*, **51,** 247–260.

Russell, P. A. (1971), ' "Infantile stimulation" in rodents: a consideration of possible mechanisms', *Psychological Bulletin*, **75,** 192–202.

Saby, J. S. and Thorpe, H. A. (1946), 'Ultrasonic ambient noise in tropical jungles', *Journal of the Acoustical Society of America*, **18,** 271–273.

Sales, G. D. (*née* Sewell) (1972a), 'Ultrasound and aggressive behaviour in rats and other small mammals', *Animal Behaviour*, **20,** 88–100.

Sales, G. D. (*née* Sewell) (1972b), 'Ultrasound and mating behaviour in rodents with some observations on other behavioural situations', *Journal of Zoology, London*, **168,** 149–164.

Schaller, F. (1951), 'Lauterzeugung und Horvermögen von *Corixa* (*Callicorixa*) *striata* L', *Zeitschrift für vergleichende Physiologie*, **33,** 476–486.

Schaller, F. (1952), 'Ultraschall im Tierreich', *Physikalish Blätter*, **8,** 543–546.

Schaller, F. and Timm, C. (1949), 'Schallreaktonen bei Nachfaltern', *Experentia*, **5,** 162.

Schaller, F. and Timm, C. (1950), 'Das Hovermögen der Nachtschmetterlinge', *Zeitschrift für vergleichende Physiologie*, **32,** 468–481.

Schevill, W. E. (1968), 'Sea lion echoranging?' *Journal of the Acoustical Society of America*, **43,** 1458.

Schevill, W. E. and Lawrence, B. (1953a), 'High frequency auditory response of a bottlenose porpoise, *Tursiops truncatus* (Montagu)', *Journal of the Acoustical Society of America*, **25,** 1016–1017.

Schevill, W. E. and Lawrence, B. (1953b), 'Auditory response of a bottlenosed porpoise, *Tursiops truncatus*, to frequencies above 100 kHz', *Journal of Experimental Zoology*, **124,** 147–165.

Schevill, W. E. and Lawrence, B. (1956), 'Food-finding by a captive porpoise (*Tursiops truncatus*)', *Breviora*, **53,** 1–5.

Schevill, W. E. and Watkins, W. A. (1966), 'Sound structure and directionality in *Orcinus* (killer whale)', *Zoologica, New York*, **51,** 71–76.

Schevill, W. E., Watkins, W. A. and Ray, C. (1963), 'Underwater sounds of pinnipeds', *Science, New York*, **141,** 50–53.

Schleidt, W. M. (1948), 'Töne hoher Frequenz bei Mäusen', *Experentia*, **4,** 145–146.

Schleidt, W. M. (1951), 'Töne hoher Frequenz bei Mäusen', *Experentia*, **7,** 65–66.

Schleidt, W. M. (1952), 'Reactionen auf Töne hoher Frequenz bei Nagern', *Naturwissenschaften*, **39,** 69–70.

Schneider, H. (1961), 'Die Ohrmuskulature von *Asellia tridens* Geoffr. (Hipposideridae) und *Myotis myotis* Borkh. (Vespertilionidae)', *Zoologische Jahrbücher. Abteilung Anatomie und Ontogonie*, **79,** 93–122.

Schneider, H. and Möhres, F. P. (1960), 'Die Ohrbewegungen der Hufeisenfledermäuse (Chiroptera, Rhinolophidae) und der Mechanismus des Bildhörens', *Zeitschrift für vergleichende Physiologie*, **44,** 1–40.

Schnitzler, H.-U. (1967), 'Discrimination of thin wires by flying horseshoe bats'. In *Animal Sonar Systems, Biology and Bionics*, ed. Busnel, R.-G., Vol. 1, pp. 69–87. Laboratoire de Physiologie Acoustique, INRA–CNRZ, Jouy-en-Josas.

Schnitzler, H.-U. (1968), 'Die Ultraschall–Ortungslaute der Hufeisen–Fledermäuse (Chiroptera–Rhinolophidae) in verschiedenen–Orientierungssituationen', *Zeitschrift für vergleichende Physiologie*, **57,** 376–408.

Schnitzler, H.-U. (1970a), 'Echoortung bei der Fledermaus, *Chilonycteris rubiginosa*', *Zeitschrift für vergleichende Physiologie*, **68,** 25–38.

Schnitzler, H.-U. (1970b), 'Comparison of the echolocation behavior in *Rhinolophus ferrum-equinum* and *Chilonycteris rubiginosa*', *Bijdragen tot de Dierkunde*, **40,** 77–80.

Schnitzler, H.-U. (1971), 'Fledermäuse im Windkanal', *Zeitschrift für vergleichende Physiologie*, **73,** 209–221.

Schnitzler, H.-U. (1973), 'Control of doppler shift compensation in the greater horseshoe bat, *Rhinolophus ferrumequinum*', *Journal of Comparative Physiology*, **82,** 79–92.

Schuller, G. (1972), 'Echoortung bei *Rhinolophus ferrumequinum* mit frequenzmodulierten Lauten: evoked potentials im colliculus inferior', *Journal of Comparative Physiology*, **77**, 306–331.

Schuller, G., Neuweiler, G. and Schnitzler, H.-U. (1971), 'Collicular responses to the frequency modulated final part of echolocation sounds in *Rhinolophus ferrumequinum*', *Zeitschrift für vergleichende Physiologie*, **74**, 153–155.

Schusterman, R. J. (1966), 'Underwater click vocalizations by a California sea lion: effects of visibility', *Psychological Record*, **16**, 129–136.

Schustermann, R. J. (1967), 'Perception and determinants of underwater vocalization in the California sea lion'. In *Animal Sonar Systems, Biology and Bionics*, ed. Busnel, R.-G., Vol. I, pp. 535–617. Laboratoire de Physiologie Acoustique, INRA–CNRZ, Jouy-en-Josas.

Schusterman, R. J., Balliet, R. F. and Nixon, J. (1972), 'Underwater audiogram of the California sea lion by the conditioned vocalization technique', *Journal of the Experimental Analysis of Behaviour*, **17**, 339–350.

Schusterman, R. J. and Feinstein, S. H. (1965), 'Shaping and discriminative control of underwater click vocalizations in a California sea lion', *Science, New York*, **150**, 1743–1744.

Schusterman, R. J., Kellogg, W. N. and Rice, C. E. (1965), 'Underwater visual discrimination by the California sea lion', *Science, New York*, **147**, 1594–1596.

Schwabe, J. (1906), 'Beiträge zur Morphologie und Histologie der tympanalen Sinnesapparate der Orthoptera', *Zoologica, Stuttgart*, **20**, 1–154.

Schwartzkopff, J. (1968), 'Structure and function of the ear and of the auditory brain areas in birds'. In *Hearing Mechanisms in Vertebrates*, Ciba Foundation Symposium, eds. de Reuck, A. V. S. and Knight, J., pp. 41–63. Churchill, London.

Scroggie, M. G. (1955), *Second Thoughts on Radio Theory*. Iliffe, London.

Scroggie, M. G. (1963), *Essays in Electronics*. Iliffe, London.

Seward, J. P. (1945), 'Aggressive behaviour in the rat. I. General characteristics: age and sex differences', *Journal of Comparative Psychology*, **38**, 175–197.

Sewell, G. D. (1967), 'Ultrasound in adult rodents', *Nature, London*, **215**, 512.

Sewell, G. D. (1968), 'Ultrasound in rodents', *Nature, London*, **217**, 682–683.

Sewell, G. D. (1969), *Ultrasound in Small Mammals*. Unpublished Ph.D thesis. University of London.

Sewell, G. D. (1970a), 'Ultrasonic signals from rodents', *Ultrasonics*, **8**, 26–30.

Sewell, G. D. (1970b), 'Ultrasonic communication in rodents', *Nature, London*, **227**, 410.

Shaver, H. N. and Poulter, T. C. (1967), 'Sea lion echo ranging', *Journal of the Acoustical Society of America*, **42**, 428–437.

Shaver, H. N. and Poulter, T. C. (1968), 'Sea lion echo ranging', *Journal of the Acoustical Society of America*, **43,** 1459.

Shaw, K. C. (1968), 'An analysis of the phonoresponse of males of the true katydid *Pterophylla camellifolia* (Fabricius) (Orthoptera: Tettigoniidae)', *Behaviour*, **31,** 203–260.

Simmons, J. A. (1968), 'Echolocation: auditory clues for range perception in bats', *Proceedings of 76th Annual Convention of the American Psychological Association*, pp. 301–302.

Simmons, J. A. (1969), 'Acoustic radiation patterns for the echolocating bats *Chilonycteris rubiginosa* and *Eptesicus fuscus*', *Journal of the Acoustical Society of America*, **46,** 1054–1056.

Simmons, J. A. (1970), 'Distance perception by echolocation: the nature of echo signal-processing in the bat', *Bijdragen tot de Dierkunde*, **40,** 87–90.

Simmons, J. A. (1971), 'Echolocation in bats: signal processing of echoes for target range', *Science, New York*, **171,** 925–928.

Simmons, J. A. and Vernon, J. A. (1971), 'Echolocation: discrimination of targets by the bat *Eptesicus fuscus*', *Journal of Experimental Zoology*, **176,** 315–328.

Simmons, J. A., Wever, E. G. and Pylka, J. M. (1971), 'Periodical cicada: sound production and hearing', *Science, New York*, **171,** 212–213.

Skolnik, M. I. (1962), *Introduction to Radar Systems*. McGraw–Hill, New York.

Smith, J. C. (1972), 'Sound production by infant *Peromyscus maniculatus* (Rodentia: Myomorpha)', *Journal of Zoology, London*, **168,** 369–379.

Stebbins, W. C., Green, S. and Miller, F. L. (1966), 'Auditory sensitivity of the monkey', *Science, New York*, **153,** 1646–1647.

Suga, N. (1960), 'Peripheral mechanism of hearing in locust', *Japanese Journal of Physiology*, **10,** 533–546.

Suga, N. (1961), 'Functional organization of two tympanic neurons in noctuid moths', *Japanese Journal of Physiology*, **11,** 666–677.

Suga, N. (1963), 'Central mechanism of hearing and sound localisation in insects', *Journal of Insect Physiology*, **9,** 867–873.

Suga, N. (1964a), 'Single unit activity in cochlear nucleus and inferior colliculus of echo-locating bats', *Journal of Physiology, London*, **172,** 449–474.

Suga, N. (1964b), 'Recovery cycles and responses to frequency modulated tone pulses in auditory neurons of echo-locating bats', *Journal of Physiology, London*, **175,** 50–80.

Suga, N. (1965a), 'Analysis of frequency modulated sound by auditory neurons of echo-locating bats', *Journal of Physiology, London*, **179,** 26–53.

Suga, N. (1965b), 'Functional properties of auditory neurons in the cortex of echo-locating bats', *Journal of Physiology, London*, **181,** 671–700.

Suga, N. (1965c), 'Responses of cortical auditory neurons to frequency modulated sounds in echo-locating bats', *Nature, London*, **206,** 890–891.

Suga, N. (1966), 'Ultrasonic production and its reception in some neotropical Tettigoniidae', *Journal of Insect Physiology*, **12**, 1039–1050.
Suga, N. (1967a), 'Neural processing involved in sonar: discussion'. In *Animal Sonar Systems, Biology and Bionics*, ed. Busnel, R.-G., Vol. II, pp. 1004–1020. Laboratoire de Physiologie Acoustique, INRA–CNRZ, Jouy-en-Josas.
Suga, N. (1967b), 'Hearing in some arboreal edentates in terms of cochlear microphonics and neural activity', *Journal of Auditory Research*, **7**, 267–270.
Suga, N. (1968a), 'Analysis of frequency-modulated and complex sounds by single auditory neurones of bats', *Journal of Physiology, London*, **198**, 51–80.
Suga, N. (1968b), 'Neural responses to sound in a Brazilian mole cricket', *Journal of Auditory Research*, **8**, 129–134.
Suga, N. (1969a), 'Classification of inferior collicular neurones of bats in terms of responses to pure tones, f.m. sounds and noise bursts', *Journal of Physiology, London*, **200**, 555–574.
Suga, N. (1969b), 'Echo-location and evoked potentials of bats after ablation of inferior colliculus', *Journal of Physiology, London*, **203**, 707–728.
Suga, N. (1969c), 'Echo-location of bats after ablation of auditory cortex', *Journal of Physiology, London*, **203**, 729–739.
Suga, N. (1970), 'Echo-ranging neurons in the inferior colliculus of bats', *Science, New York*, **170**, 449–452.
Suga, N. (1971), 'Responses of inferior collicular neurones of bats to tone bursts with different rise times', *Journal of Physiology, London*, **217**, 159–177.
Suga, N. (1972), 'Analysis of information-bearing elements in complex sounds by auditory neurons of bats', *Audiology*, **11**, 58–72.
Suga, N. and Katsuki, Y. (1961a), 'Central mechanism of hearing in insects', *Journal of Experimental Biology*, **38**, 545–558.
Suga, N. and Katsuki, Y. (1961b), 'Pharmacological studies on the auditory synapses in a grasshopper', *Journal of Experimental Biology*, **38**, 759–770.
Suga, N. and Schlegel, P. (1972), 'Neural attenuation of responses to emitted sounds in echolocating bats', *Science, New York*, **177**, 82–84.
Suthers, R. A. (1965), 'Acoustic orientation by fish-catching bats', *Journal of Zoology, London*, **158**, 319–348.
Suthers, R. A. (1967), 'Comparative echolocation by fishing bats', *Journal of Mammalogy*, **48**, 79–87.
Suthers, R. A. (1970), 'Vision, olfaction, taste'. In *Biology of Bats*, ed. Wimsatt, W. A., Vol. II, pp. 265–309. Academic Press, New York and London.
Suthers, R. A. and Fattu, J. M. (1973), 'Fishing behaviour and acoustic orientation by the bat (*Noctilio labialis*)', *Animal Behaviour*, **21**, 61–66.
Suthers, R. A., Thomas, S. P. and Suthers, B. J. (1972), 'Respiration, wingbeat and ultrasonic pulse emission in an echo-locating bat', *Journal of Experimental Biology*, **56**, 37–48.
Terhune, J. M. and Ronald, K. (1971), 'The harp seal, *Pagophilus groenland-*

icus (Erxleben 1777). X. The air audiogram', *Canadian Journal of Zoology*, **49,** 385–390.

Thorpe, W. H. and Griffin, D. R. (1962a), 'Lack of ultrasonic components in the flight noise of owls', *Nature, London*, **193,** 594–595.

Thorpe, W. H. and Griffin, D. R. (1962b), 'Ultrasonic frequencies in bird song', *Ibis*, **104,** 220–227.

Thorpe, W. H. and Griffin, D. R. (1962c), 'Ultrasonic frequencies in bird song', *Nature, London*, **193,** 595.

Treat, A. E. (1955), 'The responses to sound in certain Lepidoptera', *Annals of the Entomological Society of America*, **48,** 272–284.

Treat, A. E. and Roeder, K. D. (1959), 'A nervous element of unknown function in the tympanic organ of moths', *Journal of Insect Physiology*, **3,** 262–270.

Tucker, D. G. (1966), *Underwater Observation using Sonar*. Fishing News (Books) Ltd., London.

Turner, R. N. and Norris, K. S. (1966), 'Discriminative echolocation in a porpoise', *Journal of the Experimental Analysis of Behaviour*, **9,** 535–544.

Vernon, J. A., Dalland, J. I. and Wever, E. G. (1966), 'Further studies of hearing in the bat, *Myotis lucifugus*, by means of cochlear potentials', *Journal of Auditory Research*, **6,** 153–163

Vernon, J. A., Herman, P. and Peterson, E. A. (1971), 'Cochlear potentials in the kangaroo rat, *Dipodomys merriami*', *Physiological Zoölogy*, **44,** 112–118.

Vernon, J. A. and Peterson, E. A. (1965), 'Echolocation signals in the free-tailed bat, *Tadarida mexicana*', *Journal of Auditory Research*, **5,** 317–330.

Vernon, J. A. and Peterson, E. A. (1966), 'Hearing in the vampire bat, *Desmodus rotundus murinus*, as shown by the cochlear potentials', *Journal of Auditory Research*, **6,** 181–187.

Walker, E. P. (1964), *Mammals of the World*, Vol. II. The Johns Hopkins Press, Baltimore.

Walker, T. J. and Dew, D. (1972), 'Wing movements of calling katydids: fiddling finesse', *Science, New York*, **178,** 174–176.

Wassif, K. (1946), 'The processus muscularis and the tensor tympani muscle of bats', *Nature, London*, **157,** 877.

Webster, F. A. (1963), 'Active energy radiating systems: the bat and ultrasonic principles. II. Acoustical control of airborne interceptions by bats'. In *Proceedings of the International Congress on Technology and Blindness*, ed. Clarke, L. L., pp. 49–135. American Foundation for the Blind Inc., New York.

Webster, F. A. (1966), 'Some acoustical differences between bats and men.' In *Proceedings of the International Conference on Sensory Devices for the Blind*, ed. Dufton, R., pp. 63–87. St. Dunstan's, London.

Webster, F. A. and Brazier, O. G. (1965), 'Experimental studies on target detection, evaluation and interception by echolocating bats', *Aerospace Medical Research Laboratory report No. TR-65-172*. Wright and Patterson Air Force Base, Ohio.

Webster, F. A. and Griffin, D. R. (1962), 'The role of the flight membranes in insect capture by bats', *Animal Behaviour*, **10,** 332–340.

Wever, E. G. (1935), 'A study of hearing in the sulphur-winged grasshopper (*Arphia sulphurea*)', *Journal of Comparative Psychology*, **20,** 17–20.

Wever, E. G. and Bray, C. W. (1933), 'A new method for the study of hearing in insects', *Journal of Cellular and Comparative Physiology*, **4,** 79–93.

Wever, E. G. and Herman, P. N. (1968), 'Stridulation and hearing in the Tenrec, *Hemicentetes semispinosus*', *Journal of Auditory Research*, **8,** 39–42.

Wever, E. G., McCormick, J. G., Palin, J. and Ridgway, S. H. (1972), 'Cochlear structure in the dolphin *Lagenorhynchus obliquidens*', *Proceedings of the National Academy of Sciences of the United States of America*, **69,** 657–661.

Wever, E. G. and Vernon, J. A. (1957), 'The auditory sensitivity of the Atlantic Grasshopper', *Proceedings of the National Academy of Sciences of the United States of America*, **43,** 346–348.

Wever, E. G. and Vernon, J. A. (1959), 'The auditory sensitivity of Orthoptera', *Proceedings of the National Academy of Sciences of the United States of America*, **45,** 413–419.

Wever, E. G. and Vernon, J. A. (1961a), 'The protective mechanisms of the bat's ears', *Annals of Otology, Rhinology and Laryngology*, **70,** 5–17.

Wever, E. G. and Vernon, J. A. (1961b), 'Hearing in the bat, *Myotis lucifugus*, as shown by the cochlear potentials', *Journal of Auditory Research*, **2,** 158–175.

Wever, E. G. and Vernon, J. A. (1961c), 'Cochlear potentials in the marmoset', *Proceedings of the National Academy of Sciences of the United States of America*, **47,** 739–741.

Wever, E. G., Vernon, J. A., Rahn, W. E. and Strother, W. F. (1958), 'Cochlear potentials in the cat in response to high frequency sounds', *Proceedings of the National Academy of Sciences of the United States of America*, **44,** 1087–1090.

Whitaker, A. (1906), 'The development of the senses in bats', *The Naturalist (London)*, 145–151.

White, F. B. (1872), 'Note on the sound made by *Hylophila prasinana*', *The Scottish Naturalist*, **1,** 213–215.

White, F. B. (1877), (Untitled communication), *Nature, London*, **15,** 293.

Whitney, G., Stockton, M. D. and Tilson, E. F. (1971), 'Possible social function of ultrasounds produced by adult mice (*Mus musculus*)', *American Zoologist*, **11,** 634.

Wimsatt, W. A. (ed.) (1970), *Biology of Bats*. 2 vols to date. Academic Press, New York and London.

Witts, A. T. (1934) (7th edition 1961), *The Superheterodyne Receiver: Its Development, Theory and Modern Practice*. Pitman, London.

Worden, F. G. and Galambos, R. (1972), 'Auditory processing of biologically significant sounds: a report based on a Neuro-sciences Research Program Work Session. *Neurosciences Research Program Bulletin* **10,** No. 1, 1–119. Brookline, Massachusetts.

Worthington. L. U. and Schevill, W. E. (1957), 'Underwater sounds heard from sperm whales', *Nature, London*, **180,** 291.

Wynne-Edwards, V. C. (1962), *Animal Dispersion in Relation to Social Behaviour*. Olivier and Boyd, Edinburgh and London.

Yanagisawa, K., Hashimoto, T. and Katsuki, Y. (1967), 'Frequency discrimination in the central nerve cords of locusts', *Journal of Insect Physiology*, **13,** 634–643.

Young, D. (1972), 'Neuromuscular mechanism of sound production in Australian cicadas', *Journal of Comparative Physiology*, **79,** 343–362.

Zippelius, H. M. and Schleidt, W. M. (1956), 'Ultraschall-Laute bei jungen Mäusen', *Naturwissenschaften*, **43,** 502.

Addendum. Some additional references to cetacean sounds

Busnel, R.-G. and Dziedzic, A. (1966), 'Acoustic signals of the pilot whale *Globicephala malaena* and of the porpoises *Delphinus delphis* and *Phocaena phocaena*'. In *Whales, Dolphins and Porpoises*, ed. Norris, K. S., Ch. 28. University of California Press, Berkeley, California.

Busnel, R.-G. and Dziedzic, A. (1968), 'Étude des signaux acoustiques associés à des situations de détresse chez certains cétaces odontocètes', *Annales de l'Institut océanographique*, **46,** 109–144.

Caldwell, M. C. and Caldwell, D. K. (1965), 'Individualised whistle contours in bottlenose dolphins (*Tursiops truncatus*)', *Nature, London*, **207,** 434–435.

Caldwell, M. C. and Caldwell, D. K. (1968), 'Vocalisation of naive captive dolphins in small groups', *Science, New York*, **159,** 1121–1123.

Caldwell, M. C., Caldwell, D. K. and Evans, W. E. (1966), 'Sounds and behaviour of captive Amazon freshwater dolphins, *Inia geoffrensis*', *Contributions in Science*, **108,** 1–24.

Dreher, J. J. (1966), 'Cetacean communication. Small group experiment'. In *Whales, Dolphins and Porpoises*, ed. Norris, K. S., Ch. 23. University of California Press, Berkeley, California.

Evans, W. E. and Dreher, J. J. (1962), 'Observations on scouting behaviour and associated sound production by the Pacific bottlenose porpoise (*Tursiops gilli* Dall)', *Bulletin of the Southern California Academy of Science*, **61,** 217–226.

Fish, M. P. and Mowbray, W. H. (1962), 'Production of underwater sounds by the white whale or Beluga *Delphinapterus leucas* (Pallas)', *Journal of Marine Research*, **20,** 149–162.

Fraser, F. C. (1947), 'Sound emitted by dolphins', *Nature, London*, **160,** 759.
Lang, T. G. and Smith, H. A. P. (1965), 'Communication between dolphins in separate tanks by way of an electronic acoustic link', *Science, New York*, **150,** 1839–1844.
Lilly, J. C. (1966), 'Sonic-ultrasonic emissions of the bottlenosed dolphin'. In *Whales, Dolphins and Porpoises*, ed. Norris, K. S., Ch. 21. University of California Press, Berkeley, California.
Powell, B. A. (1966), 'Periodicity of vocal activity of captive Atlantic bottlenose dolphins: *Tursiops truncatus*', *Bulletin of the Southern California Academy of Science*, **65,** 237–244.
Schevill, W. E. and Lawrence, B. (1949), 'Underwater listening to the white porpoise (*Delphinapterus leucas*)', *Science, New York*, **109,** 143–144.
Wood, F. G. (1954), 'Underwater sound production and concurrent behaviour of captive porpoises *Tursiops truncatus* and *Stenella plagiodon*', *Bulletin of Marine Science of the Gulf and Caribbean*, **3,** 120–133.

SYSTEMATIC INDEX OF ANIMALS
(with some common names in general use)

SUBJECT INDEX

Numbers in bold type refer to pages with text figures; numbers in italics refer to pages of tables